Cold-Climate Buildings Design Guide

Second Edition

CONTRIBUTORS

Project Manager

Erich Binder, Life Member ASHRAE | Erich Binder Consulting Limited—Calgary, AB, Canada

Project Authors

Robert Bean, Member ASHRAE | Indoor Climate Consultants Inc.—Calgary, AB, Canada
Robert Bisson, PEng, Member ASHRAE | Manitoba Public Schools Finance Board—Winnipeg, MB, Canada
K. William (Bill) Dean, PEng, Member ASHRAE | Retired National Research Council Canada—Saskatoon, SK, Canada
Craig Fredeen, PE, Member ASHRAE | Cold Climate Engineering, LLC—Anchorage, AK, USA
Heather Hayne, PEng, Member ASHRAE | Carleton University—Ottawa, ON, Canada
Jennifer Holmes, PE, Associate Member ASHRAE | Design Alaska—Fairbanks, AK
Mak Kampen, PE, PMP, Member ASHRAE | ConocoPhillips Alaska, Inc.—Anchorage, AK, USA
David Lima, Member ASHRAE | Aqua Air Systems Ltd.—Calgary, AB, Canada
Cheryl McGinn, PEng, Associate Member ASHRAE | Convergint Technologies—Calgary, AB, Canada
Frank Mills, Member ASHRAE, FCIBSE, MIMechE, MIE | Frank Mills Consulting Ltd—Leyland, Lancashire, UK
Erik Ostberg, PE, Member ASHRAE | bpSolutions—Bellingham, WA, USA
Bert Phillips, PEng, Fellow ASHRAE, FEC | UNIES Ltd.—Winnipeg, MB, Canada
Justin Pockar | City of Calgary—Calgary, AB, Canada
Karen Rollins Member ASHRAE | Masters Environmental Design—Canmore, AB, Canada
Stan Yackel, Member ASHRAE | Retired Senior Designer, Scott Springfield Mfg. Inc.—Calgary, AB, Canada

Case Studies

David Lima

Illustrations, Graphs, and Schematics

David Lima, Erich Binder, Heather Hayne, Karen Rollins, Craig Fredeen, Robert Bean, Justin Pockar
DLB Associates

Jokes

Erich Binder

Updates/errata for this publication
will be posted on the ASHRAE website at
www.ashrae.org/publicationupdates.

Cold-Climate Buildings Design Guide

Second Edition

ASHRAE®

Peachtree Corners

ISBN 978-1-947192-81-2 (paperback)
ISBN 978-1-947192-82-9 (PDF)

180 Technology Parkway
Peachtree Corners, GA 30092
www.ashrae.org

Cover photograph courtesy of David Lima.

Library of Congress Cataloging-in-Publication Data

Names: ASHRAE (Firm), author.
Title: Cold-climate buildings design guide / ASHRAE.
Description: Second edition. | Peachtree Corners : ASHRAE, 2022. | Includes bibliographical references and index. | Summary: "A guide to assist designers, contractors, and engineers in constructing and maintaining buildings in cold-climate environments. Provides practical advice on the unique challenges faced when trying to maintain human comfort and energy-efficiency in cold, subarctic, and arctic climates"-- Provided by publisher.
Identifiers: LCCN 2021028090 | ISBN 9781947192812 (paperback) | ISBN 9781947192829 (pdf)
Subjects: LCSH: Building--Cold weather conditions. | Buildings--Environmental engineering. | Buildings--Arctic regions.
Classification: LCC TH153 .C594 2022 | DDC 690--dc23
LC record available at https://lccn.loc.gov/2021028090

Contents

Preface

The inspiration for this guide was the result of a working session that met at the end of the 7th International Cold-Climate Design Conference held in Calgary in 2012, cosponsored by ASHRAE, SCANVAC, and REHVA. Past Society Presidents Tom Watson and Bjarne Olesen were instrumental in the encouragement of ASHRAE hosting the Cold-Climate Design conference and conference and in ASHRAE developing this guide.

A special thanks to the City of Calgary, Aqua Air, Convergint, Uponor, Scott Springfield Manufacturing, Erich Binder Consulting Limited, Government of the Northwest Territories, Low Carbon Design, and DLB Associates for their support and effort in making this design guide a reality.

The revision of this guide is intended to improve on the first edition and expand on our knowledge of cold-climate buildings by adding additional information on residential buildings, operation and maintenance, seasonal and off-grid facilities, and as oil and gas facilities in cold climates. Also added are icons denoting extreme cold climates (snowflake) and remote locations (igloo).

GUIDE FOR RESILIENT THERMAL ENERGY SYSTEMS DESIGN IN COLD AND ARCTIC CLIMATES

Since the publication of the first edition of this book, a complementary guide has been published: *Guide for Resilient Thermal Energy Systems Design in Cold and Arctic Climates* (ASHRAE 2021c). Developed in part by the U.S. Department of Defense and the U.S. Army, this guide addresses requirements to thermal energy systems resilience, which is especially important in extreme climates. It helps to identify maximum allowable time available to correct problems with heat supply to mission-critical facilities for different facility archetypes and air leakage rates before the indoor air temperature reaches habitability or sustainability thresholds.

The guide describes best-practice examples of robust and reliable systems with emphasis on their redundancy, durability, and functionality. It also discusses the most common heating system and ventilation system approaches used in arctic climates and emphasizes the importance of a maintenance program that allows building operators to successfully troubleshoot and maintain buildings in the arctic.

Available for free download from the Whole Building Design Guide website at www.wbdg.org/ffc/army-coe/design-guides/guide-for-resilient-thermal-energy-systems-design-in-cold-and-arctic-climates, the guide is designed for energy systems designers, architects, energy managers and building operators and is a valuable resource for those who are involved in building planning and operation in cold and arctic climates.

ASHRAE Cold-Climate Buildings Design Guide kickoff "Cold Climate Rocks" at the 2015 ASHRAE Winter Conference in Orlando, FL.
From left: Bert Phillips, David Underwood (past Society President), David Lima, and Doug Cochrane

HOW COLD IS IT?

It is so cold ... people were signing up for tropical vacations in Iceland!

1 Cold Climates

WHAT IS A COLD CLIMATE?

The answer to the question "what is a cold climate?" might be as simple as "anywhere you feel cold." Of course, this feeling is different for each person and is not derived scientifically.

Historically, there have been many ways to quantify *cold* and define what a cold climate is such as lack of heat or weather cold enough to cause freezing. The Climatic Design Information chapter of *ASHRAE Handbook—Fundamentals* (ASHRAE 2021a) divides the world into climate zones based on temperature, duration of low temperature periods, and whether the zone is humid or dry. Zones 5 to 8 are listed as *cool* to *subarctic* but may not encompass all zones that have some cold-climate features. Even the hottest climates having higher altitudes may need heat a few days of the year to maintain occupant thermal comfort.

Degree days are frequently used to compare location-specific weather in terms of severity. Heating degree days are useful for calculating energy savings, but do not present a definition for cold climates. Some areas of the world have relatively few heating degree days but can get very cold for just a short time of the year. A*verage temperature* or *design temperature* are also considered in designating hot or cold climates, but some climates can experience both very hot and very cold temperatures; as such, average temperature is not always an accurate representation. Likewise, design temperature does not always clearly define the line between cold and noncold climates.

DEFINITION OF A COLD CLIMATE

A cold climate is defined by a combination of factors that together create a unique set of building design challenges. This guide touches on each of these factors and identifies strategies on how to meet the design challenges created by them. Most of these must be considered for any building, but additional consideration is needed to provide a successful building in a cold climate. If the building has a heating system, you are in a cold climate.

Cold-Climate Factors

Temperature. Many areas of the world sometimes experience temperatures below the freezing point of water. Water located throughout construction materials, support systems, and equipment is susceptible to freezing if not protected. To function reliably at low temperatures, equipment must be robust and well maintained.

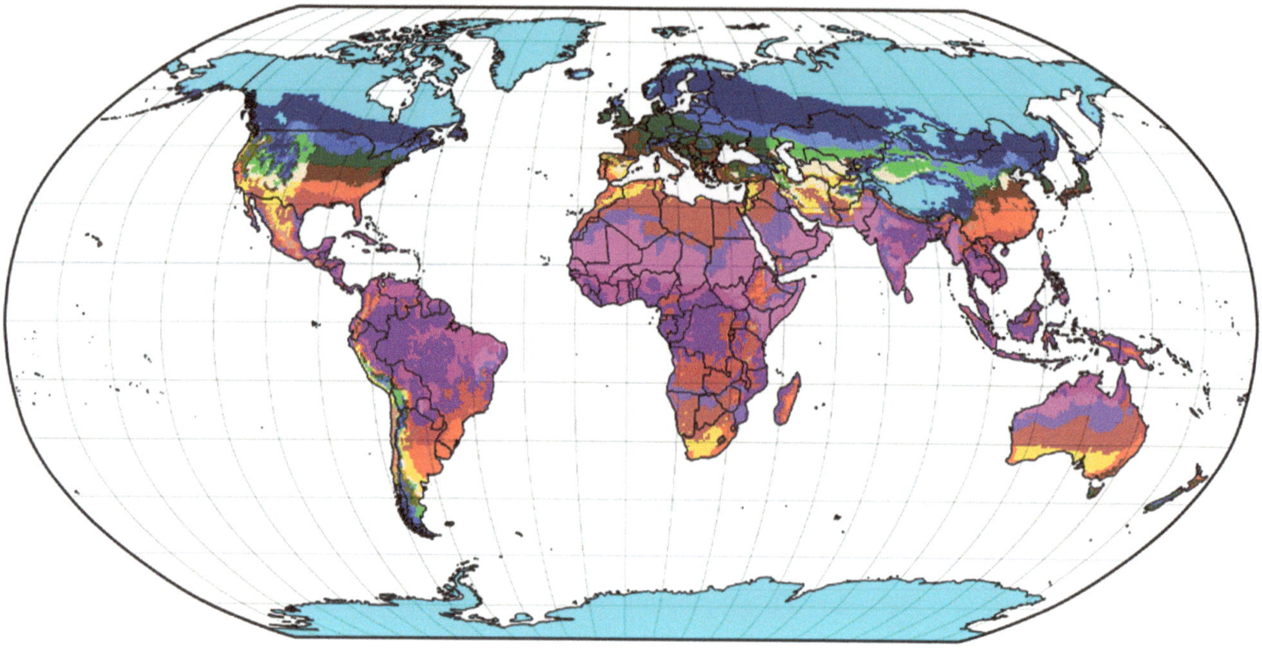

Figure 1.1 ASHRAE world climate zone map.
ASHRAE (2021c)

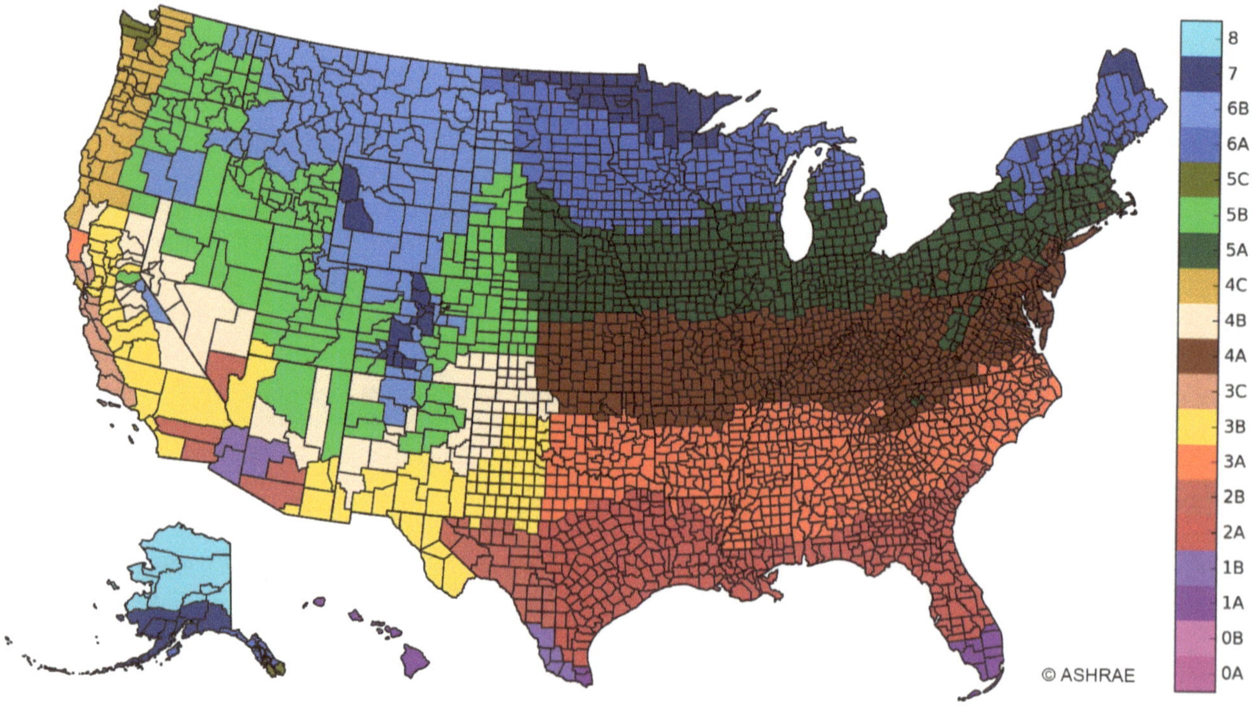

Figure 1.2 ASHRAE climate zones for states and counties in the United States.
ASHRAE (2021c)

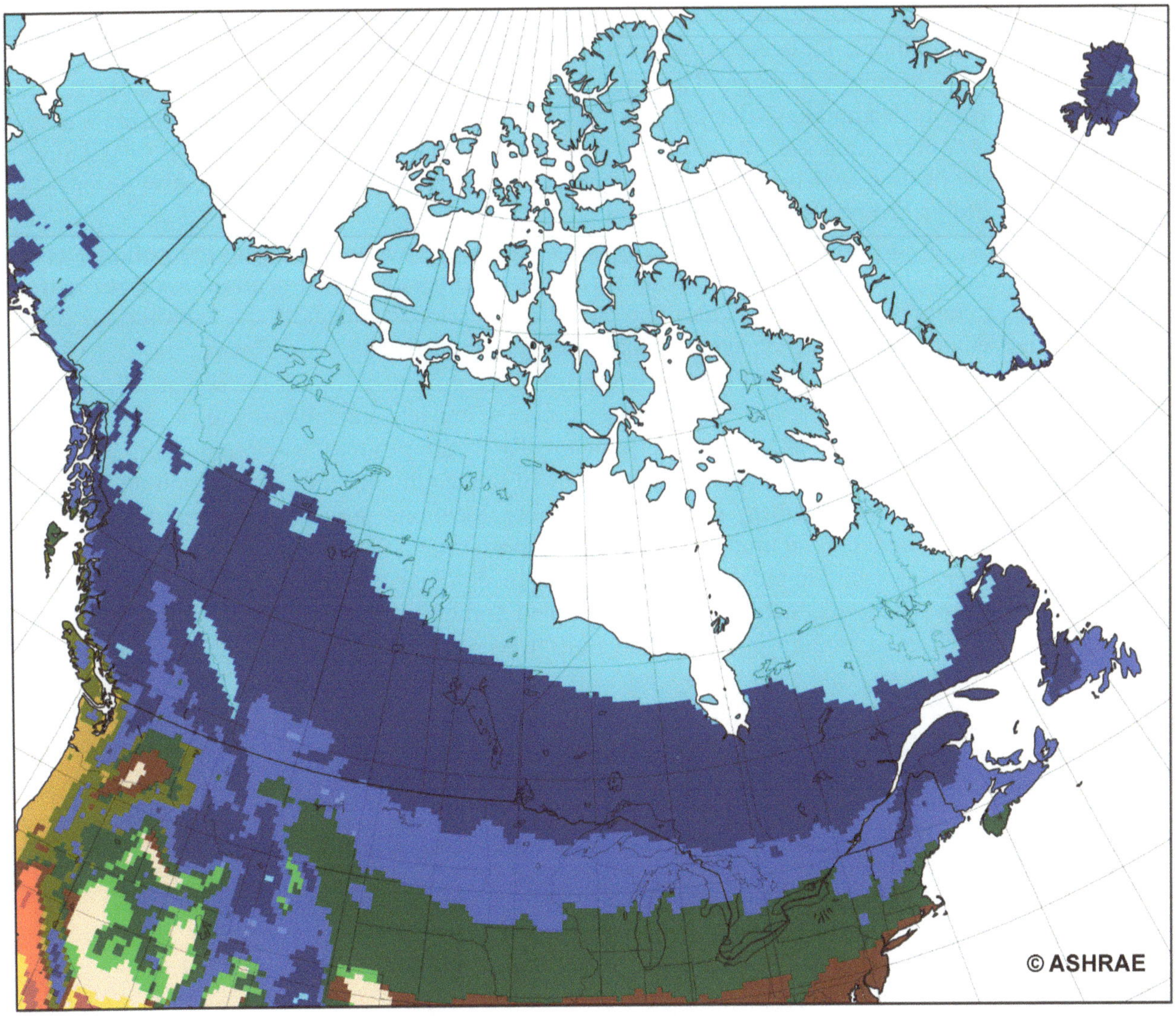

Figure 1.3 ASHRAE climate zones for provinces and territories in Canada.
ASHRAE (2021c)

Frozen Precipitation. Snow, ice, vapor, and freezing rain can wreak havoc on any building, especially when it is not designed for those conditions. Roofs need extra support to accommodate snow loads. Pitched roofs can cause miniature avalanches next to buildings. Snow can also provide additional insulation to improve the building envelope.

Wind. Pressure differentials caused by wind cause infiltration, create snow drifts, drive snow into louvers, and rip roofs and siding off buildings. Cold-climate designers must take into account air velocities and pressures in and around buildings.

Humidity. Humidity is necessary for some building occupancies in all climates, but in cold climates, higher humidities cause extra complications. Cold envelope temperatures make materials susceptible to rising relative surface humidities and surface condensation.

Thermal Comfort. Cold interior surfaces, temperature stratification, and drafts can make space conditions difficult for occupant thermal comfort. The stack effect from aggressive temperature differences across the envelope can cause infiltration, resulting in undesirable air velocities throughout the building. A combination of envelope design, compartmental strategies, and heating systems arrangement is required to establish conditions suitable for occupant thermal comfort in a cold-climate building.

Thermal Envelope (Thermal Enclosure). Thermal envelopes for cold climates are specialized and have distinct needs. Being a largely passive element, the design and thermal performance of the envelope must be integrated with the design and installation of the heating systems. Refer to Chapter 8 of this guide for more information on envelope design and performance considerations.

Maintainability. Building systems must be maintained to keep the building operational. In cold climates, this can be especially difficult when encountering freezing temperatures, wind, and frozen precipitation. Various strategies are explored to enable equipment maintenance in cold climates.

Permafrost and Frozen Ground. Frozen ground can limit the construction season, choice in construction materials and methods, impact foundations, and cause utility damage. In some areas, permafrost (permanently frozen ground) can be especially difficult to build foundations on. Refer to Chapter 8 of this guide for more information on this topic.

Remote Building Locations. Remote locations are areas where, during certain events or at certain times of the year, transportation to the building is limited or restricted. There may be limited or reduced maintenance, a higher chance of power interruption, or the building may not be regularly occupied or monitored.

Remote buildings are typically built to be more robust and stand alone. Often, critical equipment repairs must wait for staff and/or parts to be flown in. Less-critical system repairs are usually deferred or scheduled when maintenance teams and repair parts can travel to the site, so redundancy becomes important. Remote building considerations are covered throughout this guide, and particularly in Chapter 12.

If the considered building is exposed to any of the listed cold-climate factors, this book provides insight on how to deal with problems that can be encountered in the planning, design, construction, and commissioning phases of the project.

Figure 1.4 Remote Northern Alberta industrial facility.

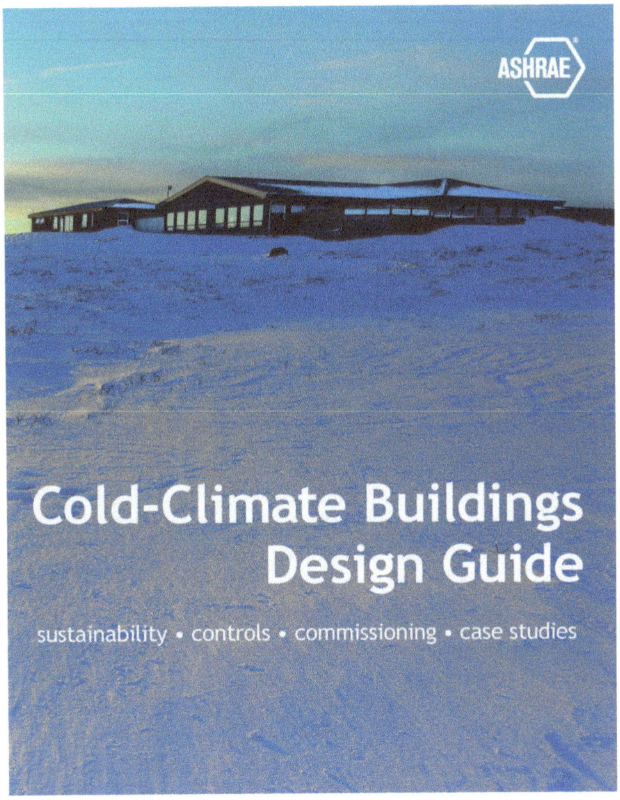

Figure 1.5 First edition cover.

Figure 1.6 Deer in the snow, Alberta, Canada.

Figure 1.7 Ice road, Northwest Territories, Canada.

HOW COLD IS IT?

It is so cold ... hitchhikers were texting pictures of their thumbs!

2 Sustainability, Resilience, and Renewable Energy

SUSTAINABILITY AND RESILIENCE IN COLD CLIMATES

Sustainability and resilience are different but complementary concepts that are important for buildings in all climates, but especially important in cold climates and remote locations. While both should be primary considerations for construction projects, either can exist without the other.

Green and Sustainable Construction

Green or sustainable construction practices are environmentally sustainable and minimize the total environmental impact of a building over its entire life cycle. Designing, constructing, and operating a sustainable building requires close cooperation among the contractors, architects, engineers, and the client/owner at all project stages, including planning and design, construction, operation, maintenance, renovation, and demolition. In sustainable designs, design decisions are guided by the objectives of protecting the natural attributes of the building site, selecting durable construction materials from sustainable sources, avoiding material and energy waste during construction, employing energy and water efficiency principles in all aspects of the building design and its systems to minimize the need for auxiliary energy, then meeting those auxiliary energy needs from renewable energy sources in as much as is practical. Thus, specific objectives of sustainable building design and construction include the following:

- Minimizing peak demand and annual total consumption of thermal and electrical energy and fresh water
- Maximizing capture, efficient storage, distribution, and use of appropriate renewable energy for required functions
- Only using fossil energy efficiently for loads that cannot be met using renewable sources

General and detailed guidance on green or sustainable construction is available from a variety of sources including ASHRAE's Building Energy Quotient (Building EQ) ratings systems, *ASHRAE GreenGuide*, and the Advanced Energy Design Guides; Leadership in Energy and Environmental Design® (LEED®) Green Building Rating System; the Building Research Establishment Environmental Assessment Method (BREEAM); the DOE's Better Buildings Challenge; United Nations Energy Program (UNEP) Energy Globe; Passivhaus; and similar certification programs.

Resilience

Resilience is the ability to survive with minimal impact, or to quickly and inexpensively recover from disaster, whether natural (such as ice storms, floods, and heavy snow falls), man-made (such as train derailments and chemical spills), or as a result of a system failure (such as running out of heating fuel, a power failure, or failure of a pump or boiler). Facilities that are able to maintain conditions supporting occupant life safety, health, and relative comfort over extended periods after such events are described as being *resilient.*

Designing resilience into buildings involves anticipating failures and disasters, then developing designs and systems to mitigate their impacts. In cold climates, particular considerations need to be given to ensuring power or heating system outages during cold weather don't result in significant damage to the building and its systems. A specific example is protecting water lines and hydronic systems and equipment from freezing. As much as is practical, equipment and systems susceptible to freeze damage should be located in the heated core of the building to maximize the time from a heating outage until these equipment and systems are exposed to freezing temperatures; measures should be included to allow pipes and equipment containing water to be easily drained in the event that freezing temperatures occur in their vicinity. Where it is impractical to drain hydronic equipment that may be exposed to freezing temperatures, the thermal fluid should be antifreeze and building operators need to know how and when to drain down systems containing water, which systems contain antifreeze, and how to correctly maintain the antifreeze protection in those systems.

Resiliency in cold climates is achieved, in part, by integrating the same energy-efficiency and conservation principles found in sustainable designs into cold-climate buildings. It must be emphasized that resiliency requires building management's committed vigilance to three seemingly mundane, yet essential, day-to-day details, before an event strikes:

1. Regular, active involvement of staff in emergency preparedness training
2. Limited stockpiling of selected supplies
3. Scheduled staff interaction sessions with local emergency services officials to review contingency planning, facility layout, and other issues of mutual importance

In resilient designs, equipment essential to the operation of the facility will be located in areas where it can be serviced during severe weather. As such, critical heating equipment must not be located in outdoor locations such as on roofs, where service personnel cannot be expected to effectively undertake repairs during a winter storm, and electrical panels and services should be located above floodwater levels.

SUSTAINABILITY, RESILIENCE, AND CLIMATE ADAPTATION

The scientific community almost unanimously believes that climate change is real and is impacting our planet and building stock. While we may not be able to accurately predict the long-term effects of climate change on the weather at a specific location, the scientific community's climate change models predict the general direction, magnitude, and rate of climate change for most locations around the globe. To be resilient to climate change and sustainable, we must adapt our designs to address probable climate change risks and effects, and to evolve our designs as the effects and predictions of the effects of climate change evolve.

Figure 2.1 Insulated and heat traced pumps, piping, and accessories.
Erich Binder

While the effects of climate change will vary from site to site and project to project, some specific situations that will require design adaptation are:

- Thawing of permafrost in some locations will require a change in foundation designs
- Changes in precipitation rates may result in water shortages and/or increased flooding
- Rising sea levels and more storms may require coastal buildings to be located at higher elevations
- More violent storms may require stronger structural tie-down of roofs, claddings, and outdoor mechanical equipment

The effects of climate change may be highly uncertain—the costs of addressing these uncertain outcomes may be very high or current practices may not be effective in solving future anticipated problems. In these cases, good designs will be able to accommodate and/or incorporate an appropriate range of upgrades in the future. While we cannot predict the future, we can design for a foreseeable range of occurrences and risks.

SUSTAINABLE ENERGY SOURCES AND SYSTEMS FOR COLD-CLIMATE APPLICATIONS

A specific objective of sustainable building design is optimizing the capture and use of renewable energy. This requires efficiently storing renewable energy between the time it is captured and the time it is needed, and efficiently distributing this energy. An underlying principle is that energy conservation be applied as much as is economically practical to reduce energy loads, and that designers focus on the whole-building impacts of each measure, not on the benefits of an individual measure in isolation of the rest of the building.

In projects in which all energy is to be provided by renewable energy sources, the cost for each potential energy conservation measure to save an additional unit of energy should be compared to the cost of generating one more unit of energy by the least-cost renewable energy source. For example, if photovoltaic (PV) panels are the least-cost renewable energy source available, and the marginal cost of generating electricity over the life cycle of the photovoltaic system for the project in question is $0.20/kWh, then roof insulation should be increased until the marginal cost of adding insulation is $0.20/kWh saved over the life cycle of the roof.

In projects where fossil energy is used to meet part or all building loads, the owner and facility designer should establish the economic criteria to be applied to each applicable energy efficiency measure to achieve cost optimization on the project. The environmental damages caused by fossil fuel spills and costs of fuel spill clean ups should be factored into economic assessments.

Energy sources and technologies that are sustainable or more sustainable than using fossil fuels include combined heat and power generation (CHP); thermal solar used for water, space, and process heating; electric generation and/or water pumping with wind turbines; electricity production using photovoltaic panels; electricity generation using small and microscale hydro; hydrokinetic technologies that generate electricity from river currents, ocean waves, and tidal flows; biomass-fueled heat and power systems; and waste heat recovery. One may also include small-scale nuclear power generation and methane recovered from landfill or sewage treatment facilities in the arsenal of sustainable energy options for cold climates.

Installing a local renewable energy system to provide power to essential equipment (e.g., emergency lighting, boiler controls, and so on) can improve resilience of a facility by keeping this essential equipment operational during extended power outages. If a renewable energy system can replace the need for a conventional emergency generator system, its economics will be significantly improved.

CHP

CHP systems have favorable characteristics for cold-climate applications. In most fuel-fired electric generation facilities, less than 40% of the energy in the source fuel is converted into electricity, the balance ending up as heat that has traditionally been rejected/wasted to the atmosphere. In smaller-scale generation facilities, electric generation efficiencies tend to be considerably lower than 40%. In CHP facilities, a significant fraction of the otherwise rejected/wasted heat is recovered and used for space, water, and process heating, potentially increasing overall plant thermal efficiency to better than 80%. In arctic climates, the largest heating loads in buildings generally coincide with times of shortest hours of sunlight, with maximum need for lighting and other electricity use. This coincidence of demand for space heating and electricity enables high average annual CHP system energy efficiencies. A CHP system that provides heat to a campus or community of buildings is referred to as a *district heating system*.

Waste Heat Recovery

Waste heat recovery from any source can be worthwhile. Cold climates offer more opportunities to capture and reuse low-grade waste heat than do warmer climates. Low-grade waste heat sources include exhaust air and domestic or process water. Applications include preheating ventilation air, preheating domestic and process water supplies, and snow melting on outdoor walks and ramps to parkades. The range of applications for low-grade waste heat can be expanded by using heat pumps to upgrade the temperature of the recovered energy.

Higher-grade waste heat can be recovered from electric generation facilities. This is covered in the section on CHP and in greater detail in ASHRAE's *Combined Heat and Power Design Guide* (2015b).

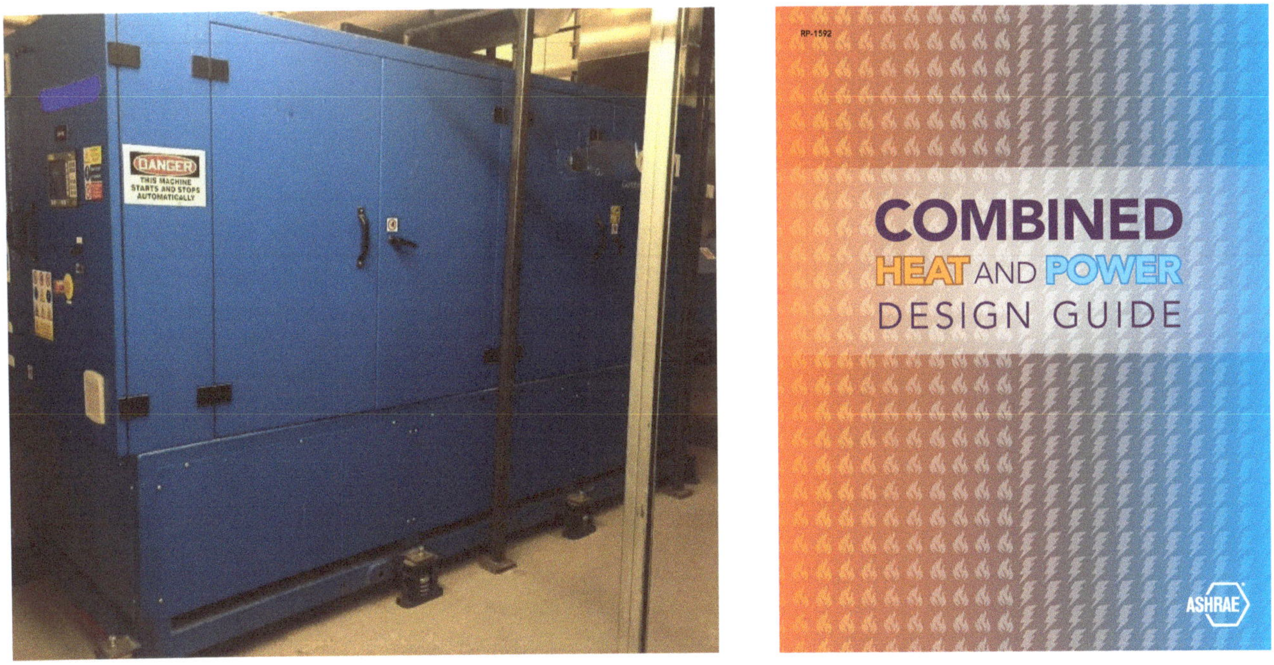

Figure 2.2 CHP unit (left) and *Combined Heat and Power Design Guide* (right).
Erich Binder (left); ASHRAE 2015b (right)

Passive Thermal Solar

Passive thermal solar design should be part of all building projects. As much as is practical, buildings in cold climates should be oriented to maximize solar energy gain through windows during the heating season.

Active Thermal Solar

Active thermal solar applications may be cost-effective in milder cold climates, but are less likely to be economically attractive in very cold climates. A promising technology for many projects is ventilation air preheat systems. Ventilation air is drawn through low-cost solar panels; incident solar radiation on the panel warms surfaces that in turn warm the ventilation air. Because the system operates near ambient temperatures, heat loss is low and heat recovery effectiveness can be quite high.

Thermal solar systems are generally not well-suited to providing space heating in cold climates because the times when space heating demands are greatest generally coincide with coldest outdoor temperatures and thus highest thermal losses from the system and lowest levels of incident solar radiation. Measures needed to protect hydronic systems from freezing complicate their operation, increase system costs, and may adversely impact operating efficiencies.

Photovoltaic

Photovoltaic panels may be very cost effective in cold-climate applications, especially remote locations in which generating electricity using conventional technologies is very expensive. Consider wall-mounting solar panels to avoid snow accumulation on collector surfaces and to keep installation costs down (Figure 2.3). As latitudes approach the Arctic and Antarctic circles, the average panel mounting angle approaches vertical. It is important to keep photovoltaic panel surfaces clear of snow, because variations in incident solar energy across the face of photovoltaic panels can have a disproportionately adverse effect on electric energy generation. Roof-mounted solar arrays can have a significant effect on snow accumulation on the roof, increasing structural requirements of roofs and supporting walls, and thus on the overall cost of the solar system array support structure.

With all solar energy collection strategies, be aware that as latitudes increase away from the Equator, the length of shadows cast by adjacent buildings and other objects increase significantly. Thus, nearby buildings, trees, and landforms can drastically affect direct solar radiation incidence.

Heat Pumps

Heat pumps use refrigeration technologies to extract low-grade heat from the environment and upgrade it to temperatures suitable for space heating and/or domestic water heating. In applications where space heating needs are met with electric energy, heat pumps can be an economically attractive way of getting more heat output for a given electric energy input. In all cases, designers must be aware that heat pump coefficient of performance (COP) is inversely related to the temperature difference between the heat energy source and the heat energy application. As the temperature difference between the source and the sink increase, heat pump performance diminishes. In most cold-climate applications, heat pumps operate with heat source temperatures well below the source temperature at which they were rated, so in-service COP may be much lower than the rated or claimed COP for the apparatus.

In cold climates, ground-source heat pumps operate primarily in heat extraction mode. Over time, soil temperatures in the region of the ground loops will decrease, increasing heat-source-to-heat-sink temperature differences, further reducing system COP. Once moist soil is cooled below freezing, its thermal capacity or *specific heat* is sig-

Figure 2.3 Wall-mounted solar panels on an apartment building in Inuvik, NWT.
Bert Phillips

nificantly reduced (the specific heat of ice is half the specific heat of water), resulting in more rapid cooldown of the surrounding ground as further amounts of heat are extracted.

Computer models that simulate annual heat-pump performance for specified heating loads and climatic conditions are available for air-source, water-source, and ground-source heat pump applications. Designers are strongly encouraged to use such programs to get better insight into long-term performance of the systems they are considering.

Wind Turbines

Wind turbines have the advantage over solar panels that they can generate electricity regardless of where the sun is. Performance may vary seasonally because wind speeds may be higher in some seasons than others, but the coincidence of energy demand and energy supply may be a better match in cold climates than is the case with photovoltaic power generation. However, there are some drawbacks to wind turbines in cold climates.

Frost and ice accumulation on turbine blades can unbalance the turbine, causing damaging vibration. Large annual temperature swings may require seasonal changes of lubricants. The summer-to-winter temperature-induced tolerance changes in bearings and gearboxes may affect performance and reliability. Extremely cold temperatures can affect the strength of some materials. All of which is to say that wind turbines developed for moderate climates may not be appropriate for cold-climate applications. Persons selecting wind turbines for cold-climate applications should specify equipment that has been tested and proven in these environments.

Wood and Biomass

Wood and other biomass fuels are considered to be a green/sustainable energy source that may be harvested specifically as a fuel or may be a byproduct from agriculture (e.g., straw or walnut shells), or the forestry industry (e.g., sawmill waste, chips from tree pruning operations) or other processes (e.g., used shipping pallets, combustible construction debris). The supply of biomass fuels generally involves some fossil-fuel usage, which varies depending on the source of biomass, the distance between the source and use, and the processing required to prepare the fuel for use. The most common use of biomass fuels is for space and water heating, although biomass can be used as the fuel source for CHP systems. In the future, thermoelectric and other technologies that directly generate electricity from flame or heat may become economically attractive for biomass CHP systems.

Several governmental organizations provide information regarding the design and operation of biomass-fueled heating and CHP systems. These include the Austrian Energy Agency, Austrian Federal Forestry Corporation, Chartered Institution of Building Services Engineers (CIBSE), U.S. Forest Service Wood Education and Resource Center, New York State Energy Research and Development Authority (NYSERDA), Vermont's Biomass Energy Resource Center, Alaska Energy Authority, the U.S. National Renewable Energy Laboratory, and the Biomass Thermal Energy Council. In addition to these, equipment suppliers also offer guidance on system design and operation. Many related publications and resources are available online.

Many renewable energy conversion technologies are relatively new and not well understood, and the industry includes many small players. Those designing biomass energy systems for the first time are encouraged to do extensive research to ensure long-term, reliable, cost-effective fuel sources are available; operating staff will have the skills and resources needed to effectively operate the facility; and that the equipment selected has a proven track record and is appropriate to the application.

Performance regulations for European biomass installations (emission limits and thermal efficiency requirements) are defined in the EU-wide rule, EN 303-5 (BSI 2012).

This standard established performance requirements for three different emission and efficiency classes of boilers. At the time of publication, there were no similar regulations applicable in North America.

There are system considerations unique to cold climates. On the positive side, gathering fuel for biomass energy systems can provide local employment in addition to the resilience and sustainability it may provide to the project. This may make biomass utilization very attractive for community sponsored facilities in smaller, remote locations.

Biomass energy systems generally are not able to respond as quickly to changes in thermal load as do gas and oil fired systems. For this reason, it may be beneficial to add some thermal storage into the system. Examples include thermal mass in conditioned spaces (e.g., concrete floors) or a hot-water thermal storage tank in the boiler loop.

Snow and other moisture in fuel supplies such as sawdust, wood chips, wood pellets, and straw pellets can lead to the fuel freezing into clumps, which can cause significant problems for fuel supply systems including hopper gates, augers, and conveyors.

Snow, moisture, and dirt in the fuel can significantly reduce the effective heat content of the fuel. It may result in cooler combustion temperatures that increase flue gas emissions. Inconsistent/highly variable fuel quality can result in inconsistent plant operation and heat generation, which may lead to user complaints.

Air-drying of biomass fuels is generally slower at cold temperatures than warmer temperatures, so it may be necessary to store up to a year's fuel supply in drying sheds to ensure acceptable moisture levels in biomass fuels. Long-term storage may also be needed where fuel supplies are only available on a seasonal basis (e.g., straw).

Many remote locations are in sensitive environments. Designers need to ensure that ash and emissions from biomass energy facilities will not create burdensome environmental problems. Note that clean ash may be considered to be a soil amendment, improving its fertility. Avoid system oversizing—harmful flue gas emissions from biomass-fuel-fired plants can increase significantly under low-fire operating conditions.

Hybrid Energy Systems

Hybrid energy systems draw upon more than one energy source to meet facility loads. This may include combining thermal solar with fuel-fired heating using a mix of biomass and conventional fuel sources. Hybrid energy systems are generally more complex, thus system operators usually require a higher skill set than do operators of single-source energy systems. For system performance optimization, operators must know which energy source is appropriate for the operating conditions.

Geothermal

High-temperature geothermal heating from hot springs or from wells is accessible at select locations around the world, most notably in Iceland. Hot springs and hot ground temperatures are commonly associated with areas with high seismic activity. Energy from this source may be collected through either open- or closed-loop systems and used for space or water heating.

HOW COLD IS IT?

It is so cold ... we had to build an ice fishing shack in the bathtub to drain it!

3 Human Comfort in the Indoor Environment

HEALTH AND COMFORT

This chapter deals with the provision of a built environment that is safe, healthy, comfortable, and productive for its occupants, visitors, and anyone nearby, and that has a minimal detrimental impact on the global environment.

In all buildings, the following issues affect the health and well-being of their occupants:

- Thermal comfort
- Air quality, including odors
- Visual environment
- Noise and vibration
- Water quality, including potable, gray, black, and solar hot water
- Electromagnetic frequencies
- Community development and infrastructure

However, in this chapter, the emphasis is on factors that are especially relevant in buildings in cold climates.

The World Health Organization defines good health as "a state of complete physical, mental, and social well-being, not merely the absence of disease and infirmity (WHO 1948)." Thus, the human factor objective in the design of spaces is to provide as good an environment as possible. And, in a cold climate, the aim is to do this throughout the year. In former times, the onset of winter saw communities draw back, seek shelter, and wait for spring when they could reemerge into the seasonal warmth and sun and light. In modern times, this is not what happens. People expect to go about their business, go to schools and colleges, and live a normal life right through the winter weather. Cold-climate building designers are expected to provide a built environment that allows them to do that.

Primarily, these buildings enable economically productive and creative settings for residential, commercial, and business activities such as interesting and stimulating spaces for education; healing environments for healthcare; invigorating and stimulating atmospheres for sports; and relaxing, pleasant regenerating spaces for leisure.

Secondly, buildings must have healthy and productive environments so these desired activities can take place successfully.

Ultimately, designers must ensure that buildings (including materials and methods of construction) and their environmental conditioning systems do not trigger or contribute to undesirable stresses.

Discomfort introduces stress, which, if maintained, can affect health (Robinson 2017). Stress-related absenteeism rises as occupants find cold conditions too uncomfortable. The hazard is the poor effect upon the person. The risk of poor health is a function of the likelihood of the effect. It is estimated in the United Kingdom that worker absenteeism costs U.K. businesses a staggering £29 Billion per year (PwC 2013). This is equivalent to every worker having an average of nine days off sick each year. Some of these are due to illnesses passed on at work through poorly performing buildings that have poor air quality and possibly sick-building features.

In a cold climate, the temptation to save energy by reducing outdoor air changes can lead to poor air quality and result in occupant sickness. Outdoor air supplies should meet levels outlined in ANSI/ASHRAE Standards 62.1 and 62.2, ANSI/ASHRAE/ASHE Standard 170, and other guidance, such as BREEAM or LEED Green Building Rating System air-quality levels.

Similar issues can occur in not providing adequate makeup air for exhaust systems. This can lead to excessive infiltration, envelope damage, or even carbon monoxide poisoning through back-drafted fuel appliances.

THERMAL CONDITIONS FOR COMFORT

The human body strives to maintain a thermal balance with its surroundings. The production of heat by the body should equal the heat losses to the environment, aiming to keep a constant core body temperature of around 98°F (37°C) (Panagoupoulos 2008).

If the body cannot achieve this balance in a cold climate, the occupant will struggle to maintain mental focus as cold discomfort interferes. If the discomfort progresses to hypothermia, extreme discomfort will result, possibly leading to illness or death depending on the severity and period of the stress encountered.

The environmental and personal parameters affecting thermal comfort, as well as the suggested range of thermal conditions to achieve this relative to the activity undertaken, are discussed extensively in ASHRAE, CIBSE, and REHVA handbooks and ANSI/ASHRAE Standard 55 (ASHRAE 2020b) and ISO 7730 (ISO 2005).

Physiological Acclimatization

People who live and work in cold climates can develop a level of acceptance known as *physiological acclimatization* that involves all the physiological processes and thermoregulatory responses to ensure adaptation in a wide range of cold environments (Mathew et al. 1981). Although not all people can feel comfortable in a very cold environment, a level of tolerance is possible.

People living in cold countries are more likely to be accustomed to cold and can function at temperatures below the optimum, whereas those who live in hot countries tend to prefer warmer temperatures and have low tolerance to cold. This forms, in part, the rationale of the adaptive approach to thermal comfort, described in ANSI/ASHRAE Standard 55 (ASHRAE 2020b) and European adaptive standard. BS EN 16798-1 (BSI 2019), indicating that there is a tolerance to a range rather than a specific temperature. ANSI/ASHRAE Standard 55 shows how this applies as *adaptive comfort* and the level to which it can apply; however, prevailing mean outdoor temperatures below 50°F (10°C) are not covered by the adaptive method described in the standard.

Cold Discomfort/Stress

In the human body, a drop in the core body temperature results in vasoconstriction, in which blood circulation to the peripheral parts of the body, particularly hands and feet, is reduced to restore heat balance by reducing heat loss. The body prioritizes the head, torso, and all central functions but starts to lose use of peripheral functions. Of equal importance is the hypothalamus/pituitary/thyroid response and production of T3 and T4 hormones to involuntarily increase the body's metabolic rate. If no remedial action is taken and cold exposure persists, body temperature will decline. Air and mean radiant temperature, humidity, air velocity, activity level, and the clothing of the individual all affect the rate and total amount of heat loss from the body.

Working under cold conditions can adversely affect performance. Although pain and decreased motivation induced by coldness play a part, performance in the cold deteriorates largely for physiological reasons. Manual dexterity is affected. Hand and finger sensitivity decreases, resulting in sensations of numbness and a loss of feeling and possibly grip. Strength and speed of movement (particularly flexion and extension of fingers) are similarly affected. Slow performance and increased mistakes are likely.

To offset the effects of cold temperatures, workers use heavily insulated coats and gloves. This also reduces dexterity and free movement. Any mechanical systems located outdoors that may require maintenance during cold temperatures should take this into account; an example is the use of oversized fasteners that are both easier to handle and to manipulate with gloved hands.

Acceptable Temperatures

Building services must provide heating to achieve a operative temperature range that is appropriate when considering occupant clothing and activities. In some cases, there may be several differing and even conflicting requirements. For example, in hospitals, patients, medical staff, and visitors will all desire different temperatures.

Unacceptable Temperatures

Figure 3.1, which is based on CENELEC Guide 29, "Temperatures of Hot Surfaces Likely to be Touched" (2007), indicates risk of burning when touching hot surfaces for lengths of time. This is an issue where heating equipment and distribution systems are accessible and likely to be touched by occupants.

While considering the issues associated with hot surface temperatures in rooms, another area that should also be noted is the use of solar control glass that will present hot glass surface temperatures inside rooms during high solar gain periods, particularly low-sun angles in spring or autumn, as the glass absorbs the sun's rays and limits the amount of heat entering the room. However, the glass itself heats up to temperatures that may exceed 158°F (70°C). Cold climates in northern altitudes are most at risk as the sun angle can be low and direct solar gains cause such problems.

Heating Surface Temperatures

Where heating systems use hot surface panels such as radiators that could be touched by occupants, it is necessary to ensure that the temperature of the surface does not cause injury through surface burns if people come into contact with them.

The safe surface temperature differs for different users. Older people and children are more susceptible to burns and so lower temperatures are desirable. This translates to larger surface areas at a lower fluid temperature for the same output. For example, whereas office areas may be heated using conventional 180°F/172°F (82°C/78°C) com-

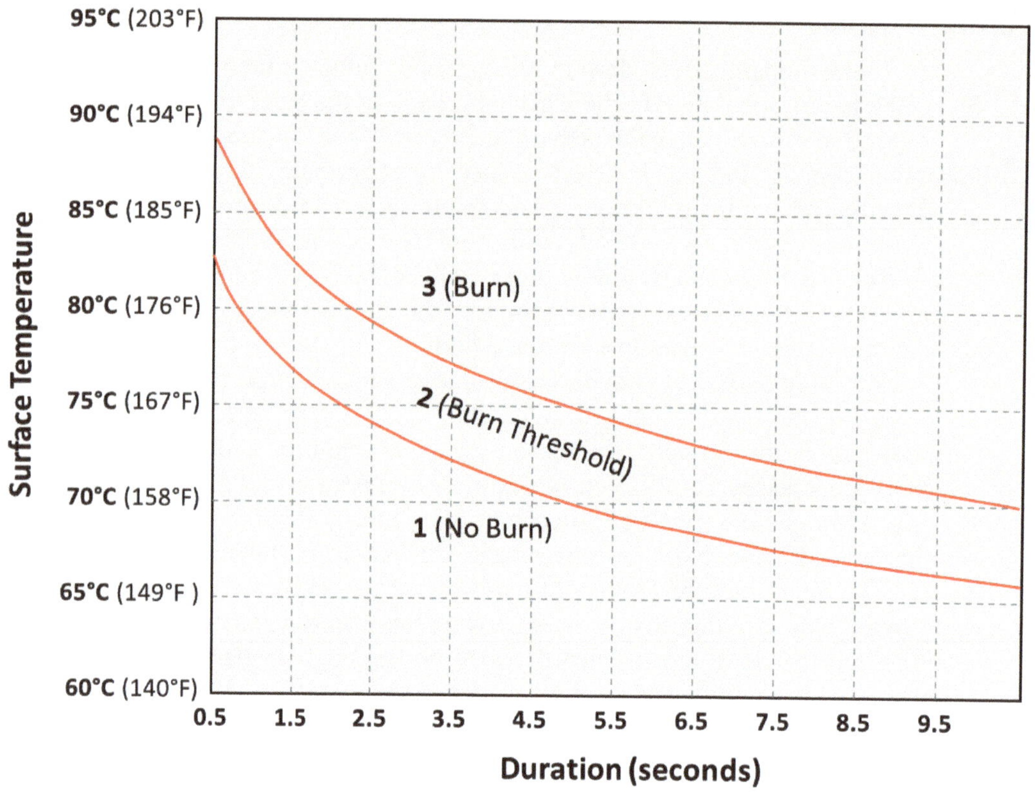

Figure 3.1 Burn threshold spread when the skin is in contact with a hot smooth surface made of ceramics, glass, and stone materials.

Courtesy DLB Associates

pensated radiator heating circuits, schools require surfaces to be limited to 110°F (43°C). This may be achieved by fitting low-surface-temperature covers over standard radiators or operating at 110°F/90°F (43°C/33°C) flow and return temperatures. Besides the safety benefit, such lower temperatures reduce pipe distribution losses and improve plant efficiency.

Care homes and hospitals usually maintain all exposed heater surfaces at or below 110°F (43°C). Temperatures above and around 110°F (43°C) will begin to cause skin damage if the skin contact is of sufficient duration (CIBSE 2015).

Patient Reaction Times

Human reaction times to stimulus have been the subject of research since the middle of the 19th Century (Kosinski 2013). There is sufficient research and agreed knowledge on the topic that the following table has been derived and included as normative advice within Annex B of BS EN ISO 13732.1:2008 (BSI 2009).

Age, arousal, and state of health are factors that can influence reaction times. These are considered to be the most relevant influences within a hospital environment.

Age. The influence of age is something that engineering design must consider in hospitals, care homes, residences, and anywhere that older people may be present. Simple reaction time shortens from infancy into the late 20s, then increases slowly until the 50s and 60s, and then lengthens faster as the person gets to age 70 and beyond (Kosinski 2013).

Table 3.1 Causes of Touching a Hot Surface (BSI 2009)

Contact Period of Up To	Examples of Touching a Hot Surface	
	Unintentional	Intentional
0.5 s	Touching a hot surface and fastest possible withdrawal following pain sensation without restriction of movement	NA
1 s	Touching a hot surface and quick withdrawal following pain sensation	NA
4 s	Touching a hot surface and extended reaction time	Activation of a switch, pressing a button
10 s	Falling against a hot surface without recovery	Prolonged activation of a switch, slight adjustment of a handwheel, valve, etc.
1 min		Turning a handwheel, valve, etc.
10 min		Use of control elements (controls, handles, etc.)
8 h		Continuous use of control elements (controls, handles, etc.)

NA = Not applicable

Children have a longer reaction time to stimulus. For design, a contact time of 4 s is considered appropriate for young children, and a contact time of 1s is the minimum that should be considered for older children. A contact period of 15 s is considered appropriate for children under the age of 24 months (Table 3.1) (BSI 2009).

Using these contact times, it is clear from the graph that unintentional contact with a hot surface by children results in a burn threshold temperature of approximately 56°C for a bare metal surface, and higher temperatures for ceramic, glass, and coated metal surfaces.

Arousal. *Arousal*, or state of attention, is a factor influencing reaction time. Reaction time is fastest with an intermediate level of arousal, and deteriorates when the subject is either too relaxed or too tense (Kosinski 2013).

Health Impact of Underheating

Hypothermia occurs if the core temperature of an individual falls to approximately 95°F (35°C) from the normal body temperature of 98.6°F [37°C]). A person with hypothermia will often not realize they are in trouble, act counter to common sense, and further precipitate cooling (CIBSE 2015). They may also become aggressive, making treatment difficult as they become detached from reality.

The World Health Organization indicates that no hypothermia risk is expected between 64.4°F and 75.2°F (18°C and 24°C) for the average person, although 68°F (20°C) is the minimum recommended temperature for the very old and the very young. Temperatures below 53.6°F (12°C) may also pose a health risk for preschool children and the elderly, the sick, and the disabled, particularly when bedroom temperatures fall at night (WHO 1987).

A review of relevant literature concluded that conditions below 60.8°F (16°C) and above 65% relative humidity impose additional hazards, particularly from respiratory diseases and allergic responses to molds, fungi, house dust, and allergens from domestic animals (CIBSE 2015).

HUMIDITY AND HUMAN COMFORT

Problems with High Humidity

At cold temperatures, high relative humidity is not an issue as the body seeks to conserve heat. Though there many reasons to keep relative humidity below 60% (e.g., microbial, hydrolysis, and/or material preservation), higher relative humidity reduces evaporative cooling from the skin and is beneficial to mucous membranes, including the eyes.

The thermal insulation of clothing and bedding is strongly influenced by the moisture content of the materials used. Damp materials associated with high humidities lose much of their thermal insulation (CIBSE 2015).

High humidity in cold climates can also cause significant issues if the building envelope and systems are not designed properly. As noted previously and in more detail in Chapters 8 and 9, the condensation from humidified spaces can cause significant damage to building envelopes as well as promote mold growth, which can create significant indoor air quality (IAQ) issues.

Problems with Low Humidity

In cold climates, absolute moisture can be very low even though relative humidity can be high. In these situations, complaints of dryness are often raised. Investigations show that there is a correlation between low room humidity and symptoms associated with dryness and irritation of mucous membranes. Low room moisture content may increase evaporation from mucous membranes and can produce microfissures in the upper respiratory tract that may act as sites for infection (Leinster and Mitchell 1992).

At design temperatures for sedentary occupancy, room relative humidity should be above 40%. Lower humidity is acceptable for short periods. Relative humidity of 30% or below may be acceptable, but precautions should be taken to limit the generation of dust and airborne irritants and to prevent static discharge from occupants (CIBSE 2006).

Note also ANSI/ASHRAE/ASHE Standard 170 (ASHRAE 2017) revisions regarding low humidity levels in medical facilities. These advise that humidity can be as low as 30%.

Static Electricity

Static electricity shocks are unlikely with humidities above 40% rh, or at lower humidities if particular materials are selected to prevent the buildup of static electricity. Static electricity can lead to shocks when occupants are not adequately grounded via their shoes and the floor covering. The incidence of electrostatic shocks depends on the electrical resistance of the floor covering and the speed and distance walked. The resistance is a function of the floor material and its moisture content (CIBSE 2015).

Some types of carpet can become highly charged at low room humidity, creating an environment for electrostatic shocks. Typical body voltages are shown in Figure 3.2 as a function of room percentage saturation.

The CIBSE Guide for Environmental Design reports that women are more sensitive to shocks than men. Carpeted buildings with underfloor heating can have particularly dry carpets and require humidity to be above 55% saturation to avoid electrostatic shocks (CIBSE 2015).

Carpeted spaces heated exclusively with air still have floors exposed to short-wave sunlight, which will easily raise the carpet temperature higher than if the floor had been heated with embedded pipes. Most heated floors in modern high-performance buildings in cold climates can have temperatures 20°F (12°C) less than core body temperature, meaning they are actually cooling the floor when exposed to short-wave radiation. It is likely that the same carpet in the same environmental conditions but heated exclusively with air would have the same issues.

Electrostatic discharges can be dangerous in the presence of hazardous gases or explosive substances. In these conditions, an atmosphere above 65% rh is considered safe (CIBSE 2015). In cold to severe-cold climates (> –40°F [–40°C]) such high levels, though electrostatically safe, would cause aggressive condensation even on high-performance glazing systems.

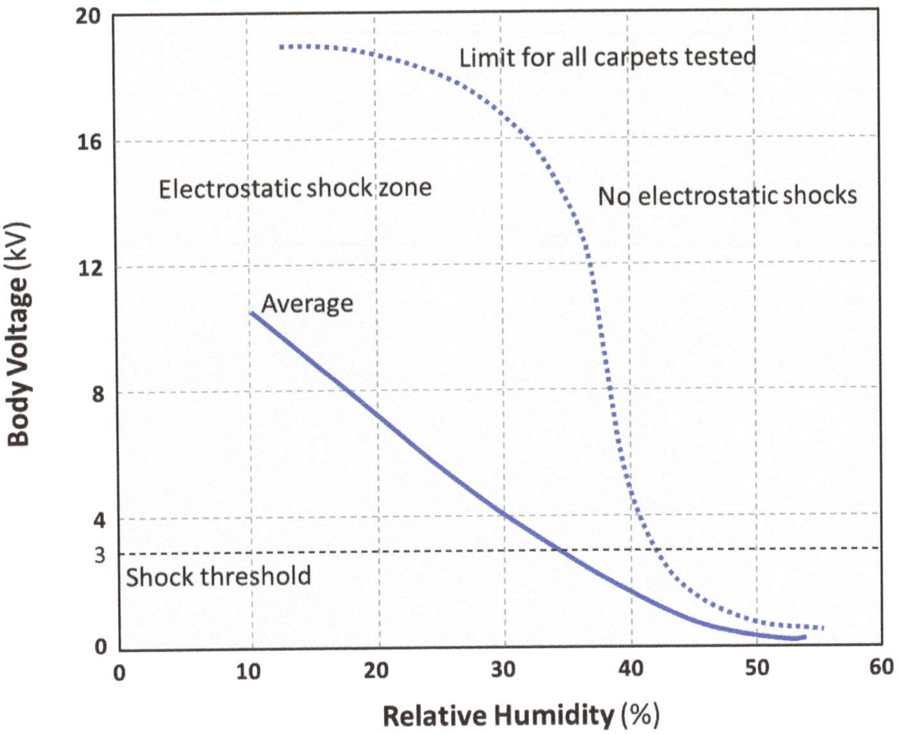

Figure 3.2 Relationship between relative humidity and electrostatic shocks.
Courtesy DLB Associates

Manufacturing machinery may encounter electrostatic problems at relative humidity below 40%. Humidification can be used to eliminate this effect and protect both the machinery and its operator.

EXTREME DESIGN CONSIDERATIONS

In extreme cold climates, it is recommended to locate all equipment and serviceable devices within the conditioned space rather than outside to ease the human factors of maintenance, as well as prolong equipment life. Snow and ice can create hazardous conditions for accessing the equipment.

Humidification in extreme cold climates and remote locations is not recommended unless specifically required for an industrial application such as a data center or in medical applications such as an operating room. Where provided, the design team needs to carefully detail the building envelope design. Additional extreme design considerations for humidification are covered in Chapter 9.

HOW COLD IS IT?

It is so cold ... we had to kick a hole in the air just to breathe!

4 Utilities in Buildings

This chapter only focuses on cold-climate concerns for utilities close to the building, including electricity, fuels, natural gas, propane, water, and sewer services.

ELECTRICITY

Electricity in cold climates can be very expensive because of the difficulties in generating and transmitting power. Many remote, cold locations are not on a grid and rely solely on a local station for power generation. The most common generating means are hydroelectric and diesel. Typically, the costs for environmental cleanup at diesel generation sites is on par with fuel costs; however, at remote sites the costs could be significantly higher and should be considered when selecting a power source. Natural gas and propane are becoming more viable and cleaner options for heating and power generation as long-term storage options improve.

Wind and photovoltaic generation can be considered in certain cases, especially where costs to generate electricity are very high. A rapid fall in panel prices and improvements in control/converter technologies have significantly changed the economics of these technologies. These types of power generation are generally used to aid in powering a single site, not an entire community. A life-cycle cost analysis should be performed before selecting wind or photovoltaic. The electrical loads in a facility should be trimmed to minimum before selecting a solar- or wind-powered system. For example, use propane or natural gas instead of electricity for stoves, water heaters, and clothes dryers.

Power quality voltage fluctuations and outages are typical. Ice, snow, wind, and wildlife can damage distribution lines, resulting in power failures. Sensitive electronic equipment may need to be carefully protected. Not all types of service (e.g., 347/600 volt) may be available in certain communities.

Consider the reliability of electrical power supply when designing a building. If building heating is entirely electric, a backup heating system is needed to prevent freeze-up. Even if a building is heated by oil or gas, if critical components or activities occur in the building, a backup diesel generator sized for part or full load may be required.

It is important to know the correct current electricity rate in the community. There may not be any demand charges, merely a usage rate. The cost for electricity in small communities can be up to ten times the rate in cities, so minimizing electrical consumption is desirable. However, and counterintuitively, lowering usage may not always result in cost savings, because the local generating station may still need to keep the same number of generators running.

Figure 4.1 Power line failure caused by ice storm.
Credit ziggy1, iStockphoto.com

Emergency power should be provided where required by code or necessity. However, if there is a generator, standby power is required so that the generator can start in case of power outage. Consider shunt tripping the building load from the generator when/if the fire pump has to start up while maintaining only the essential building loads. To make generator testing easy, it may be worth considering an artificial load bank. In lieu of a generator, consider a CHP unit, which can be used for construction power and then incorporated into the constant building power system.

Lighting in cold-climate buildings must be designed for the darkest days. However, daylighting can be used during times of the year when more light is naturally available. The lighting system should be designed so that lights not required during bright times can be shut off. Do not use photovoltaic panels to provide electricity for lighting when daylight can be used directly through windows, light tubes, or skylights. However, remember that skylights and light tubes are not suitable for use in sleeping rooms in summer unless they have some type of adjustable shading.

Consider equipping pumps and fans of more than 1 hp (0.75 kW) in size with variable-speed drives to save power. Maintenance costs on fractional horsepower variable-frequency drive equipment can negate any operational savings.

Electrical alarm systems are useful in remote and cold locations for fire, security, and mechanical breakdown. Alarms issue a warning, preferably before any major damage occurs, and allow maintenance staff to address the problem before it worsens. Typical alarms include fire alarm, low building temperature, high sewage level (in the building sewage tank), and low domestic water (in the domestic water tank).

Power supply to buildings is generally provided through overhead wiring. In some cases, however, safety considerations require buried service and teck cable to be used. In non-permafrost areas, the cable should be installed below the frost line, especially if the soil is susceptible to frost. In permafrost areas, or where the cable must be placed in the active layer, the cable should be surrounded by non-frost-susceptible soil and the screened sand bed be surrounded with gravel to provide good drainage.

In many parts of the world, grounding of electrical systems is made by connection to municipal water mains. This is not always possible in remote, cold locations. If there are community water lines, they will be insulated, with metal not in direct contact with the ground. The electrical resistance of the ground in arctic or subarctic areas is extremely variable. Other grounding options include welding to the piles of steel foundations, rod electrodes, Ufer (concrete encased electrode) grounds, or plate electrodes.

Notwithstanding the cost of electricity, there are several applications peculiar to cold climates that may require electricity:

- Electrically heated plumbing vent stacks with self-regulating cables used to prevent freezing of vents in extreme cold
- Electric heating jackets for propane tanks to ensure that sufficient vapor pressure is maintained to supply the building
- Tracing of sanitary or water lines
- Artificial heating of foundations to speed freezeback cycle
- Heating cables under steps, walks, and drives for safety and accessibility, or emergency evacuation routes

FUEL

The type of fuel available, ease of transport and storage, and cost are key considerations in design of building mechanical systems. Common fuels include natural gas, propane, fuel oil, or biomass.

The availability and cost of these fuels can vary with location. Be sure to check with the building owner and local utilities on the availability of fuel types and pricing early in the design process.

The availability and associated cost of these energy sources should not only be evaluated on a cost-per-therm (cost-per-megajoule) basis, but also as part of an overall equipment life-cycle cost analysis that takes the overall efficiency of the equipment into account. An example of this would be a situation where the local propane may be more costly than fuel oil per therm (megajoule), but the ability to have the heating plant as a 95% efficient propane-fired condensing boiler compared to a 87% efficient oil-fired boiler may make propane the more desirable long-term choice.

Another consideration in fuel choice is greenhouse gas emissions. This may drive the fuel choice from a fuel oil to propane or biomass.

Choice of fuel can be a critical decision for those projects that need to meet energy performance requirements. Different fuels have significantly different properties, efficiencies, and equipment requirements:

- High-efficiency condensing equipment is readily available for natural-gas- and propane-fired equipment. Though there are some fuel-oil-fired condensing pieces of equipment, they tend to be smaller in size and require more regular maintenance than their gas-based equivalents.
- Fuel-oil-fired equipment in general requires more maintenance than gas-fired equipment.
- Biomass is becoming more popular and is discussed more in Chapter 2. The larger equipment and supporting storage-space requirements for biomass needs to be taken into account early in the project space planning stage.

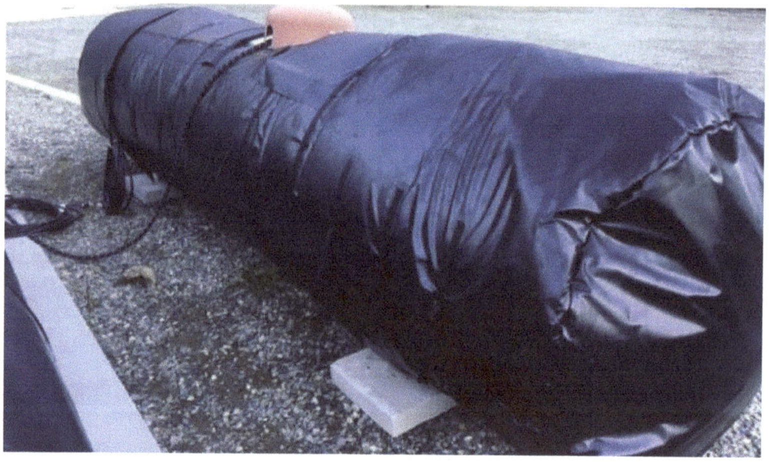

Figure 4.2 Insulated covers on exterior propane tanks.
Erich Binder

In some situations, backup heat sources are required if the primary fuel supply fails or becomes unavailable. This is more common in high-seismic locations where natural-gas distribution may be disrupted; for example, critical infrastructures such as hospitals normally fueled by natural gas may require backup tanks of fuel oil or propane. This may also impact the selection of fuel-fired equipment that can be quickly and easily converted to the backup fuel when the need arises.

Cold-Climate Considerations for Fuel

The following sections discuss cold-climate design considerations that should be taken into account for different fuel systems.

Propane and Natural Gas

Propane liquefies at –44°F (–42.2°C). In locations where outdoor temperatures can drop below this temperature, the tanks can be buried or insulated and heated to keep the propane in its gaseous form.

Regulators are commonly used in both propane and natural gas systems to reduce the pressure before it is supplied to a piece of equipment. The moisture level in these gases can vary from utility to utility.

In extremely cold temperatures, the moisture can condense when exposed to the colder aboveground temperatures and accumulate at the regulator, causing operational issues. Therefore, pressure-reducing regulators may need to be installed inside the heated envelope or be heat traced.

Fuel Oil

Fuel oil is generally available in grades No. 1 and No. 2. No. 1 fuel oil is more refined, less viscous, and has less heating capacity per unit volume than No. 2 fuel oil. At low temperatures, any fuel oil can have viscosity issues. No. 2 fuel oil is better for low-temperature use and may be the only fuel delivered in some regions during the winter months. Additional chemicals may need to be added to the fuel in extreme arctic conditions. The designer should be aware of the various performance impacts on equipment using fuel other than No. 2 fuel oil. The designer must verify performance differences with the manufacturer for different grades of fuel oil. This can include a derating of heat output in heating equipment and power generation in generators.

Figure 4.3 Natural gas service coupled with pressure-reducing regulators.
Erich Binder

Figure 4.4 Insulated and heated outdoor fuel oil storage tank.
Erich Binder

Having cold fuel oil delivered to equipment can be detrimental to equipment operation. It is recommended practice to allow fuel oil to heat up before arriving at the fuel burner. Strategies to accomplish this include the following:

- Installation of a day tank in the mechanical room to allow the fuel to mix with room temperature fuel. Consider routing fuel oil return lines back to the day tank to assist in preheating incoming fuel.
- Use of single-pipe deaerators at the burners that mix warm returning fuel from the burner with the new, cold fuel oil at the device.
- Installation of looping segments of pipe in the warm mechanical room to increase heat transfer prior to the burner. Note that moisture will condense on these surfaces if there is much humidity in the mechanical room.
- Preheating incoming fuel with fuel heaters. This is more commonly seen on high-capacity, industrial applications such as incinerators. Care must be taken to not overheat the fuel oil.

Fuel oil storage tanks can also accumulate water condensation through their atmospheric vents. Water will settle in the bottom of the tank, where the fuel supply drop tube terminates. Water in fuel oil not only impacts on the burner firing efficiency but may also lead to freezing of fuel lines or fouling of devices in the fuel delivery system. Regular maintenance is required to check for and remove water in storage tanks. Consideration should also be taken of fuel filters that remove water from the fuel oil prior to the equipment.

Sizing of Fuel Storage Systems

When fuels other than natural gas are used, and fuel is delivered and stored at site, the proper sizing of storage is important. Storage quantity is based on system cost, frequency of delivery, storage area size, and need for extra allowance in case of weather or other emergencies.

Propane, oil, and biomass are fuels often used in cold climates. Although these may be stored in bulk at a community depot or tank farm, a certain amount of fuel is required on site, usually up to two weeks' supply. A two-week supply, calculated at continuous maximum operating load (including heat and standby power), is a reasonable minimum capacity to be provided in buildings where a standby generator is to be installed, when a long disruption in heating would require relocation of residents, or when an outage could damage essential equipment. A one-week supply, calculated on continuous maximum operating load, is about the minimum amount of fuel storage for a nonessential building, where the building could be closed and plumbing drained down if needed (GNWT 2013).

WATER

The danger to water and sanitary lines and equipment in cold-climate locations is freezing and splitting from the expansion of water as it freezes.

Potable water supply may be supplied from a town or central system through buried or aboveground piping. In smaller communities with no piping system, water may have to be hauled by residents themselves or trucked to tanks in individual buildings.

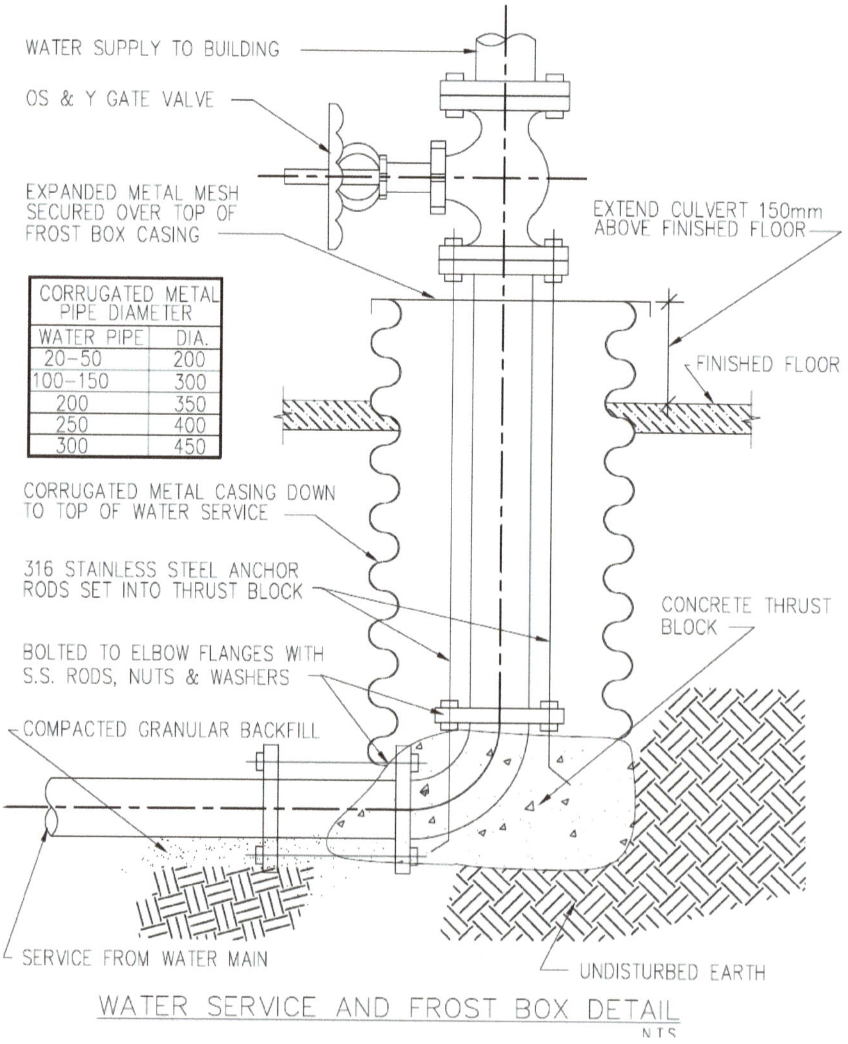

Figure 4.5 Underground water service detail.

Courtesy Erich Binder

Service water heating may be performed by the heating boilers, if the service water load is not required all times. When calculating the amount of service water required, bear in mind that even recent water estimates may be on the high side, and that some remote communities (traditionally) use less water than southern ones. For example, school showers in remote communities are generally not used at all except to store hockey gear.

Consider the use of low-flow fixtures for water conservation, except at end of runs where it is desirable to have regular-flow water closet(s) to ensure good water flow. When sizing water heaters, check the lowest annual anticipated incoming water temperature, which can be quite low.

If the building is likely to be unoccupied or rarely used during very cold times (for example, a school over winter holidays), it may be necessary to provide some water flow to prevent freezing pipes.

Piped Water

Water Supply Freeze Protection

In piped systems, freeze protection can be provided by heat tracing, insulation, heating of water, recirculation, or water wasting. Although aboveground piped water and sanitary systems housed in utility corridors (or *utilidors*, raised and insulated housings built to contain water, sanitary or other utility piping) have been used in the past, buried piping is now more common.

Buried piping is not subjected to extremes of ambient temperature, is safer from vandalism, and does not interfere with roads and property lots. However, if a utilidor must be used, it is better to run individually insulated lines than use a box utilidor (in which sev-

(a)

(b)

Figure 4.6 (a) Indoor storage tank and (b) wall fill cap.
Erich Binder

eral lines are housed together), because the box has too large a surface area to be insulated economically.

In extremely cold regions, permafrost may exist. *Permafrost* is defined as soil that is at or below the freezing point of water (32°F [0°C]) for two or more years. Permafrost may be expected if the mean annual soil surface temperature is below 32°F (0°C). It is important to keep permafrost frozen to retain integrity of the building foundation and any services attached to it, such as water, gas, or electrical lines, which could be damaged or ruptured during building movement.

To prevent freeze-up in any piped system, energy available above freezing must be greater than heat loss from piping system (NMIC 2017).

$$Q_t = Q_{rw} + Q_{as} + Q_{au} + Q_{ldt} + Q_{ldu} + Q_{lc} + Q_{lwwt} + Q_l$$

where

Q_t	=	heat loss or gain
Q_{rw}	=	heat of raw water
Q_{as}	=	heat addition at source and in distribution
Q_{au}	=	heat input by users
Q_{ldt}	=	heat loss during treatment and distribution
Q_{ldu}	=	heat loss during use
Q_{lc}	=	heat loss during collection
Q_{lwwt}	=	heat loss during treatment and disposal
Q_l	=	heat for change of state

For ice to form in flowing water, the water must be supercooled to several degrees below freezing. However, ice will form soon at 32°F (0°C) on any pipeline features such as valves.

Design freeze-up time is time before a system water reaches freezing point. As many cold climate residents know, smaller diameter pipes freeze first. (To thaw a frozen water line, steam or hot water is used.)

To prevent freezing, the following measures are often used:

- Water is heated at the treatment plant (because incoming water may be very near 32°F [0°C])
- Town water piping systems are designed to keep water moving via looping mains
- Building incoming domestic water lines are provided with a recirculation line and pump to move water back to the main
- Piping is insulated
- Piping may be heat traced
- Water may be bled from lines (however, this is wasteful and costly)

If there is no permafrost, and piping can be installed below the frost line, freeze protection may not be required except where piping rises through the active layer.

Heat tracing is often used on water and sewer piping, especially on piping to and from raised buildings. The heat tracing supplies enough heat to offset losses and maintain a minimum temperature in the fluid so it does not freeze. Hydronic heat tracing is generally more economical than electric, but if there is no hydronic system, electric self-regulating temperature heat tracing may be used. When using heat tracing on polyethylene piping, ensure the piping is rated for that use.

On water recirculation lines back to the street, use heat tracing as backup, activating on loss of flow, in case of power outage. This prevents freeze-up when the circulation pump fails. It is also a good idea to have a pilot light to show when heat tracing is operating.

For aboveground piping design, use the lowest anticipated air temperature. For buried piping, use the minimum ground temperature. Ground temperature is best determined from actual measurements, but can be estimated using soil properties and climatic data. Mean annual ground surface temperature is generally from 4°F to 15°F (2°C to 8°C) warmer than the annual air temperature. However, the ground temperature on various surfaces varies with the air temperature and type of surface.

Equations for determining heat loss in steady-state condition buried insulated or uninsulated pipe are dependent upon the geometry of the piping and soil properties. They can be quite complex. A very good reference is *Cold Regions Utility Monographs, 3rd Edition* (Bernard 1970), which gives equations and examples for a number of situations.

It is important to note that the actual ground temperature has a "delay" factor compared to the air temperature. Average air temperature varies as a sinusoidal function. So does the ground temperature, but it is delayed by weeks or months, depending upon the type of ground. A site may have its coldest air temperatures in February but the ground may be coldest in May.

Figure 4.7 shows a typical "trumpet" curve for annual ground temperature variation with depth in a permafrost environment. At some point, ground temperature is constant all year (depth of zero annual amplitude), but that temperature is below zero. If the location were not in permafrost, the depth of zero annual amplitude would be above freezing. The trumpet curve is actually an envelope including all the "whiplash" curves that show soil temperature at a particular time (see Figure 4.8).

Depths of freeze and thaw can be calculated using modified Berggren equations (Rice 2008).

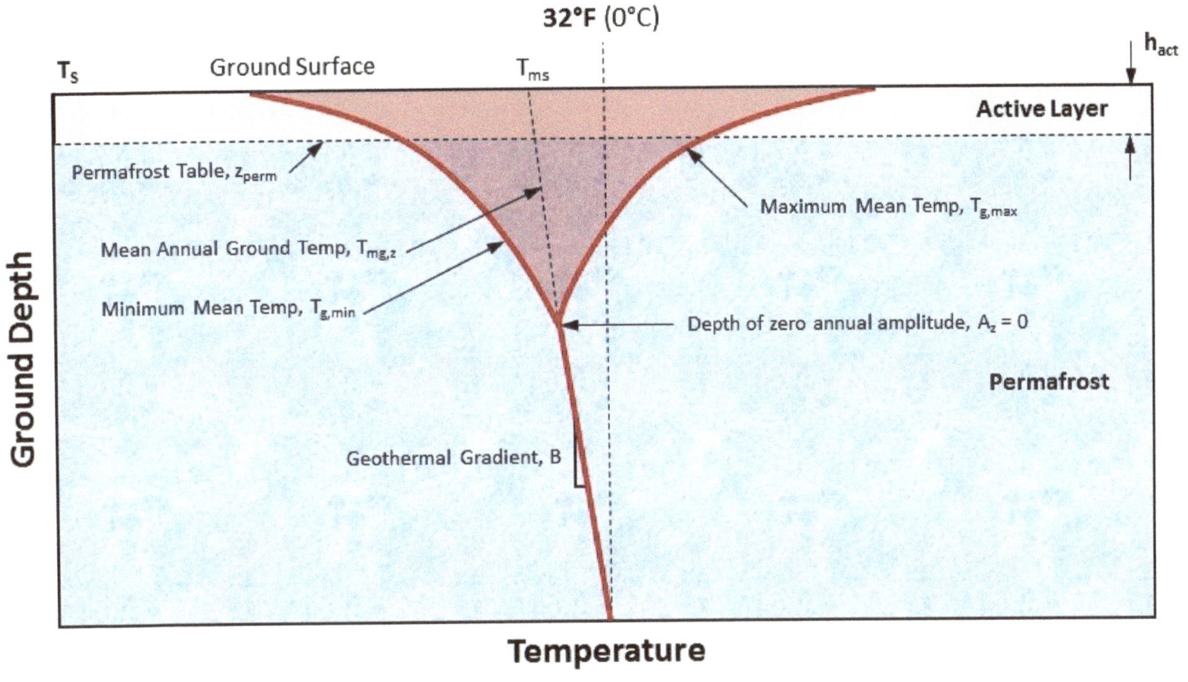

Figure 4.7 Trumpet curve for ground temperature in permafrost region.
Courtesy DLB Associates

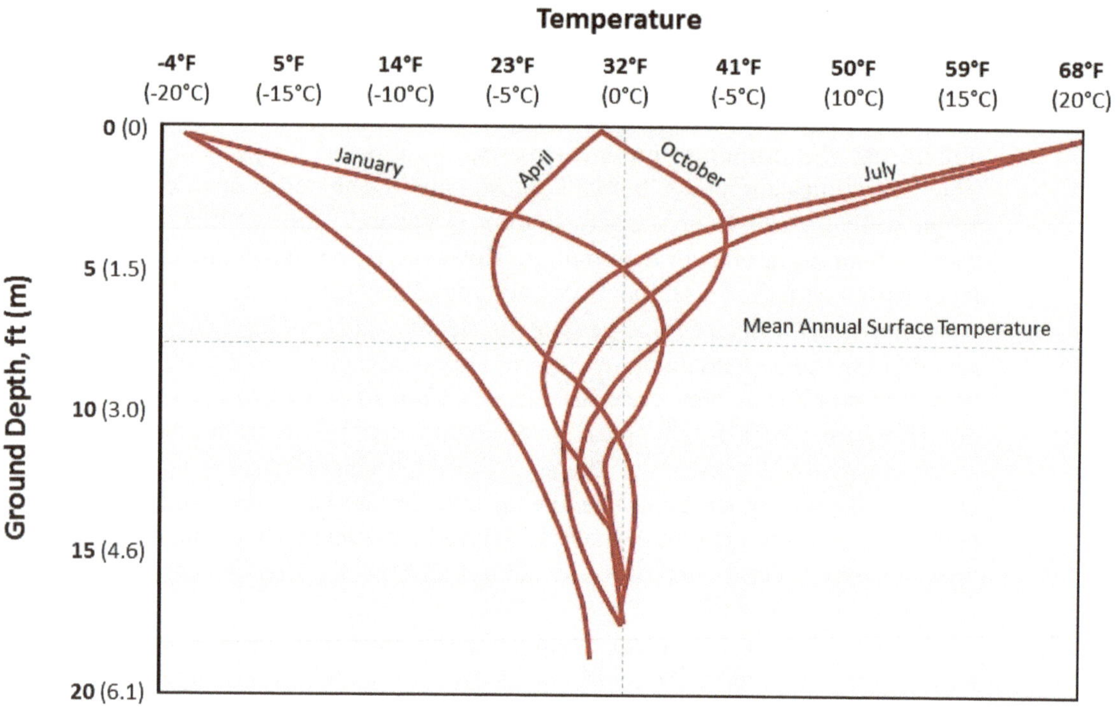

Figure 4.8 Whiplash curves of ground temperature.
Courtesy DLB Associates

Sizing of Water Supply

Minimum domestic water quantities and pipe sizes are determined by code; however, a larger water supply should be allowed for buildings classified as emergency evacuation centers or those buildings in which water shortages would require inhabitants to be relocated. Note that some buildings, such as hospitals, may require two sources of potable water by code.

For a piped water system, size the cold-water recirculation pump and recirculation line back to main. To prevent pipe freeze-up, water temperature should not drop below 33°F (0.5°C), which allows a safety factor above the actual freezing temperature. Check ground temperature curves before designing.

Note that minimum ground temperatures may occur weeks or months after the coldest times of year; the recirculation pump should not be turned off arbitrarily based upon outdoor temperature.

If a water, recirculation, or sewer pipe does freeze up, the heat tracing alone will take a long time to thaw the ice. (Homeowners may resort to use of a hair dryer with good success.) Do not use electric welders or other open-flame devices, because they can cause a fire. A good solution is to use a thaw pump or an electric thaw device. Steam may also be used. In cold climates, there is usually a local company or two with a steam or hot-water thawing system.

Example: Insulated Pipe in Air.

$$q = Q / L = \Delta T / R_p$$

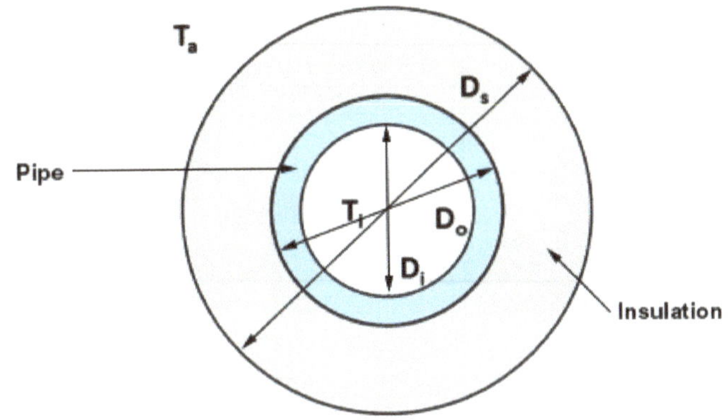

Figure 4.9 Insulated pipe in air.

Courtesy DLB Associates

where

q	=	heat transfer from pipe per unit length
Q	=	heat transfer from pipe
L	=	length
ΔT	=	temperature difference between warm and cold side of pipe
R_p	=	combined thermal resistance of inside film, pipe, insulation, and outdoor air film

For practical purposes, the thermal resistance of the inner film is very small, as is the resistance through the pipe wall, and both can be ignored.

$$q = \pi(T_f - T_a) / [(\text{lend}_s - D_o) / 2k) + 1 / (h_s D_s)]$$

The surface heat transfer coefficient h_s is a function of the surface temperature of the outside insulation surface, which is not known.

Therefore, the outside surface temperature must be assumed. That value is used to calculate other properties. The calculation is repeated substituting in the new value (of surface temperature) until the answer and the assumption coincide.

Spreadsheets or calculators perform this task easier than by hand.

Example of Parameters for a Pipe Heat Loss Calculation.

D_s	=	outer diameter of the insulation
D_o	=	inner diameter of the insulation
D_i	=	inner diameter of the pipe
Pipe outside diameter D_o	=	4 in. (0.1016 m)
Fluid temperature T_f	=	140°F (60°C)
Ambient temperature T_a	=	0°F (–17.8°C)
Insulation thickness	=	$D_s - D_o$ = 0.0635 m (2.5 in.)
Insulation outside diameter D_s	=	9 in. (0.2286 m)
Insulation thermal conductivity	=	0.023 Btu/h·ft·°F (0.0398 W/m·K)
Surface heat transfer coefficient h_s	=	2.13 Btu/h·ft^2·°F (12.1 W/m^2·K)
q	=	24.1 Btu/h·ft (25.1 W/m)

Calculations of heat loss from insulated buried pipe are more complicated due to the need to know soil conductivity and coldest temperature. It is recommended that computer analyses be used (NMIC 2016).

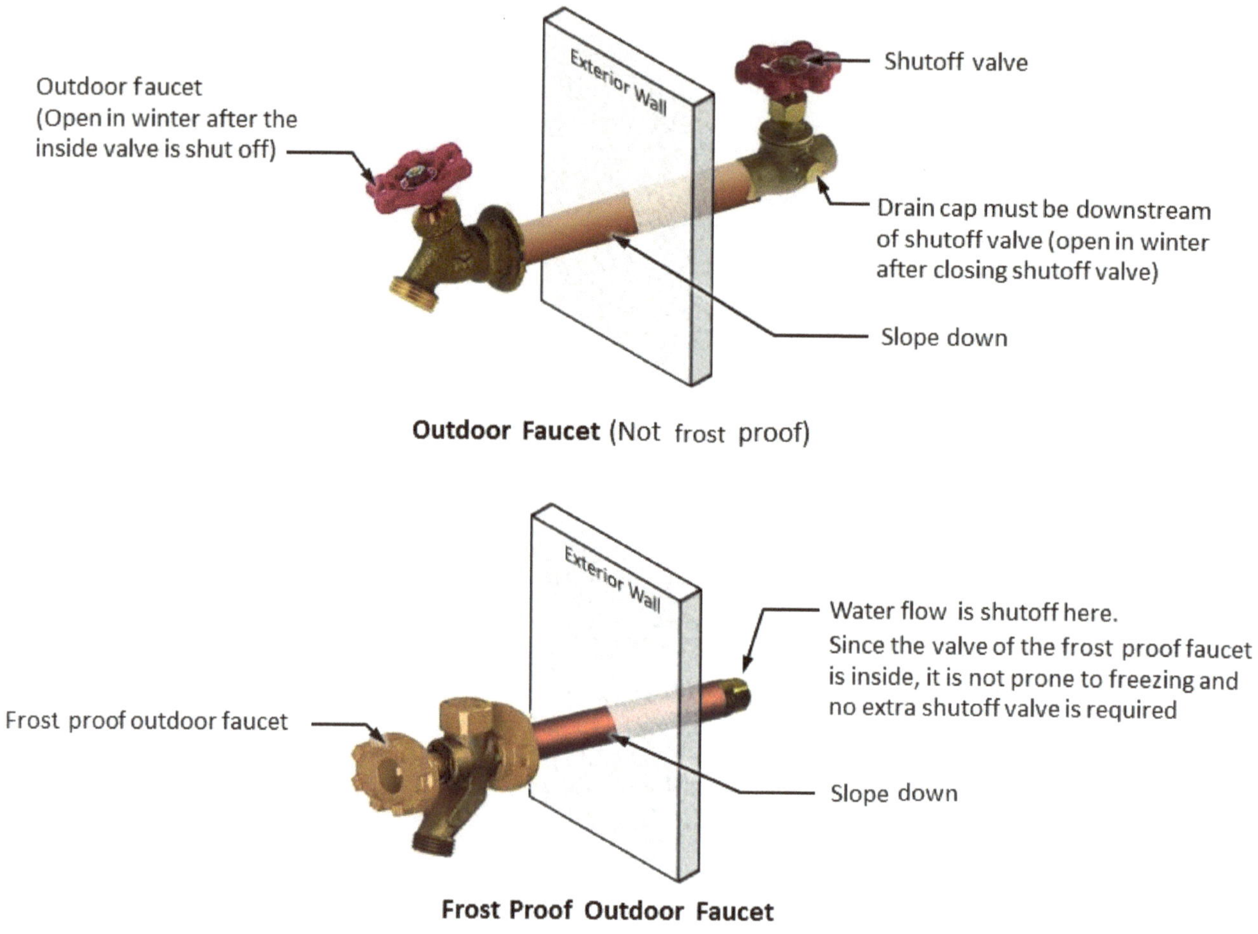

Figure 4.10 Shutoff valves for outdoor faucets.
Courtesy DLB Associates

Trucked or Hauled Water

In many smaller communities, each building has its own water tank. Individuals or water trucks fill the tank at intervals. The water fill location should be such that the water fill is well separated from and the person filling the tank does not have to pass the sewage tank pump out. Both connection locations should be clearly labeled. It is best to provide two separate vents on the water tank in case one ices up. The water tank should also have an overflow pipe to the outdoors.

Freeze Protection. The water tank must be located in a heated area and not have cold air entering nearby (e.g., combustion air). However, water tanks should not be placed in a hot location, such as a boiler room, to keep water fresher.

System Sizing. Water systems can be sized based on past water consumption records, if available, or sized based on occupancy type and frequency of water delivery. The following is an example of some water supply sizing rationale (GNWT 2013):

- Daily usage rates for residential occupancies: 24 gal (90 L)/resident/day (6.6 gal [25 L]/staff/day is sufficient for nonresident staff)
- Daily usage rates for nonresidential occupancies: 6.6 gal (25 L)/person/day

Figure 4.11 Fire hydrant locater—snow flag.
Erich Binder

Figure 4.12 Frosty dry pipe fire sprinkler line and valve.
Erich Binder

To determine water tank size, multiply daily usage rates by the usual number of days between water deliveries in the community. If an infrequent delivery is assumed, the tank will be oversized and more costly. But, more importantly, the tank water will not be refreshed as frequently and may stagnate.

If a building using trucked water requires a wet sprinkler system, a separate firewater tank must be provided. Tankage can take up a lot of floor area. It should be sized early in the design process so it can be considered in the architectural and structural design. Firewater tank capacity must provide the duration of sprinkler water supply required by code or local authority.

Fire Protection—Hydrants

Fire hydrant locations need to be specially marked or flagged, especially in high snow or drifting snow locations.

Fire Protection—Sprinklers

Passive fire suppression options should be considered, especially in remote locations.

Even if building code analysis indicates that a sprinkler system is not needed, safety concerns may prompt the designer or building owner to install one. The local authority having jurisdiction (AHJ) may also require one.

Freeze Protection. Firewater tankage must be located in heated areas of the building. Wet-system sprinkler system piping and heads must be located in heated areas. Individual dry sprinkler heads (on a wet system) should be used in entrance foyers. Glycol loops can be installed in a water wet system in locations where the piping may be subject to freez-

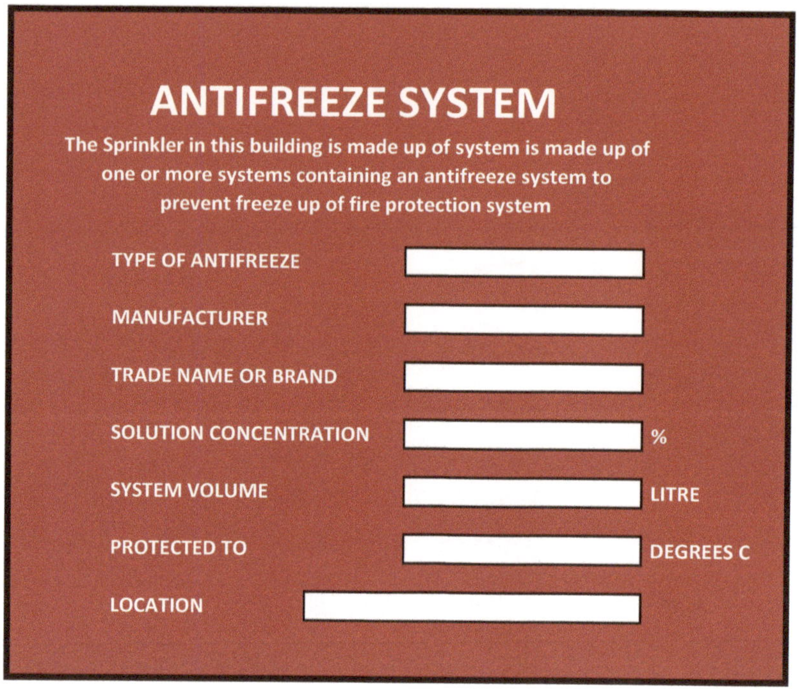

Figure 4.13 Antifreeze sprinkler line tagging.

ing, such as crawlspaces, ambulance drive-in bays, correctional facility entrances, and others.

Entirely dry systems are used in places that are subject to freezing temperatures, such as maintenance garages and warehouses and, even then, only if the size of the building makes the provision of glycol loop(s) excessively expensive. The maintenance requirements for dry systems are more intensive than for wet systems.

System Sizing. The fire protection engineer will design and size a sprinkler system, but the building services engineer will need to design incoming piping or storage tankage. The sprinkler system may not yet be completely designed at this point, so some initial estimation is required.

A piped water system may not require a fire pump, if city water flow and pressure are sufficient. However, a flow test of city water should be performed to double-check at a hydrant nearest the site.

A tanked water sprinkler system will always require a fire pump. Whenever a fire pump is provided, an emergency generator is also needed. In cold climates where construction is expensive, the generator can be housed in a shipping container or other insulated structure separate from the building.

Potable water and firewater tanks must be separate, with a backflow preventer installed so that firewater cannot flow into the domestic water system.

Sanitary

Freeze Protection. Sanitary system pipe slope and required drop are calculated using code slope values. But higher-than-minimum slopes may be desirable to keep sewage flowing and prevent freeze-up in buildings that may have sporadic or little use for long periods of time (such as schools during winter holidays).

Figure 4.14 Insulated aboveground septic tank cold-water pump out connections.
Erich Binder

On the other hand, designers may wish to use the minimum slope in order to avoid installing a lift station in cases where the building system must connect to an existing aboveground utilidor.

System Sizing. For buildings using trucked water, sewage tank size should be 1.5 times water tank size (small buildings), or equal to the water tank size (large buildings).

HOW COLD IS IT?

It is so cold ... we didn't clean the house, we defrosted it!

5 Strategic Design

As with any design, the design team needs to fully understand the guiding parameters of the project. This includes the traditional design guidelines such as weather data, codes, and energy performance goals. In cold-climate designs, additional conditions such as construction logistics, available fuels, and predominant wind direction can be just as important. Though this book is predominantly focused on the mechanical portion of the design, this chapter covers design considerations that affect all members of the design team.

WEATHER CONSIDERATIONS

All aspects of the local weather conditions at the particular project site need to be taken into consideration. Chapter 1 better defines these challenges and their associated design impacts.

When planning your project, the following specific weather-related items should be considered:

- **Geography.** Mountains, plains, and bodies of water can affect the wind conditions of a site, its annual snowfall, relative humidity, and cold-climate-specific conditions such as temperature inversions, which can create significant air quality issues.
- **Wind.** Windblown snow getting inside of building openings or creating snow drifts is a significant design consideration for all cold-climate projects. The predominant wind direction should be identified for all months of the year. Locate mechanical outdoor air intakes, exhaust/relief air discharges, and combustion exhaust with these in mind. This is discussed in more detail later in this chapter.
- **Sun Exposure.** When a facility is located at the more extreme latitudes, sun exposure and its positive physiological effects becomes a high priority in the orientation and layout of the facility. For the mechanical systems, sun exposure also translates to solar gain, which increases the cooling load and overall energy usage during the summer months. Balancing these requirements is an important consideration.
- **Temperature.** Some mechanical systems (e.g., air-to-air heat pumps) do not work effectively when exposed to extreme cold temperatures. Chapter 9 covers mechanical systems, their associated cold-temperature limitations, and tips on how to make them work in colder climates. Cold temperatures can also be detri-

mental to some material properties such as brittleness. Extreme temperatures can create unique design challenges, for example, hoarfrost on HVAC intakes.

- **Heating Degree Days.** The amount of heating and cooling required depends on several factors other than outdoor temperature: how well-insulated a particular building is, the amount of solar radiation reaching the interior of a building, the number of electrical appliances running (e.g., computers that raise their surrounding temperature), the amount of wind outdoors that increases infiltration loads, and what space temperature the occupants find comfortable. Important thermal comfort factors include relative humidity, air and mean radiant temperature, and air velocity. External variables such as precipitation, cloud cover, heat index, building albedo, and snow cover can also alter a building's thermal response.
- **Humidity.** Indoor relative humidity levels have pros and cons when it comes to designing buildings in cold climates. Mechanically increasing relative humidity can have a positive impact on user comfort as well as process considerations such as the reduction of electrostatic discharge. However, higher humidity levels in an improperly designed HVAC system and building envelope can result in surface condensation, mold growth, and failed building materials. This is explained in more detail in Chapter 3. Reference Chapter 8 for building envelope considerations.

CODES

In addition to the different codes that a jurisdiction may use, the designer should know what year of each code is currently adopted by the local AHJ. This can vary widely, even within the same region. The designer needs to contact the local AHJ and identify which codes are currently enforced as well as any local amendments to those codes that may be in place. It is very common for cold climates to amend national and international codes for the local climactic conditions. An example of this is increasing the minimum bird screen size to address frosting of intake screens.

Verify with the AHJ the anticipated codes that will be in place at the time of your finished design. Unexpected code adoptions at the end of a project can result in countless hours of rework, or worse, rejected designs or costly field change orders.

An owner or owner's representative may also contractually require the design team to follow additional design guidelines or codes. This may include

- design guidelines from organizations such as ASHRAE or REHVA;
- owner-developed design or material standardization guidelines;
- sustainability certifications such as LEED or Green Globes;
- occupancy-specific guidelines such as the Facility Guidelines Institute's guides for health care facilities;
- owner's insurance company requirements such as FM Global;
- other design principles such as PassivHaus, NetZero, or the Living Buildings Challenge; and
- additional analysis or activities such as energy studies, life-cycle cost analyses, and commissioning.

Be sure to verify all requirements with clients during the fee negotiation phase and have them either included or specifically excluded from contracts. These requirements can significantly impact design budgets and construction costs as well as the direction taken in selecting the mechanical systems and construction materials for the project.

Local politics and cultural traditions should be taken into consideration. Though local politics are not limited to cold climates, some cold-climate regions have strong cultural identities and traditions that can influence the design of the building. An example of this is the inclusion of saunas in senior housing buildings in indigenous communities for traditional cleansing. As always, the design team should work closely with the owner, local community, and key stakeholders during the planning stage of the project to ensure all opinions are listened to and appropriately addressed.

Anticipated usage of the facility has a significant impact on system design and energy usage. It is not uncommon for community buildings to be used extensively by the public outside of normal business hours.

In small communities, it is common for the school's gymnasium to be in use almost every weekend and week night until as late as midnight. The designer should take this into account when zoning their mechanical systems so that unoccupied areas of the facility can either be turned off or placed into night setback temperature operation. Refer to Chapter 10 for more tips on night setback strategies. The most energy-efficient system is one that is turned off.

ENERGY PERFORMANCE

There are several design guidelines available that provide benchmarks for energy performance, most notably ANSI/ASHRAE/IES Standard 90.1 (ASHRAE 2019b) and *International Green Construction Code*® (ICC 2018). The benefit of these standards over others is that the energy performance goals are specific to a climate region. The energy priorities of a warm climate are different from a cold climate, and so must be the design philosophies and energy benchmarks.The specific strategies for designing for energy conservation in cold climates are covered more in depth in other chapters of this book.

As noted, owner requirements and sustainable certifications such as LEED may require that the building performance exceed the base model performance of these standards by a set percentage. Identify what percentage improvement the team will be achieving at the beginning of the project.

PROJECT SCHEDULE

A complete project schedule must be completed at the concept stage of design and include the following:

- Design
- Bidding
- Permitting
- Material submittal review and procurement
- Construction

Construction drives the design schedule in cold climates. Design and procurement schedules that affect material delivery methodologies or push building envelope construction into the winter can significantly drive up the costs of construction and delay a project until the next build season. Oftentimes, an owner will set the construction window and then work backwards through material procurement, bidding, and permitting with the remainder being the amount of time the team will have to complete their design.

The contract delivery methodology for a project affects the design, bidding, and procurement schedules. The design methodology for a project in a cold climate as well as a more in-depth review of delivery methods are defined further in Chapter 6.

A traditional design-bid-build delivery method will include separate design and bidding schedules, while an alternative delivery method such as design-build or construction-manager-at-risk can go from design right into construction, sometimes overlapping the design and construction schedules through a process known as *phased permitting*.

Phased permitting is where the design team completes an early design package that includes the site, foundation, and all subgrade systems, including plumbing waste and storm water piping. This allows the contractor to begin working on the site in early spring after the ground has thawed. The remainder of the aboveground design can then proceed and be permitted on a schedule based on aboveground material procurement requirements.

Early procurement material packages for long-lead items such as generators and custom air-handling units can also be done to ensure they are on site before the envelope if finished. Building structural steel packages are another common early procurement package.

In addition to the design scope and deliverable milestones, the design team should keep in mind preconstruction regulatory requirements in the schedule including the procurement of environmental permits for site development, utilities applications, and local AHJ building permits. Note that sometimes there are multiple entities that have AHJ control over a project site.

MATERIAL ORDERING AND DELIVERY LOGISTICS

A project can be located in the middle of a large city or in an extremely remote location such as the South Pole. The time for both manufacturing and delivery of materials and equipment can vary significantly by product and project location. The key is to get the materials and equipment to the job site when it is needed to maintain the construction schedule.

All equipment and materials must be delivered to the site. This includes the tools and equipment to complete early site prep and foundation installation. This is a straightforward mobilization exercise for those sites that are accessible via the road system. However, many cold-region projects are not on the road system. Heavy equipment and building materials need to be transported to these more remote locations via water, air, and even seasonal ice roads. Often, it is a combination of these methods.

In today's global economy, it is common for goods to be manufactured thousands of miles away from the project site. The farther the job site is from the factory, the longer it will take to get the materials to the site. This delivery time, which can be several weeks, must be added to anticipated equipment and materials lead time.

For some remote, cold-climate areas, ships and barges are regularly used to get materials and equipment to a job site. This might include oceangoing cargo container ships or even small barges pulled up rivers hundreds of kilometers/miles. Ships and barges have departure dates that are set months in advance. All materials and equipment must be manufactured and delivered to the dock ahead of time to ensure their inclusion in that trip. These delivery dates may be limited to the summer months when rivers and waterways are free of ice. Some communities only receive a couple of barge deliveries a year. If the departure dates are missed, construction and materials transportation costs grow exponentially.

If time is more important than money, many materials can be airlifted to the site. Size and weight will limit what can be transported this way. The designer will need to verify the cargo size and weight capabilities of the local aircraft as well as the airport's ability to accept aircraft of varying cargo capacity. The cost to airlift materials to a site can be very expensive, so the designer should rely on this methodology sparingly, if possible.

Figure 5.1 A truck hauls a prefabricated building module to a remote cold-climate worksite.
Courtesy Matt Stone

In extreme cold climates, ice roads are a common mode of transportation, oftentimes to industrial sites. They allow transportation of materials over soft ground that would normally be impassible during warmer temperatures and should be designed by civil engineers experienced in cold-climate design. Ice roads have limited operational windows that can change based on seasonal conditions, so the designer should verify the anticipated window of operation with the entity in charge of the road.

SITE PREPARATION

Prior to construction of a facility, the site must be cleared and prepped. This may include early design packages or a phased construction process as noted. Preparing the site for construction may be done in early spring or even the prior summer/fall season. In spring construction, the thawing of the frozen ground can be accelerated by clearing overburden and using steam or hydronic heat tubes covered by insulated blankets. Once a layer is thawed, the ground can be excavated and the process repeated until the proper depth is achieved.

In locations that require driving piles into the ground as part of the foundation, this can be completed during the winter months. This is especially true in regions that have wet or exceptionally soft ground conditions, where it is logistically easier to maneuver and operate the pile-driving equipment on the frozen ground.

EXTERIOR ENVELOPE CONSTRUCTION LOGISTICS

Cold-climate job sites typically have a limited "summer construction" window where as much site and envelope construction is completed as possible. It is the most economically viable scenario for a contractor to have a building enclosed before cold

Figure 5.2 Amundson Scott 2009 South Pole construction.
Courtesy Daniel Leussler via Wikimedia Commons

winter temperatures set in. This normally includes the installation of foundations, structural elements, and substantial completion of the exterior envelope, preferably with insulation and vapor barriers.

This allows the interior construction to be completed in a conditioned space during the winter. Many adhesives, mastics, concrete, and other construction materials require specific temperatures to be maintained during their application and curing.

There is also a significant reduction in labor efficiency when working in colder temperatures. This can be caused by restrictive clothing, worker comfort, and equipment operation. These effects are explained in more detail in Chapter 3.

If the building is not enclosed prior to winter, then the contractor might need to tent the structure and provide temporary heat. Tenting a structure commonly includes enclosing the project with vast sheets of heavy plastic, which can be very susceptible to damage from high winds. The plastic also has little to no insulating value, so the cost of heating the facility during this transitional period can be very expensive.

Bear in mind that there is a hazard in using direct-fired fuel heaters in a building that has been enclosed. The large amount of water vapor produced, combined with poor heat distribution through the construction site, can lead to large amounts of condensation forming within the exterior envelope surfaces. More than a few buildings have had early mold and envelope issues created because of this construction process. The use of indirect-fired space heaters are preferred if logistically possible.

The contractor also has thc option of abandoning the site until the following spring. This used to be a more common practice, but current construction techniques allow construction to continue through the winter months. The old adage that "time is money" is especially true in cold-climate construction.

Figure 5.3 Site preparation.

Erich Binder

OTHER CONSTRUCTION LOGISTICS CONSIDERATIONS

Mechanical equipment can have a very long manufacturing lead time, sometimes upwards of 12 to 16 weeks plus transportation. The design team will need to consider the logistics of getting this equipment to the site as well as into the building. If it is anticipated that the equipment will show up later than the exterior envelope is installed, the design team will need to make provisions for their installation, such as removable wall sections, postponing the installation of roof sections until equipment can be lowered in, or providing permanent operable openings that the equipment can be moved in through.

One other methodology is to have that equipment purchased during the design phase by the owner and installed by the contractor (sometimes referred to as *owner-provided-contractor-installed*). Startup and warranty service needs to be included in the contract as this can sometimes be a point of contention with the contractor on material they did not purchase.

The design team should also keep in mind warranty periods for equipment. Traditional warranties begin at the time of equipment startup. If the heating equipment is being used for temporary construction heat, the factory warranty may be mostly expired by the time of occupancy. Extended warranties may need to be included in the contract if a full year of post-occupancy coverage is expected.

Figure 5.4 Tented construction that creates a conditioned space for workers and material curing.
Craig Fredeen

MAINTENANCE AND OPERATIONAL CONSIDERATIONS

Designing around the long-term maintenance and operation of a system is critical to cold climate environments. Chapter 16 covers this topic in depth, but a summary of some of the major points of consideration include the following:

- **System Simplicity.** A simpler system is easier to maintain as well as operate. Match the system complexity to the skills of the operators the owner will employ to run the building.
- **Preventative Maintenance.** Systems should be designed to provide ready access to equipment to easily complete preventative maintenance and annual inspections. Facilitating these activities will increase the equipment life and reduce the chance of sudden failures that can significantly affect the indoor environment.
- **Redundancy.** Replacement parts and service may not be readily available for a project. Redundancy, including standby power sources, should be designed into a project to ensure the facility does not freeze.
- **Property Damage.** When systems fail in cold climates, significant property damage can occur from pipes freezing and bursting. When moisture, whether in gaseous or in liquid form, enters a building envelope, it can create mold problems or permanently damage the envelope insulation. The strategic use of drip pans and area floor drains can minimize the damage from leaks in mechanical

and plumbing systems. Providing alarms for high building pressure conditions will minimize the driving of moisture into the exterior envelope system.

EQUIPMENT LOCATION

The location of equipment can be of great consideration for the long-term operational costs of a facility. In extreme cold climates, the designer should try to locate all mechanical equipment and piping inside the heated envelope. Equipment that is located outdoors will typically have a shorter life span because of the exposure to extreme temperatures. Also, it is much easier to fix equipment and conduct regular preventative maintenance activities inside a conditioned space than outside in the dark, snow, cold, and wind. It is common in very cold regions for client design guides to not allow equipment to be installed outdoors.

The designer should consider how the equipment and its components will be accessed for regular maintenance and replacement. Envelope integrity is crucial to energy savings as well as preventing damage from frozen pipes. Care should be taken not to require the removal of envelope components, such as wall or soffit panels, for regular maintenance procedures or even equipment replacement. Experience shows that envelope components are never sealed up to the same condition as the original construction, if resealed at all.

EQUIPMENT SELECTION

For some owners, the choice of equipment is very simple, because they have set manufacturers that they use. Sometimes this includes specific equipment model numbers. This reduces the amount of parts storage and simplifies staff training, which, in turn, results in improved maintenance and shorter down times when there is a mechanical failure.

When designing around multiple vendors, care should be taken to specify equipment that has replacement parts readily available in the region. The availability of trained technicians should also be taken into account. Remember that in remote locations it can be a considerable cost just to get a technician to and from a project site.

It is common for highly energy-efficient, yet complicated systems to be operated incorrectly and actually use more energy than a simpler system. This can be mitigated to a certain degree with proper owner training and project close-out documentation. This is explained more in Chapter 11.

Provisions can be made to reduce the impact of equipment failures or extend the time until repairs are needed to be completed. This may include the following strategies.

Sizing and Redundancy

A system can have 100% redundancy, such as a set of pumps piped and controlled in a primary/backup configuration. Or, redundancy can be spread over multiple pieces of equipment, such as two boilers each sized at 66% of the total heat load, or three at 50% total heat load. Both methods allow one piece of equipment to be off-line for maintenance or repair while still providing the capability to maintain the building above freezing temperatures at full design conditions.

One strategy for sizing redundancy is to consider the heating load associated with the ventilation system. If one boiler goes down, temporarily turning off the ventilation system in the building is oftentimes all that is required to maintain space heating temperatures with the remaining boiler capacity.

Equipment selection is based on the rare 0.3% or 2.7% design-day conditions. System efficiency and minimum equipment modulation/turn-down capabilities need to be considered for operation during the rest of the year. Providing multiple smaller pieces of equipment, such as three boilers sized at 50%, might provide a more year-round efficient system compared to a single 66% design-day sized boiler. Similar strategies should be taken into account when sizing heating distribution equipment.

Spare Parts

Select equipment so that the same model can be used in multiple locations. This will allow spare units to be kept on hand and readily available for emergency replacement. This may be critical for the heating system, but is also useful for all mechanical, plumbing, and electrical system components. This strategy is also recommended for reducing the number of unique consumables, such as filter sizes and types of light bulbs.

For remote locations, it is not uncommon to require entire pieces of equipment, such as variable-frequency drives and pumps, to be provided as additional materials in the contract documents.

It is recommended that close out documents, such as operation and maintenance (O&M) manuals, include equipment parts lists as well as contact information on where spare parts can be ordered.

Alarms and Notifications

Use alarms to immediately notify personnel of a system failure. This can be done either through the building automation system (BAS) or through stand-alone controls. These types of alarms might be for cold temperatures in a room, a failed piece of equipment, or the identification of a liquid leak from a pipe failure. Consider using automatic texting and remote monitoring services for critical alarms. These strategies are covered in Chapter 10.

MATERIALS

In addition to maintenance considerations for equipment, the geographical location of a site may impact the materials used on a project. The availability of local building materials should be verified. Some remote locations do not have ready access to gravel, so the use of concrete for housekeeping pads or exterior fuel tank pads can be very expensive. The use of steel stands indoors and treated wood timber pads on the exterior may be more cost-effective.

The spatial and weight limitations of shipping can also impact the selection of construction materials and equipment. An example of this is ductwork. Traditionally, spiral ductwork is preferred over rectangular ducts for its efficiency of installation and performance. However, to conserve shipping costs, many materials are packed as tight as possible. Prefabricated spiral ducts and fittings are limited in their ability to nest inside each other and can take up a lot of room in transit containers. It might be more cost-effective to ship rectangular ducts that are broken down into two or more pieces and nested for shipment. Another option would be to use round ducts that have a longitudinal seam. These types of ducts are limited to smaller diameters but can have significant quantities nested for shipment. The shipping costs would need to be weighed against labor installation costs to find the most cost-effective combination.

FUEL

A key aspect of design strategy depends on the availability and cost of energy for the project site. In addition to electricity, this includes fuel sources such as natural gas, propane, fuel oil, and various biomass fuels. The availability and cost of these different fuels can be geographically based. Be sure to check with the owner and local utilities early in the design process regarding the availability of fuel types and pricing. See Chapter 4 for further discussion of fuel choice considerations.

PREDOMINANT WIND DIRECTION

When laying out the mechanical support spaces, keep in mind wind direction(s) that the facility will see during the entire year. The predominant wind direction may change

Figure 5.5 Wind-blown snow in an air-handling unit.

based on the season. For any location with winds, the designer should be mindful of the locations of environmental exhaust, plumbing vents, and products of combustion discharge points (e.g., boiler flues) as they relate to the outdoor air intakes. Care must be taken to ensure that the contaminated air is not reintroduced into the outdoor air intakes either directly via the wind or indirectly through air current eddies. Some facilities install carbon monoxide detectors in the outdoor air intakes to guard against this occurring.

The consideration of wind-driven snow plays a significant role in cold-climate design. Wind-driven snow can enter air intakes, create drifts that cover up mechanical openings, or make mechanical space entrances inaccessible. Architectural and structural concerns with snow drifting should be taken into consideration as well.

It is preferred to not have air intakes facing into the predominant wind direction. Snow can be very light, even granular in extreme cold climates. The snow can easily be entrained in the wind and blown into louvers and hoods when located on the predominant wind direction side of a building. High winds can also create turbulent wind eddies in corners and on the backsides of building surfaces that may result in snow getting inside hoods and louvers. Refer to Chapter 9 for additional information about designing for wind and cold-climate specific outdoor air intake design solutions.

TEMPERATURE INVERSIONS

Cold temperatures can create inversions that limit the natural convection of these pollutants up and away from their sources. This can increase the distance a pollutant will remain at the same elevation, increasing the likelihood of being pulled into a outdoor air intake. The designer should consider separating intake and pollutant sources as far away as possible, both horizontally and vertically.

Figure 5.6 Temperature inversion. The mist from the aerial coolers at this Northern Ontario facility was drifting over the nearby freeway, creating difficult driving conditions.

EXTREME DESIGN CONSIDERATIONS

As the project conditions become colder and the locations more remote, the construction logistics have a much greater effect on the construction budget and schedule. This can easily push a project completion out years after the design is completed.

An important part of construction logistics for all projects is labor. The contractor must be able to hire skilled workers in each trade or bring them in from other communities. The process of acquiring room and board for workers in a remote community can be a significant logistical challenge. On larger projects, modular camps are setup by the contractor to house their workers and construction offices.

Extreme cold and project remoteness makes the issues noted in this section exponentially more important:

- Energy efficiency is very important because of the significant amounts of energy used as well as the typically higher costs of fuel.
- Project schedules are hampered by shorter construction seasons.
- Winter construction costs are higher because of the inefficiency of working outdoors in the cold as well as the cost of providing temporary heat conditions.
- Remote sites affect material delivery logistics and contribute to overall higher costs of construction.

The importance of locating all equipment within the heated envelope becomes more important, to the level where many owners require it within their design standards. Most equipment is not rated for operation in extreme cold temperatures. If equipment needs to be located outdoors, even if it sits idle during the winter, electric strip heaters may need to be added to various parts of the equipment to ensure that the electrical components such as circuit boards do not break under extreme cold temperatures.

Redundancy and ease of maintenance are critical as the time between equipment failure and occupant comfort impacts, or even burst water pipes, is a matter of hours rather than days.

Other physical properties of materials in extreme cold climates should also be taken into consideration. This can include items such as the brittleness of PVC or the poor operating performance of some compact-fluorescent lighting in cold temperatures.

HOW COLD IS IT?

It is so cold ... someone spilled hot coffee on me and I thanked them!

6 Building HVAC Design Process

As with all engineering projects, the first step in the design process is to identify the end goal. For HVAC designs, this means identifying and prioritizing the owner's definition of success. This can vary from client to client and also from project to project. Some examples of parameters that the designer should work with their client to prioritize on a project include the following:

- Performance
- First cost
- Design and construction schedule
- Energy efficiency/sustainability
- Long-term operational costs
- Operational workability
- Design simplicity
- Durability/resilience

The next step of the design process is to identify how the project design and construction projects will be delivered. Examples of types of project delivery methods include the following:

- Traditional design-bid-build
- Design-build
- General contractor/construction manager (also known as *construction manager at risk*)
- Public-private partnership

The role of this book is not to provide an in-depth discussion of each type of delivery method and the associated contractual relationships. For more information about these different types of delivery methods, along with pros and cons, reference *The Project Resource Manual: CSI Manual of Practice* by the Construction Specification Institute (2005).

Under each delivery method, the designer's role and responsibilities may change. The priorities for the design may also shift as well. For instance, a project for which the owner has decided to use design-build as a project delivery method might have more emphasis on a faster schedule and lower first costs.

ROLES AND RESPONSIBILITIES

The roles and responsibilities of each of the design team members should be clearly identified early in a project.

Key Stakeholders

This is a catchall phrase encompassing those individuals who will have input on the design or will be impacted by the end product. This category includes the owner and users of a facility, but also may include other entities such as the public, financial backers, and code authorities.

Owner

This entity is usually the one paying for the project. The owner may elect to have a project management team be their owner's representative that speaks for them during the design and construction phases.

User

This is the occupant of the facility. Sometimes, but not always, this is also the owner. There may be multiple groups in a facility having a voice in the design decisions. This is sometimes referred to as a *user group.*

An important group of users are those individuals who will be maintaining the mechanical systems. They will define the level of complexity, as well as preferences on system design. This group should be included in the project throughout the design process including ensuring that they have an opportunity to review and comment on the mechanical design during the delivery milestones. In the end, they are the ones who will have to maintain the designed system for its expected life (Lewis 2004).

The Design Team

Each of the following entities have their own unique design challenges and considerations when it comes to cold-climate design. It is important that all members of the design team have cold-climate design experience or at least have a consultant on the team who can review and advise them of the design requirements.

Architect. The architect is responsible for the building layout, envelope design, and finishes. One of the architect's important roles, especially in cold-climate designs, is that of defining the size of and access to mechanical spaces. As noted elsewhere in this book, the mechanical systems in cold climates take up a significantly larger footprint in a facility than they do in milder climates, and the use of roof-mounted equipment can be detrimental to the long-term operability and maintenance of a cold-climate facility.

The mechanical systems must be coordinated with the architect at the concept level of the project in order to fully integrate them into the building. There are several cold-climate mechanical system considerations that can be very difficult to implement in the later stages of a project. These are discussed in Chapter 9.

Civil Engineer. The civil engineer defines the site, utilities, location of fuel tanks, and other items that impact the mechanical design. In addition to the coordination of site utilities and storm drainage discharge, the design of the site plays a significant role in the development of ground-source heat pumps and laydown areas for biomass and fuel storage.

Geotechnical Engineer. Due to the varying ground conditions among cold climates locations, the geotechnical engineer plays a critical role in deciding the building foundation type (such as slab-on-grade, pile foundations, or post and pad). Foundations for frost-susceptible and permafrost ground conditions commonly have elevated structures or the use of thermosiphons or active freezeback systems, which affect the mechanical system design.

Structural and Electrical Engineers. The roles, interaction, and coordination of the mechanical design with the team's structural and electrical engineers follow the traditional requirements of other climates.

Commissioning Provider. The use of commissioning in cold-climate designs is very important. It is also important to have the commissioning provider identified and involved early in the design process. Commissioning is covered in Chapter 14.

Specialty Designers. Additional specialty consultants can be involved in a project, based on the usage and complexity. This group may include renewable energy and sustainability specialists, hazardous material/environmental engineers, acoustics engineers, construction cost estimators, and a computational fluid dynamics modeler (flue gas and fume hood plume dispersion, envelope hygrothermal studies, and snow drifting). Other specialty consultants might be unique to a type of facility, such as a lab, medical, or kitchen consultant.

PROJECT DEFINITION

As designers know, every project starts with a significant number of variables. To help focus the system selection and move a design team toward a solution that meets the owner's definition of success, the first step is to define the most common variables.

Five major variables must be defined with the owner:

1. **Budget.** Is there a set construction budget? Or will the needs of the project define the budget, such as in a conceptual planning project?
2. **Schedule.** When does the project need to be constructed and operational? This is a significant variable for cold climates because of time windows associated with construction logistics. For more information on construction logistics, see Chapter 5.
3. **Usage.** What is the building going to be used for and what spatial requirements will be needed to provide that function? Are there future renovations or expansions planned? What is the anticipated daily, weekly, and seasonal usage of the facility? The mechanical systems may need to be zoned to properly address these variables in an energy-efficient manner.
4. **Indoor Environment.** What are the parameters for comfort? These may include room temperatures, IAQ, humidity, and other parameters specific to a project. The requirements for a warehouse are vastly different from those for a hospital.
5. **Life Expectancy.** It is not uncommon for cold-climate buildings to be erected with the understanding they will be relocated, disassembled, or abandoned after a fixed time.

It is highly recommended to work through the project variables with the owner as early as possible. This will reduce owner worries as well as designer's budgets. There are several different mechanisms available to the design team to facilitate these discussions; one of them is a design charrette.

A *design charrette* is a sit-down meeting with the entire design team, the owner, and all of the key stakeholders in a project. For mechanical systems, it is recommended to have representatives from the facility staff and operators who will be maintaining the system present. If the owner has a renewable, energy, and/or sustainability specialist, that individual should also be present. The meetings are formal, run based on an agenda, and followed with meeting minutes to document the direction and design criteria the design team will use in the development of the facility. It is helpful to forward the agenda along with questions ahead of time to allow the attendees to coordinate with stakeholders prior to the meeting.

For the mechanical designer, topics to cover with the owner may include the following:

- What is the owner's comfort level on system and control complexity?
- Are there any specialty systems that are required, such as compressed air and fuel systems?
- What are the owner's lessons learned on past projects in that region? What are their "hot buttons"?
- What design guidelines and codes, as noted in Chapter 5, will be applied to the project?

The design team will need to identify if the client is looking to the design team to select the systems based on the known parameters or if the client would like to be active in the decision process. Where the client wants to take an active role, the design team should review the different solutions available and note the pros and cons of each system along with how they address their design priorities. This often leads to a life-cycle costs analysis comparing the best options before a final system solution is selected.

By the end of the charrette process, the design team and owner should have a clear scope of work for the project and have identified the systems that will be installed.

SITE SELECTION

Whether the designer has the opportunity to have input on the site selection for a project or is given a preselected site, the following items are good to keep in mind.

Utilities

Recovered heat distribution systems from a utility or exterior heat source are one of the least expensive forms of energy. If heat recovery is not developed on the project site, it may be cost effective to extend the heat from a nearby source, such as a power plant. The length and complexity of pipe routes influence whether this is cost effective. Modern buried pipework can travel long distances with little heat loss (about 1.8°F/mi [0.62°C/km] or less), but costs can be high, typically $1 million per mile ($620,000 per kilometer) or more, depending on site factors.

Utilities and fuel availability are an important part of selecting mechanical systems. Extending utilities to a project site can be very expensive and take a large part of an available project budget. Ensure that utilities are sized for any planned expansions of the facility to avoid future disruptions.

The availability of domestic water and sanitary sewer utilities can also have a significant effect on project costs. Where these are not available from a municipal utility, on-site wells and septic systems may be required. Information on surrounding wells will help the design team to anticipate available flow and water quality. If a fire suppression system is required, the lack of a municipal water source may require the installation of a water storage tank and fire pump assembly.

Site Layout

As part of the development of the site, the design team should take into account access to laydown areas for biomass fuel or access to stand-alone systems such as fuel storage tanks or flush-and-haul plumbing systems. The development of roads and parking areas should also take into consideration snow removal and on-site snow storage. Guidance on biomass deliveries and storage can be found in the CIBSE Application Manual AM15 (2014) and in Chapter 2 of this Guide.

Sound should also be considered for adjacent property lines, especially if installing a generator or other significant noise-producing equipment. Some jurisdictions have codified limitations on the acceptable sound levels at property lines.

When developing multiple facilities on a site, the designer should consider whether it is more economical to have a central heat plant that serves all facilities or have a distributed system with separate heat plants in each building. If development of the site is phased, the infrastructure discussions should occur as part of the installation of the first facility. However, building a large energy center at the start can be costly and inefficient, and phased growth must be considered where the energy supply keeps pace with the development buildout.

Mixed-use development is becoming more popular. There are many design challenges and opportunities involved in the design of mixed-use campuses. Because these developments can be very different from each other depending on their size and usage, and typically having a long-term development timeline, which changes as market factors dictate, it is important that a flexible strategy is developed using a sustainability tool such as BREEAM UK Communities or LEED for Neighborhood Development (BREEAM 2015; USGBC n.d.), both of which help engineers to develop schemes allowing for future growth and changes. Energy and water strategies are two examples of aspects they cover, but there are many more issues that engineers must consider.

Future-Proofing

All designs need to take into account known additions and/or renovations on a project, and also make provision for future developments and changes that will inevitably take place. A scheme that is so bespoke that it cannot be updated is not sustainable and may lead to a building that becomes unusable in time. A simple example of this is an office building in the United Kingdom that relies solely on natural gas as its fuel source. When U.K. gas runs out—in two years based on current estimates—the building may have no fuel or may have to pay high costs to have it imported. Designers should allow for future changes to allow other fuels or renewables to be used.

This is known as *future-proofing* the design. The anticipated changes may come in the form of additional square footage, changes in potential programmatic usage, or a phased approach to renovations to existing systems.

To facilitate future-proofing of a facility, the designer needs to take into consideration several scenarios. If the changes are going to require additional HVAC infrastructure, space can be left in a mechanical room for future equipment, such as additional boilers or pumps. Distribution systems should be sized for the additional load. These future equipment locations and points of connection for future extensions should be clearly identified on the construction documents so that the contractor does not inadvertently place a piece of equipment in an open spot, as well as letting future designers know the original designer's intent of how to incorporate the additional infrastructure.

Underground systems, such as interior buried sanitary sewer piping, should be installed at an elevation that anticipates the future expansion. This location and depth should be clearly identified in the documents for future locating and use. The use of building information modeling will ensure that embedded data will identify space set aside for future development.

The design should also be robust enough to allow for unanticipated future use or changes in the concept when the future funding is finally achieved. Technology is a great example of an ever-changing variable that is difficult to anticipate.

If the owner wants to incorporate renewables or alternative energy systems into a facility at a later date, the design of the original facility should be completed with this in mind. A good example is designing the heating terminal units and distribution system for a lower supply temperature that will be available from a future recovered heat system or ground-source heat pump. The space for adding equipment to support the renewables (pumps, expansion tanks, fans, and other equipment) should also be identified.

Another consideration for future-proofing of a facility design is the anticipation of future energy costs in life-cycle cost analysis calculations. Long-term energy costs are very difficult to predict on a monthly basis, let alone years in advance. Considerations should include extensions of lower-cost fuel and energy infrastructure upgrades that may reduce energy costs or make one fuel more desirable over another for equipment selection.

Long-term planning reports may be available from the local utility to assist in these estimates. The best form of planning for the future is developing robust documentation of the final facility systems, including operation and maintenance data for each piece of equipment, as-built drawings noting all changes made during construction, and good documentation of the installed controls sequences. Any provisions for future expansions should be clearly noted in the documentation.

HOW COLD IS IT?

It is so cold ... we had to push the house down the street to jump-start the furnace!

7 HVAC Design Calculations

After the design team has established the goals and parameters to suit the owner's objectives as outlined in Chapter 6, one of their primary tasks is to produce design calculations to determine the building heating, cooling, and ventilation loads to meet either the comfort of the occupants or the maintenance of environmental conditions to suit the requirements of a process or product within the space.

These design calculations also form the basis to size equipment, avoid condensation on building elements, achieve certain space conditions, or predict energy usage. The level of accuracy and calculation method chosen depends on many things: stage of design, project size or complexity, budget, and other factors. Manual calculations or estimating methods based on engineering principles may give good results for simpler calculations, such as estimating the heating load of a single-family home or sizing a pump in a small commercial building. But for larger or complicated buildings, load- and equipment-sizing calculations should be done with spreadsheet or computer analyses.

The science behind design calculations is the same in any climate. ASHRAE, CIBSE, REHVA, SCANVAC, and other groups (e.g., equipment manufacturers) publish handbooks and guidebooks with detailed calculation procedures and methods. Commercial and freeware computer programs are widely available to determine loads, size equipment, and model building conditions during every hour of the year. Although the equations and methods are universal, there are certain things one must consider with regard to calculations for extreme cold design.

These programs can help designers

- ensure that parameters or constants used in rule-of-thumb or equations for moderate climate calculations reflect actual cold-climate conditions;
- determine and manage the presence and migration of moisture indoors and throughout the envelope, avoiding condensation within or on the envelope;
- determine the combined ventilation and infiltration heating loads (which may, in some cases, be larger than the envelope heating load) and ensure infiltration and ventilation loads are minimized;
- consider the long hours of sunlight (or darkness) encountered in far northern or southern regions; and
- give thought to the need for heating plant redundancy or backup, taking into account also the location and logistics involved to repair or replace systems or components.

This chapter provides an overview of HVAC design calculation considerations for cold climates. Because the majority of calculation methods are the same in any climate, it does not review the commonalities but instead focuses on special considerations for cold climates.

PSYCHROMETRIC CHARTS AND PERFECT GAS LAW

How Air Behaves at Different Conditions

Many common HVAC equations and rules of thumb are expressed for standard conditions. However, it is important to be sure exactly what those "standard" conditions are, as definitions vary. The definition may also vary between standards published by the same group or industry. For example, the U.S. standard atmosphere, referred to in *ASHRAE Handbook—Fundamentals*, Chapter 1 (ASHRAE 2021a), is 59°F (15°C) at 14.696 psia or 101.325 kPa. In Chapter 18, another value for standard air is given as having a mass of 0.075 lbm/ft^3 (1.2 kg/m^3). This could apply to saturated air at about 60°F (15.5°C) or dry air at 69°F (20.6°C). AMCA (Air Movement and Control Association International) defines standard air in ANSI/AMCA Standard 210 (AMCA 2016b) as air with a density of 0.075 lbm/ft^3 (1.2 kg/m^3), which are the approximate properties of air at 68°F (20°C), 50% rh, and 14.696 psi (101.325 kPa). Some fan manufacturers define standard air as clean, dry air at sea level and 70°F (21.1°C).

In cold climates, high altitudes, or other nonstandard conditions, air properties are sufficiently different from standard that the familiar constants change.

The perfect gas law $pV = nRT$ applies with good accuracy to dry air, water vapor, and mixtures of the two in the range of temperatures and pressures commonly found in building science calculations. A psychrometric chart can be built for any pressure condition or temperature range. Chapter 1 of *ASHRAE Handbook—Fundamentals* (ASHRAE 2021a) describes psychrometrics and perfect gas relationships for air.

The psychrometric chart is useful in determining the properties of air at certain conditions. It is especially useful for cold-climate calculations to show air dew point in order to anticipate and prevent condensation or frosting in or on the building elements. Once the interior air conditions are pinpointed on the chart, the constant humidity line through this point can be traced across to the saturation temperature line. Examples follow on how to locate the dew point in a building envelope.

Sea level 14.7 psi (101.325 kPa) standard atmospheric pressure charts are published for a variety of temperature ranges.

Cold-weather conditions may also occur at high altitude, where air density is lower and standard-pressure psychrometric charts no longer apply. ASHRAE publishes charts for 2460 ft (750 m) altitude 13.4 psi (92.66 kPa), 4920 ft (1500 m) altitude 12.26 psi (84.54 kPa), and 7380 ft (2250 m) altitude 11.17 psi (77.04 kPa).

Air Density Variation

If the density of air is (much) different from that at standard conditions, this must be considered in calculations. For example, as cold makeup air or combustion air enters into a building, it expands as it warms. This volume change can be substantial and will affect equipment selection. Fans are normally rated at standard conditions. Fans that operate in cold locations with cold air should be selected with consideration of these nonstandard conditions.

Note in Table 7.1 that at certain conditions (e.g., –20°F and 5000 ft [–29°C and 1524 m]) the temperature and altitude corrections cancel out.

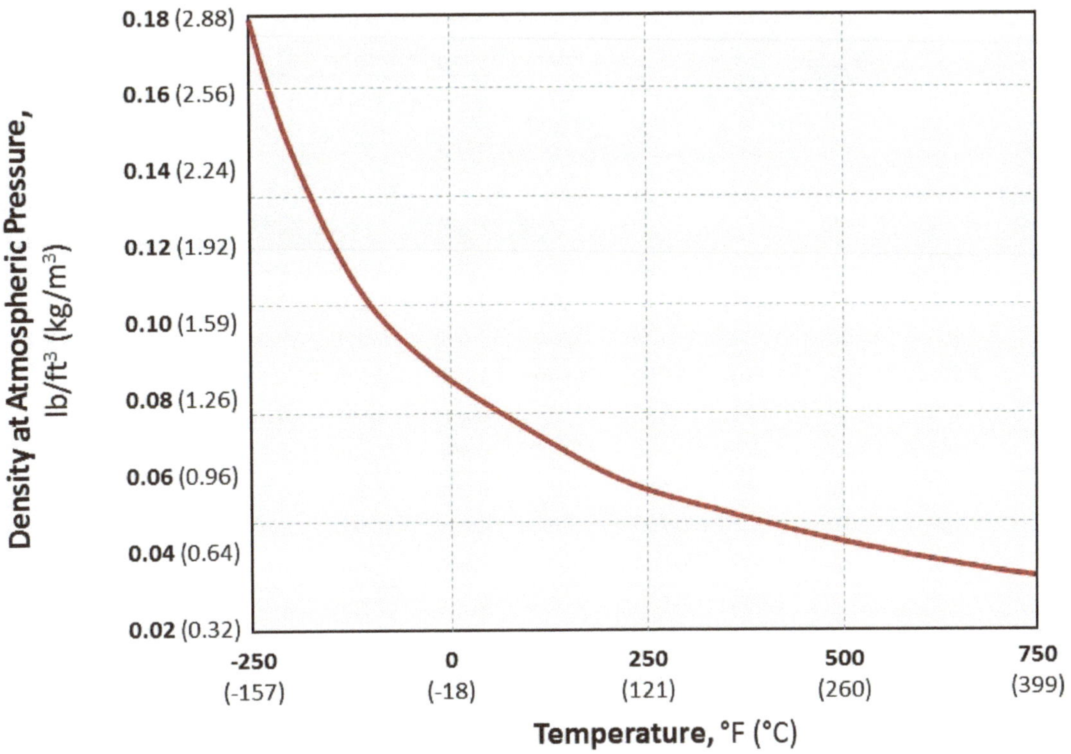

Figure 7.1 Air density variation with temperature.
Courtesy DLB Associates

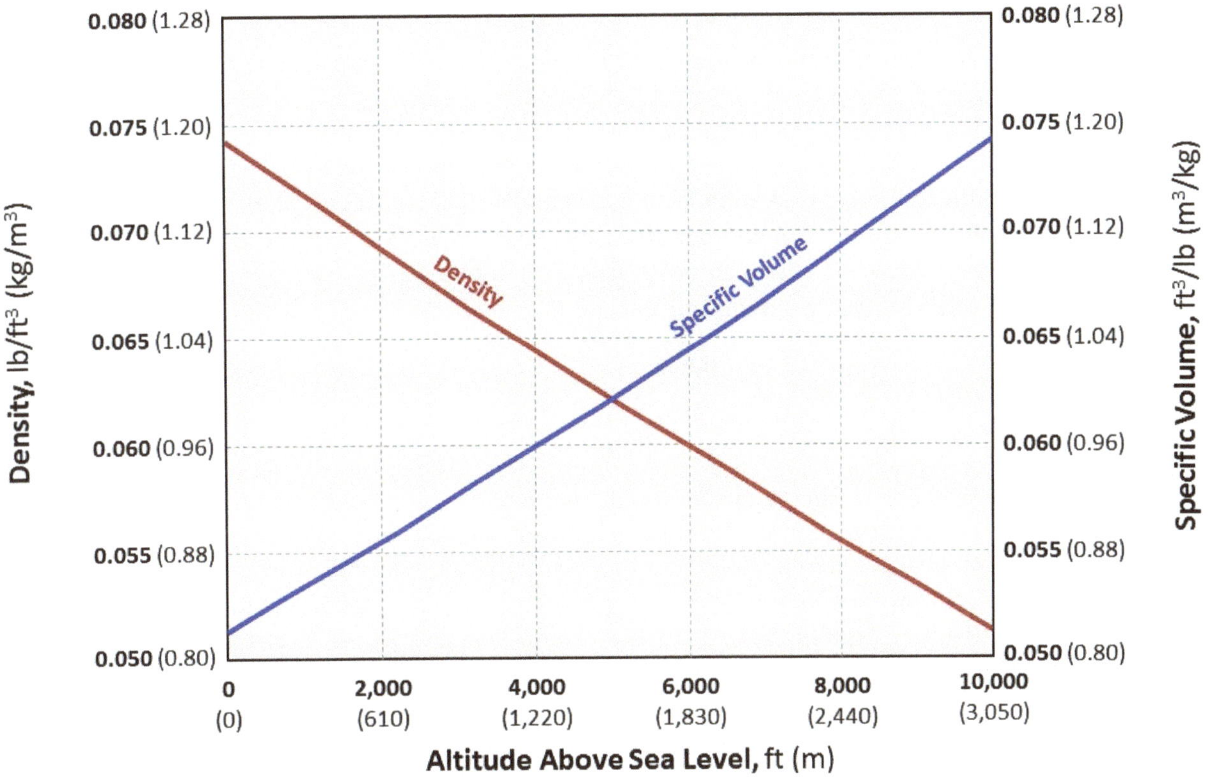

Figure 7.2 Air density variation with altitude.
Courtesy DLB Associates

Table 7.1 Air Density Variation (Temperature and Altitude)

Air Temperature, °F (°C)	Altitude Above Sea Level, ft (m)										
	0 (0)	305 (1000)	610 (2000)	915 (3000)	1220 (4000)	1525 (5000)	1830 (6000)	2135 (7000)	2440 (8000)	2745 (9000)	3050 (10 000)
	Barometric Pressure, in. Hg (kPa)										
	101.32 (29.92)	97.73 (28.86)	94.21 (27.82)	90.79 (26.81)	87.50 (25.84)	83.61 (24.69)	81.21 (23.98)	78.19 (23.09)	75.25 (22.22)	72.33 (21.36)	69.69 (20.58)
−50 (−46)	1.293	1.247	1.202	1.159	1.117	1.076	1.036	0.998	0.960	0.924	0.889
−40 (−40)	1.262	1.217	1.174	1.131	1.090	1.050	1.011	0.974	0.937	0.902	0.868
−30 (−34)	1.234	1.191	1.150	1.105	1.066	1.028	0.987	0.956	0.914	0.883	0.847
−20 (−29)	1.204	1.160	1.120	1.080	1.040	1.000	0.965	0.930	0.895	0.860	0.828
−10 (−23)	1.178	1.136	1.095	1.056	1.017	0.980	0.944	0.909	0.875	0.842	0.810
0 (−18)	1.152	1.100	1.070	1.030	0.995	0.958	0.923	0.888	0.856	0.822	0.792
10 (−12)	1.128	1.088	1.049	1.011	0.974	0.938	0.904	0.870	0.838	0.806	0.776
20 (−7)	1.104	1.065	1.027	0.990	0.954	0.919	0.885	0.852	0.820	0.789	0.759
30 (−1)	1.082	1.043	1.006	0.969	0.934	0.900	0.867	0.835	0.803	0.773	0.744
40 (4)	1.060	1.022	0.986	0.950	0.916	0.882	0.849	0.818	0.787	0.758	0.729
50 (10)	1.039	1.002	0.966	0.931	0.898	0.865	0.833	0.802	0.772	0.743	0.715
60 (15)	1.019	0.983	0.948	0.913	0.880	0.848	0.817	0.786	0.757	0.729	0.701
70 (21)	1.000	0.964	0.930	0.896	0.864	0.832	0.801	0.772	0.743	0.714	0.688
100 (38)	0.946	0.912	0.880	0.848	0.818	0.787	0.758	0.730	0.703	0.676	0.651
150 (66)	0.869	0.838	0.808	0.770	0.751	0.723	0.696	0.671	0.646	0.620	0.598

Air Volume Increase with Temperature

Sometimes it may be necessary to calculate the volume change of cold air as it warms up upon entering the building.

For example, a makeup air unit is to supply 15,000 cfm (7079 L/s) of outdoor air at 70°F (15.4°C) to a building. How much outdoor air at –30°F (–34°C) should the fan bring into the building? Elevation is approximately 1000 ft (305 m) above sea level.

From Table 7.1, indoor air at these conditions has a density ratio of 0.964 compared to standard air. The outdoor air has a density ratio of 1.191.

The volume of outdoor air needed is

$$15{,}000 \text{ cfm } (7079 \text{ L/s}) \times (0.964/1.191) = 12{,}141 \text{ cfm } (5730 \text{ L/s})$$

Combustion Air Supply

Both combustion and ventilation air are required in boiler rooms. The ideal amount of air needed for complete combustion depends on the fuel being burned. If there is too little air, incomplete combustion results and carbon monoxide will form. In practice, sufficient air must be provided not only to support complete combustion but to keep the boiler room positively pressured. Excess air is wasteful and decreases boiler efficiency. In addition, too much cold air entering can freeze piping and/or equipment.

To determine the amount of combustion air required at standard rated conditions, consider the fuel combustion equation. For example, the complete combustion of natural gas (which is basically methane [CH_4]), can be represented as

$$CH_4 + 2\,(O_2 + 3.76\,N_2) = CO_2 + 2\,H_2O + 7.52\,N_2$$

Expressing this equation in molecular weight, the stoichiometric air/fuel ratio is 17.2 lb air/lb natural gas (17.2 kg air/kg natural gas). Similar calculations for oil and coal give ratios of about 14:1 for fuel oil and 12:1 for coal.

To calculate the volume of air required for combustion of natural gas, convert the air and fuel ratios to volumes.

At standard conditions (60°F [15.5°C], sea level), natural gas (methane) weighs about 0.042 lb/ft^3 (0.671 kg/m^3), and air about 0.0765 lb/ft^3 (1.225 kg/m^3).

One pound (unit) of methane occupies a volume of 1/0.042 or 23.8 units volume, compared to 17.2 pounds of air or 225 units.

This is approximately a 10:1 volume ratio, but in practice, extra air is provided to ensure complete combustion.

However, as noted previously, these air quantities are based at 60°F (15.5°C). To adjust for cold climates, the required amount of cold air must be calculated.

For example, at 30°F (–1.1°C) and sea level, the volume relationship between 30°F (–1.1°C) entering air and standard 60°F (15.5°C) air is

$$1.234{:}1.022 \text{ (or } 1{:}0.833)$$

Therefore, the volume of outdoor air needed is only about 83% of the standard condition air volume. For modulating combustion equipment, outdoor air openings can be reduced in size accordingly, because at warmer temperatures, the combustion equipment will have lower firing rates and thus require less outdoor air (Halley 1998). Ensure that the AHJ of the project concurs with the design.

Fan Sizing

When sizing fans and motors to move air at nonstandard conditions, the different air density must be considered. For example, as air density increases (for example, when temperature decreases), the required static pressure and fan power increase. Refer to Table 7.1 for the increase factor.

For example, a fan required to deliver 15,000 cfm (7080 L/s) at 3 in. w.g. (747 Pa) at –30°F (–34.4°C) at sea level would require static pressure rating (and motor fan power) correction (increase) by a factor of 1.234. (Refer to Table 7.1 for the increase factor.)

Air Heat Flow Equations

Heat flow equations for air are determined at standard conditions. The familiar constants used in calculations at moderate temperatures may not be accurate enough for calculations at cold temperatures. Constants used commonly in HVAC equations are based upon air characteristics (for example, specific heat and specific volume) at a certain temperature and pressure (approximately room temperature). If the actual conditions of the calculation are at vastly different temperature or pressures, these differences must be accounted for.

BUILDING HVAC CALCULATIONS

Calculations for Heating Systems Design

In cold climates, building heating requirements are typically the dominant design criteria impacting the HVAC systems. Although buildings can require cooling too, space heating loads are generally larger and exist for longer periods during a typical year compared to buildings in other climates. It is generally desirable to minimize the building heating load (and hence, plant size and energy consumption) to realize cost savings, energy savings, and greenhouse gas emission reduction.

Heating equipment operates most efficiently at full load and lowest possible return temperatures, so it is best not to oversize the plant. Of course, redundancies, safety factors, or use of CHP generation must be taken into account.

For any building, it is important to make good design decisions early in the planning and design process. There are usually other drivers during design besides the building mechanical system performance, and compromises may have to be made with the architect. Many of these factors are described in Chapters 5 and 6 of this guide.

Getting an initial estimate of building heating load is often useful. This estimate may start with a scaling of heating consumption from another similar building. The estimate can then be refined using a spreadsheet or load calculation software based on type of building, conditioned area, and typical envelope construction, with an estimated infiltration rate and preliminary ventilation load. Later, more detailed calculation iterations will probably give lower heating loads once the floor plans and systems are closer to being finalized. Initial building heating and ventilation load estimates provide guidance during the schematic design stage to help develop mechanical room sizing and ceiling space requirements for ducts.

As buildings become more complex and with the increasing emphasis on energy efficiency, many designers are choosing energy modeling as their methodology to develop the heating and cooling loads, which in turn can be used to model the energy performance of various building envelope and mechanical and electrical system alternatives. Energy modeling is covered in greater detail in *ASHRAE Handbook—Fundamentals*, Chapter 19 (ASHRAE 2021a).

Outdoor Design Conditions

Outdoor design conditions for many areas, even quite remote communities, are available. *ASHRAE Handbook—Fundamentals*, Chapter 14 (ASHRAE 2021a), now has weather data for 9237 locations around the world available via Handbook Online. Data categories include the following:

- Annual heating and humidification design conditions
- Annual cooling dehumidification and enthalpy design conditions
- Extreme annual design conditions
- Monthly climatic design conditions

For locations where there are no existing values available, designers may have to estimate and err on the conservative side. Be aware that microclimates exist, and adjacent locations may have quite different temperature regimes. Seek out local knowledge and history about the weather and site.

The primary consideration in cold-climate design is outdoor dry-bulb temperature. Selection of an outdoor design temperature occurrence, whether 99% or 99.6% of the time, implies that for the other 1% or 0.4% of the year, the indoor temperature would drop below design conditions if the heating plant were sized exactly to meet the load. Typically allowances of between 20 to 25% are given for a combination of warm-up and heating safety, depending on the climate, building use, and type of construction. If even closer control over the indoor temperature is required, *ASHRAE Handbook—Fundamentals*, Chapter 14 (ASHRAE 2021a), contains a method to estimate recurrence of extreme weather events.

Outdoor air humidity in cold climates must be considered, but the actual quantity of moisture the air can hold is reduced as temperature drops.

Wind speeds and direction are also very important for infiltration calculations. They are presented in climate tables; they do not necessarily occur at the same time as the coldest temperatures, but a conservative calculation might assume so.

Indoor Design Conditions

The building owner must decide on desired interior temperature and humidity and the tolerance for the building condition to vary from those conditions.

Desired space conditions of temperature and humidity are determined by the occupancy. ANSI/ASHRAE Standard 55 (ASHRAE 2020b), ISO-7730 (ISO 2005) and other standards provide a range of values for human comfort. Industrial processes or special buildings (museums, laboratories, etc.) may require special considerations.

Normally, building codes require an indoor design dry-bulb temperature of 71.6°F (22°C), but this value represents only one of many factors considered for thermal comfort. Even if all factors are satisfied, it is possible to still have 10% occupant dissatisfaction.

Humidity control is very important to avoid condensation in or on building elements, so the dew point of indoor air must be noted and monitored. The amount and type of glazing can impact the dew point limits. If added humidification is not specifically required by the client or by the process, it may not be necessary. Many cold-climate buildings do not have humidification.

In cold climate areas, the ventilation requirements necessitate the introduction of cold, dry outdoor air, resulting in relatively low space humidity levels. This condition may result in an increased level of static electricity discharge in areas such as carpeted offices, which is detrimental to electronic equipment such as personal computers. A balance must be achieved between maintaining an acceptable space relative humidity without exceeding a level that will compromise the building envelope. High-performance glazing may be required to maintain a minimum relative humidity level of about 30%. Refer to Chapter 3 for more details on the interaction between humidity levels and static electricity.

Envelope Heat Losses

Transmission heat losses occur throughout all surfaces of a building. *ASHRAE Handbook—Fundamentals* (ASHRAE 2021a) presents the methodology for estimating heat losses from buildings. ANSI/ASHRAE/IES Standard 90.1 (ASHRAE 2019b) gives prescriptive thermal insulation values for various types of building envelope assemblies.

Heat loss from a building envelope can be decreased by adding insulation (and cost) but there is an optimum economic point. As insulation levels are increased, the incremental reductions in conductive heat loss and energy savings diminish.

When calculating envelope transmission losses, use realistic values. For example, use overall window U-factor or door value (not center-of-glass) for calculations of transmission heat loss through glazing. Likewise temperature-adjusted R-values should be used and weighting be applied to assemblies.

Heat transmission through insulating materials that are compressed, wetted, or aged, is less than the published "ideal" value. For example, the long-term thermal resistance of polyisocyanurate insulation reduces as gases in its pores seep out and is replaced by air.

Envelope conduction losses can be significantly affected by thermal bridging. Up to 50% of conductive heat loss through the opaque areas of buildings is associated with details often overlooked (Roppel and Marif 2011) such as construction details or poor workmanship.

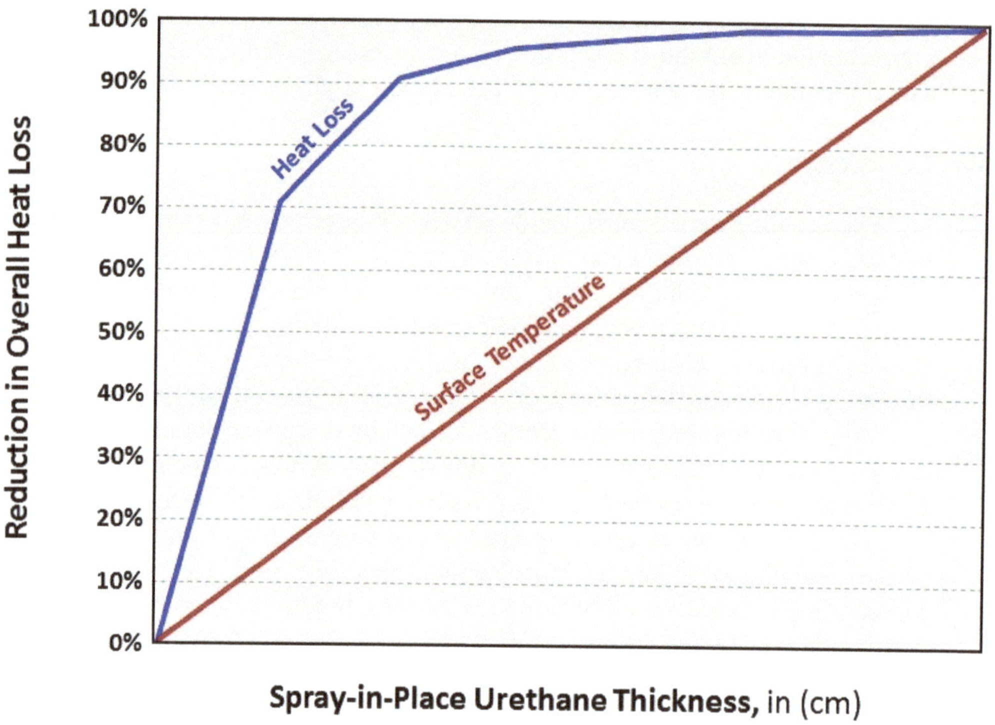

Figure 7.3 Heat loss and surface temperature variation with insulation thickness.
Courtesy DLB Associates

To account for thermal bridging in simple (non-computer-assisted calculations) the envelope U-factor may be calculated by one of three methods:

1. Isothermal planes that are one dimensional, with no thermal bridging effects considered (for components with layers of materials, where U = sum of 1/RSI (R-value).
2. Isothermal planes with parallel path, in which it is assumed that heat transfers through thermal bridge parallel to the heat flow through insulation, as in wood-stud walls.
3. Metal frame assembly method (or zone method [ASHRAE 2013]) where heat flow through thermal bridges is not parallel to the heat through insulation.

These methods are explained and examples given in *ASHRAE Handbook—Fundamentals* (ASHRAE 2021a) and in *ANSI/ASHRAE/IES Standard 90.1-2016 User's Manual* (ASHRAE 2017). These methods do not necessarily address complexities in three-dimensional construction. It is recommended that, for complex cases, computer modeling be used; for example, the ASHRAE equivalent wall method (Carpenter 2012).

RP-1365—Thermal Performance of Building Envelope Details for Mid- And High-rise Buildings (Roppel and Marif 2011) provides thermal performance data for 40 common building envelope assemblies and gives procedures to use for modeling. This report describes modeling that was done with 3-D time transient finite element analyses, and results calibrated against measured thermal performance data.

Calculations alone, even if derived from very accurate computer simulations, cannot account for all installation or operational situations. For example, dead air spaces that

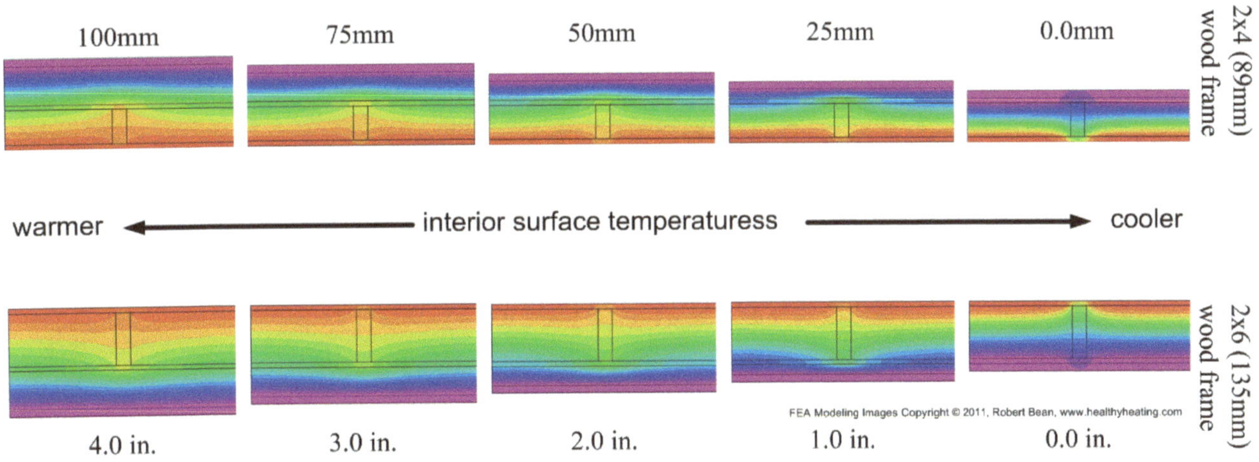

Figure 7.4 Finite element analysis modeling of exterior wall with incremental increases for exterior insulation.

Courtesy Robert Bean, Indoor Climate Consultants Inc.

receive minimal air movement or heat can readily become cold and condensation can occur.

Doors, access hatches, and window and skylight frames act as thermal bridges and must be considered in the envelope effective heat transfer calculations. Many buildings in cold-climate locations are set on steel pilings joined to steel frames. Heat transfer into the building should be considered and minimized by using insulated pile caps or another method.

Dew Point Calculation

For example, the following shows one method of calculating the dew point in an assembly.

A wall with no vapor barrier is constructed as shown in Figure 7.5. Outdoor temperature is –31°F (–35°C), and interior temperature is 64.4°F (18°C) at 30% rh. Determine whether/where moisture will condense within or on the assembly.

Note that this simple example does not account for infiltration or thermal bridging. For more accurate calculations, use computer analyses.

Obviously, the dew point location is within the mineral wool batt. Its location may be calculated using the temperatures at the interior side of batt (59.0°F [15.2°C]), the total temperature drop through the batt, (59.0°F – (–18.6°F) = 77.6°F [15.02°C – (–23.99°C) = 39°C], and the width of the batt (2.48 in. [63 mm]). With these variables, we can solve for the location of the dew point.

$$L = [(15.2 - 0.5) / 39] \times 63 \text{ mm} = 24.7 \text{ mm from inside surface of batt}$$

I-P units:
Temperature of interior side of batt = –23.99°F
Temperature drop through batt = 59.0°F – (–18.6°F) = 77.6°F
Width of batt = 2.48 in.

$$L = [(59.0 - 33) / 77.6] \times 2.48 \text{ in.} = 0.83 \text{ in. from inside surface of batt}$$

In this case, the mineral wool insulation offers little resistance to water vapor. During steady-state conditions when water vapor is constantly diffusing through the permeable

Table 7.2 Conditions throughout the Example Wall (Figure 7.5)

	Temperature, °F	R per Layer (°F·ft²·h /Btu)	R/R Total	(R/R total) ·(ΔT)	Temperature Increase, °F	Layer Thickness, in.	Cumulative Thickness	Temperature at Right Side of Layer, °F
Outdoor air temperature	–31	—	—	—	—	—	0	–31
Exterior air film, 0 in.	—	0.17	0.013	1.236	1.24	0	0	–29.7
Fiberboard siding, 0.37 in.	—	0.57	0.043	4.145	5.38	0.37	0.37	–25.6
Particle board, 0.63 in.	—	0.97	0.074	7.053	12.43	0.17	0.54	–18.6
Air space, 1 in.	—	1.02	0.078	7.417	19.85	1	1.54	59.0
Mineral wool batt, 2.48	—	9.65	0.736	70.168	90.02	2.48	4.02	59.4
Sand plaster, 0.51 in.	—	0.06	0.005	0.436	90.46	0.51	4.53	64.4
Interior air film	—	0.68	0.052	4.945	95.40	0	—	—
Indoor air temperature, °F	64.4	—	—	95.40	—	—	—	—
Column totals	—	**13.12**	—	—	—	**4.53**	—	—
ΔT total, °F	95.4	—	—	—	—	—	—	—

	Temperature, °C	RSI per layer, m²·°C/W	RSI/RSI total	(RSI/RSI total) ·(ΔT)	Temperature Increase, °C	Layer Thickness, mm	Cumulative Thickness	Temperature at Right Side of Layer, °C
Outdoor air temperature, °C	–35	—	—	—	–35	—	—	—
	—	—	—	—	—	—	0	–35
Exterior air film, 0 mm	—	0.03	0.01	0.69	–34.31	0	0	–34.3
Fiberboard siding, 9.5 mm	—	0.1	0.04	2.29	–32.02	9.5	9.5	–32.0
Particle board, 16 mm	—	0.17	0.07	3.90	–28.12	16	25.5	–28.1
Air space, 25 mm	—	0.18	0.08	4.13	–23.99	25	50.5	–24.0
Mineral wool batt, 63 mm	—	1.7	0.74	39.00	15.02	63	113.5	15.0
Sand plaster, 13 mm	—	0.01	0.004	0.23	15.25	13	126.5	15.2
Interior air film	—	0.12	0.05	2.75	18.00	0	126.5	18.0
Indoor air temperature, °C	18	—	—	53.00	—	—	—	—
Column totals	—	**2.31**	—	—	—	**126.5**	—	—
ΔT total, C°	53	—	—	—	—	—	—	—

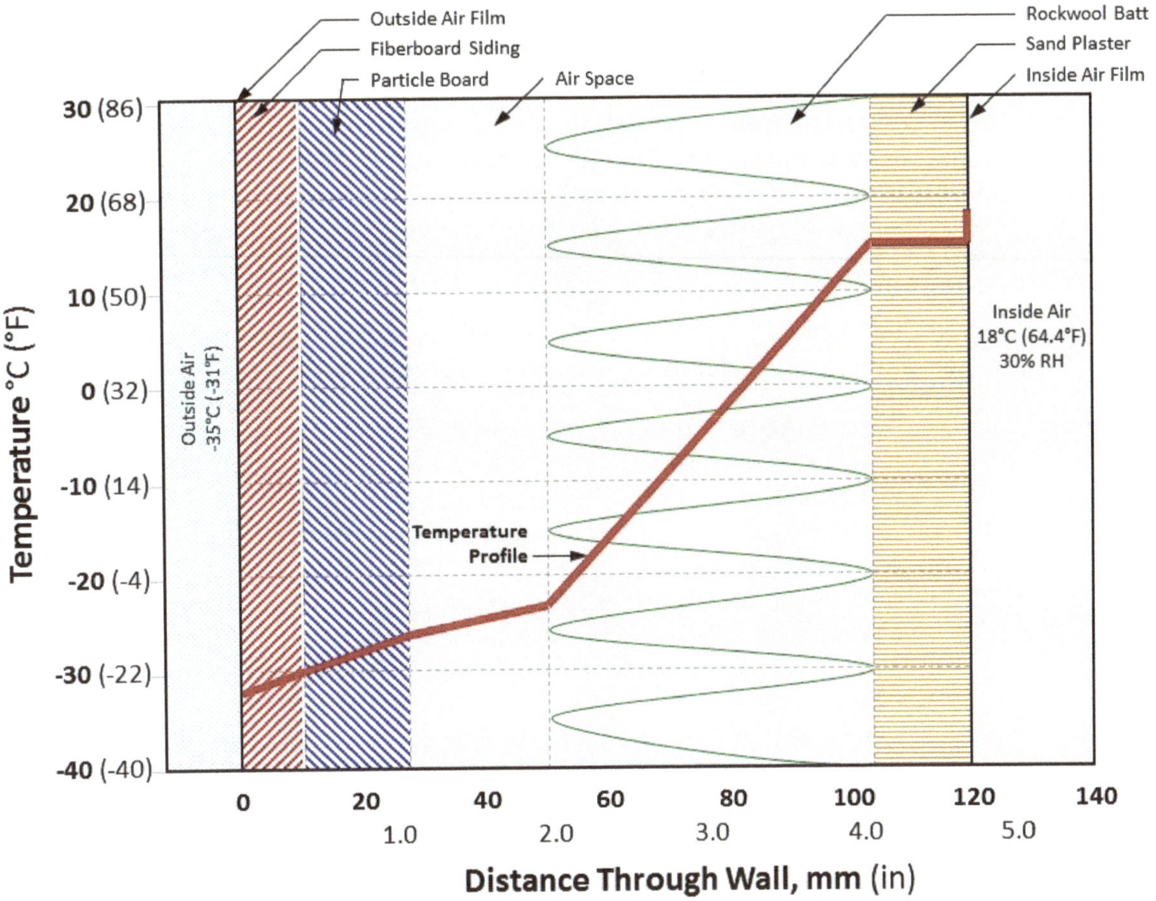

Figure 7.5 Temperature profile through the wall assembly.
Courtesy DLB Associates

insulation, the vapor condenses on the next colder interface (the inside of the particle board). During non-steady-state conditions, the vapor condenses in the insulation.

This example illustrates the need for a vapor barrier to be located so that its temperature is always above the dew-point temperature. Determine where moisture will condense inside or on the assembly. Obviously, the 32°F (0°C) location is within the mineral wool batt.

Infiltration and Ventilation

Outdoor air can be exchanged with indoor air within a building in two different ways: infiltration and ventilation. Ventilation can also be subdivided in either natural or mechanical means. All of these outdoor air exchange methods interact with each other and affect occupant thermal comfort and IAQ.

Uncontrolled infiltration is generally detrimental to human comfort and building energy usage. Design and construction practices aim to reduce it or at least to control it, whereas ventilation is required and designed. Infiltration depends on factors such as wind, temperature difference, tightness of construction, stack effect from building height, and overall building pressurization. In general, it is desirable to have a slight positive pressure on a building. This minimizes air infiltration and moisture migration.

Building pressure through a building envelope varies with time, season, and other factors, and depends on four pressure fields: exterior, interior, interstitial, and air system. *ASHRAE Handbook—Fundamentals*, Chapter 16 (ASHRAE 2021a), deals with infiltration in great detail. Computer analysis is generally required to solve the complex transient calculations pertaining to infiltration. In cases with uniform indoor pressures, the combined effect of wind, stack effect, and mechanical systems can be added.

In practice, simple air-change method estimations are often used to calculate infiltration. Air-change rates in various parts of the building are estimated (e.g., 1 to 2 air changes per hour in entrance vestibules).

Another method is to assume rates of infiltration depending on the type of building and condition, as in the following example scenarios:

- 0.03 cfm/ft^2 at 0.01 psi (0.15 L/s·m^2 at 75 Pa) infiltration rates in buildings with low indoor air relative humidity less than 27%, and low to moderate part-day occupant loads and few sources of water vapor, or buildings that have dehumidification systems (e.g., warehouses, large-volume stores, low-occupant-load offices, vehicle storage, and repair facilities)
- 0.02 cfm/ft^2 at 0.01 psi (0.10 L/s·m^2 at 75 Pa) infiltration rates in buildings with a normal indoor air relative humidity between 27 and 55%, and low to moderate part day occupant loads: (e.g., schools, community halls, libraries, and large offices)
- 0.01 cfm/ft^2 at 0.01 psi (0.05 L/s·m^2 at 75 Pa) infiltration rates in buildings with an indoor air relative humidity greater than 55% (e.g., pools, spas, student residences, and humidified portions of hospitals)

A third method is the crack method, in which the air leakage rate (obtainable from door and window manufacturers) is multiplied by the length of sash crack or door area.

Another method is to use an assembly tested for air infiltration using one of the test standards, either ASTM E2357 (ASTM 2018) or CAN/ULC S742 (SCC 2011).

Infiltration also occurs over roofs and through floors and basement walls, because of pressure differences. Even if buildings are pressurized during operating hours, infiltration will occur when fans are turned off at night.

Envelope penetrations such as elevator shaft exhaust openings can increase the impact of stack effect by providing a path for the air to exit.

Calculations for Ventilation Systems Design

The heat load of airflow in or out is determined by the sensible heat flow equations discussed in the previous section. If the conditions of air are not near standard, then the constant in the terms must be adjusted (e.g., for a building is at high altitude).

Traditionally, small buildings and homes had no mechanical ventilation systems and depended on leakage or windows that could be opened for ventilation air. Newer designs include provision of heat-recovery ventilators or energy-recovery ventilators to provide outdoor air and makeup exhaust air quantities. Heat recovery of ventilation air in cold climates is almost always economical.

Chapter 16 of *ASHRAE Handbook—Fundamentals* (ASHRAE 2021a) provides information in greater detail for ventilation and infiltration calculations.

Ventilation Requirements

Ventilation air requirements are determined by codes and standards, such as ANSI/ASHRAE Standard 62.1, *Ventilation for Acceptable Indoor Air Quality*. Because heating and/or cooling air and moving it around costs money and uses power, ventilation air quantities should be minimized while maintaining acceptable IAQ. In winter, preheat of air should be used. Economizers can provide free cooling during spring, summer, or fall. Another option to reduce ventilation air quantities is disinfection of return air through the use of technologies such as ultraviolet filtration.

Required ventilation rates for humans, animals, and processes are found in many references (for example, ANSI/ASHRAE Standards 62.1 and Standard 62.2 and ANSI/ASHRAE/ASHE Standard 170). However, air quantities given in these standards assume perfect mixing of air within spaces, which is not always true in real life. To take into account the uneven mixing, ASHRAE Standard 62.1 has incorporated a correction factor for outdoor air calculations called the *ventilation effectiveness* for various air distribution types. Placement of the supply inlets and return/exhaust outlets are also considered in the ventilation effectiveness calculations.

If no air moves past cold surfaces, dead-air spaces can readily develop condensation or mold. Examples of dead-air spaces include windowsills piled high with books or floor-level heating vents blocked by furniture.

In addition to providing the occupant outdoor air for acceptable IAQ, a ventilation system may be required to offset infiltration or outdoor pressure effects, or to exhaust contaminants such as radon or volatile organic compounds.

Ventilation is required in crawlspaces, dead-air spaces, and attics to prevent condensation and ice. If the crawlspace is heated and enclosed, it should be ventilated as a conditioned space; if it is unheated, it can be vented to the outdoors without mechanical means, in the same manner as an attic.

Building Pressures

A common way to keep a building under positive pressure is to provide 5 to 10% extra outdoor air in the ventilation system relative to the exhausted air volume. The actual amount of air needed would be difficult to calculate; hence, this rule of thumb can be used.

Some buildings, such as laboratories or hospitals, require certain differential pressures be maintained between spaces for infection control, contamination control, or other reasons. This can become complicated when the entire building has to remain at a slightly positive pressure at all times.

Makeup air must be provided for exhausts for a building, such as for a garage, washroom, workshop, fume hood, or kitchen. Care must be taken with variable exhaust quantities (for example, multiple fume hoods) so that the quantity of makeup air supply matches the air being exhausted. For shop equipment such as in a woodworking shop, it may be more economical to select a self-contained filter unit with recirculation back into the conditioned space so that no makeup air is required.

Ventilation Air Heating

For ventilation heating loads, select the indoor and outdoor design conditions for heating. For cooling, use cooling indoor and outdoor design temperatures and humidity levels.

Ventilation air will require preheat to make sure that the air is above freezing before entering the air handler. Frosting of filters will occur if below-freezing air enters; it is best to remove filters on the preheat coil in winter to prevent them frosting up. If a mixed-air system is used, winter ventilation air will be calculated based on required outdoor air for occupants and may only be 15% or so of the total airflow. The heating coil must be sized to meet winter heating load but also to allow all air to go through in summer (or add a bypass to the coil).

Size ventilation air intakes at low velocity 500 fpm (2.5 m/s), because higher velocities will entrain snow and rain (Armstrong 2012).

In cases where occupancy and outdoor air requirements vary during the day, a demand-control ventilation system using a CO_2 sensor may be a good choice. However, if any of the building envelope heating is provided through the air system, it is important to check for sufficient heating capacity at times of low occupancy.

Heat Recovery

Heat recovery ventilators or other means of reclaiming energy from exhaust air, such as glycol runaround loops, heat wheels, or others are definitely desirable in cold climates. In buildings with larger, assembly-type occupancies, heat recovery can play a large part in reducing the building heating and cooling loads, plus will have a significant impact on the owner's energy costs. ANSI/ASHRAE/IES Standard 90.1 also requires heat recovery on air-handling equipment based on climate zone and percentage of outdoor airflow rate at design conditions (ASHRAE 2019b).

There will be a reduction in efficiency because of defrost cycles. This loss of efficiency may be difficult to calculate, but the manufacturer may be able to give information. Refer to Chapter 9 for further details on heat recovery methods, equipment, and design considerations.

Heating System Sizing

The heating load of a building is the sum of all the loads that would normally occur. Service water heating may or may not be done by the main heating plant. If it is to be provided through indirect-fired heaters, the amount of overlap has to be determined—the entire hot-water load will not generally be added to the building heating load (except in situations with large and constant service water loads).

Once a heating load is determined, safety factors or redundancy must be determined. Sizing a heating system may or may not require redundancy in equipment depending upon the location and service. For example, hospitals require redundancy. For buildings with no inherent need for heating system redundancy, consider maintenance and reliability requirements of the system. Sometimes, two types of boilers can be used: (1) a biomass boiler to take the base load (80% of the time heating plants operate under 50% capacity) with (2) a peaking oil or gas boiler to address the peaks.

Full redundancy in a building may not be required. Some facilities are sized with two boilers at 60% load so that some loads such as ventilation can be trimmed temporarily should one of the boilers go down. This also allows both boilers to be in service at a reduced firing rate throughout most of the season, thus reducing idle time and standby losses.

Oversized systems should be avoided as they can cycle and do not operate at best efficiency. Ways to reduce heating system size include using indirect-fired service water heaters with hot-water priority and implementation of emergency plans (e.g., if one boiler fails or if heat loss exceeds heating plant capacity, reduce noncritical ventilation and heating loads).

For non-condensing heating appliances, equipment operates most efficiently at full load but is subject to minimum return temperatures to avoid corrosive condensing of the flue gases. Condensing heating appliances are typically not subject to the minimum return temperatures and operate most efficiently at higher temperature differentials and lowest possible return temperatures. However, there are many types of condensing appliances with varying design flow and temperature requirements, so it is best to consult with the manufacturer representative for these parameters. Of course, redundancies, safety factors, or use of CHP generation must be taken into account.

Although heat recovery is frequently used to preheat incoming air, the designer must use judgment as to whether to downsize the heating plant accordingly. If the heat recovery unit fails, the plant may not be able to maintain temperature without reducing ventilation rates.

Another reason to minimize the heating plant size is local legal requirements for classes of operating engineers for various plant sizes. It can be very difficult to find the proper class of operating engineers in small, remote communities.

Radiant floor temperatures should not be too hot. ANSI/ASHRAE Standard 55 and ISO Standard 7730 both state that, for comfort, floors should be at a maximum surface temperature of 84°F (29°C) for heating, with 75°F (24°C) being the optimum temperature (ASHRAE 2020b; ISO 2005).

Glycol System Calculations

Glycol solutions are often used instead of water in cold-climate heating systems to prevent freezing, especially for coils that have contact with outdoor air. Glycol has a different density and specific heat than water. These properties vary with temperature and percent glycol solution. Ethylene glycol has better antifreeze properties than propylene glycol, but should not be used if there is a hazard to children, food, or water. Propylene glycol has a much higher viscosity than ethylene glycol (or water) and thus requires more pumping power at cooler temperatures.

Selection of the percent glycol depends on whether freeze or burst protection is required. Freeze protection is more stringent and requires higher concentrations of glycol. Usually, this is the concentration used in cold-climate heating piping or coils when fluids may need to be pumped near their freezing temperatures such as for systems that would be at risk if power or pump failed. Using too high a percent glycol solution will increase pumping power and energy usage.

Example: Selecting Glycol Solution

What percent glycol should be used for freeze protection on an outdoor heating coil to be used in a building in Moscow, Russia?

For safety, choose to use propylene rather than ethylene glycol. The 99.6% heating dry-bulb design temperature in Moscow is –9.1°F (–22.8°C) (ASHRAE 2013).Therefore from Table 7.3, 42% propylene glycol should be used. In practice, it is usually easier and safer to use premixed glycol mixtures than to mix at the site.

Example: Choosing Pump and Motor for Glycol System

Assuming a premixed solution of 45% glycol is used and that the coil is rated for 50,000 Btu/h (14.653 kW) at entering water temperature of 140°F (60°C), at 20°F (12.5°C) ΔT, and pump flow of 5 gpm (0.32 L/s), what is the actual coil heat output, and what is the corrected coil flow and increase in system pressure drop?

Using correction factors from Table 7.4, the actual coil heat output will be 50,000 Btu/h (14.653 kW) × 0.928 = 46,400 Btu/h (13.598 W).

The adjusted pump flow is 5 gpm (0.32 L/s), (0.32 L/s) × 1.077 = 5.385 gpm (0.34 L/s).

The system pressure drop is increased by a factor of 1.139.

Calculations for Cooling Systems Design

Cooling Loads

Although cold-climate buildings are heating-predominant, many require some form of cooling, even if only during summer months. Depending on the building geometry, there may also be some cooling load at during spring or fall during parts of the day.

Table 7.3 Propylene Glycol Freeze and Burst Properties

Propylene Glycol Temperature, °F (°C)	Freeze Protection, % by volume	Burst Protection, % by volume
20 (–7)	18	12
10 (–12)	29	20
0 (–18)	36	24
–10 (–23)	42	28
–20 (–29)	46	30
–30 (–34)	50	33
–40 (–40)	54	35
–50 (–45)	57	35
–60 (–51)	60	35

Table 7.4 Correction Factors for Propylene Glycol Mixtures

% by Volume	Heat Transfer Correction	Pump Flow Correction	Pressure Drop Correction for 140°F (60°C) Solution	Pressure Drop Correction for 100°F (38°C) Solution
20	0.987	1.013	1.067	1.098
25	0.978	1.022	1.078	1.120
30	0.969	1.032	1.089	1.141
35	0.957	1.045	1.106	1.168
40	0.944	1.059	1.122	1.196
45	0.928	1.077	1.139	1.228
50	0.912	1.096	1.156	1.261
55	0.893	1.120	1.172	1.293
60	0.873	1.145	1.189	1.326

During the summer, the long hours of sunlight do not allow the building much time to cool down overnight. Of more concern is overheating during spring or fall, especially in buildings with high thermal mass or improperly designed and controlled radiant floors.

In the more northern latitudes, solar angles will be lower and may contribute to a higher cooling load and uncomfortably high space temperatures. Shading of fenestration must be considered and used wisely.

Economizers can provide free cooling during spring, summer, or fall.

Selecting Indoor and Outdoor Design Conditions. Usually, indoor design conditions for summer allow slightly higher indoor temperatures than in winter. However, the owner must verify the desired conditions. For industrial processes, very strict limits may be imposed year-round for indoor temperature and humidity.

Calculating the Building Cooling Load. Cooling loads can be calculated as per *ASHRAE Handbook—Fundamentals*, Chapter 18 (ASHRAE 2021a), which identifies two load calculation methods called the *heat balance method* and the *radiant time series*. Other methods and load calculation software have also been developed by manufacturers and software developers.

Be careful about adding safety factors, because an oversized system will cycle. This is inefficient and may also cause problems with humidity; when water has condensed on a cooling coil and the compressor is turned off, water will reevaporate into the air.

Humidification and Dehumidification. Recommended humidity conditions for buildings in other climates may not be appropriate for those in cold climates, except in critical locations such as health care facilities or museums.

Be cognizant of the temperature of building elements in relation to indoor air dew point (e.g., window frames, window glazing, or cold-water piping in humid locations). The problems of condensation in and on the building envelope are covered in detail in Chapter 8.

Watch out for condensation on building structural members, uninsulated valves, or air handler inlet flanges, particularly in heated penthouses in buildings with high interior humidity levels.

Ventilation, humidification, or dehumidification loads can be calculated by equation or by using the psychrometric chart.

Dew Point Avoidance

Example: Dew Point Determination on Building Structure. A hospital penthouse is located in a cold climate at sea level. Hospital air is heated and humidified to 77°F (25°C) and 50% rh. In winter, external temperatures are at –4°F (–20°C). The penthouse air is essentially at those conditions, due to air supply into the room and the presence of stairs and elevator shaft. Although the air intake plenums are insulated, a section of metal flange across the width of the top of the plenum was not covered. At what temperature will condensation form on the flange?

Looking at the psychrometric chart in *ASHRAE Handbook—Fundamentals*, Chapter 1 (ASHRAE 2021a), the dew point for the penthouse air is about 57°F (14°C). Assuming the metal flange is at about the same temperature as the outdoors, condensation on the flange can be expected when outdoor temperatures fall below 57°F (14°C), and frost will form on the flange when outdoor temperatures fall below freezing.

Condensation on Glazing. Condensation occurs on (or inside) window glazing or frame assemblies when the surfaces are at or below dew-point temperature. It is wise to ensure that this does not occur. However, the calculation can be complicated, because the point R-values are different at different places on the window assembly, so a number of calculations must be completed, taking into account the actual R-value at edge of the glass and on the window frame.

ENERGY MANAGEMENT

Building energy consumption is the sum of heating, cooling, ventilation, air conditioning, and service water heating loads plus lighting, plug, and process loads. Because energy use and costs are typically higher in cold climates, it imperative to sensibly minimize this energy consumption.

Degree-day methods can be used to provide a very rough approximation of fuel usage, but for more accurate predictions of heating and cooling loads and equipment sizing, spreadsheet or computer calculations are required. Energy modeling software is available online to simulate the building heating and cooling loads and energy requirements.

For energy use estimation, use the best and most recent rates for power and fuel. Electricity costs can vary hugely between communities, from mere cents per kilowatt hour in cities to over $2.50/kWh or more in remote locations. Contact the local utilities if in doubt. With development of or the closing of natural gas wells, communities' fuel choices may change quickly. For example, recently, in Inuvik, Northwest Territories, Canada, the closing of a natural gas well required substitution of more expensive synthetic natural gas, almost doubling the cost to consumers. Since then, a less-expensive liquid natural gas facility has come on-line (NTPC 2014).

In small, cold-climate communities that are diesel-powered, there may be no actual cost incentive to reduce community building energy consumption because the same generator would have to be run in any case.

Demand charges may not apply in some locations. Pumping power and fan power are expensive; designers should use variable-flow equipment where possible. If demand charges are not a concern, consider using instantaneous water heaters.

The cost of a recirculation pump power on domestic hot-water lines may be sizeable; consider adding in a night shut-off to lower costs.

Also, consider any added costs necessitated by the cold climate. For example, if propane is used for heating, a vaporizer, or electric blanket is required in deep cold; this will have electrical cost to run.

When designing a system and calculating energy savings, use common sense and err conservatively. Consider the situation in which the equipment or system will exist. In a large center with trained operators, energy savings from the use of heat wheels, night setbacks, or variable-air-volume system are probably able to be realized. However, a missed energy saving opportunity occurs if, for example, a hospital fan system for office areas, occupied only during the day, is (by mistake or from lack of training) run 24/7 like the fan system for the patient care areas, which obviously need constant ventilation.

HOW COLD IS IT?

It is so cold ... that when I turned on the shower I got pelted with hail!

8 Building Envelope

INTRODUCTION

The relationship between active heating systems and the passive heat loss resistance provided by the building envelope is one of balance. Providing heat is good. Providing the right amount of heat is better. Providing the right amount of heat at the right time in an envelope that prevents that heat from escaping to the atmosphere is best. This good-better-best logic becomes one of the most critical questions of building design.

To be reasonably efficient, durable, and comfortable, the envelope must meet certain performance criteria. These criteria are interrelated with the design and performance of the heating and ventilation systems and must account for everything from local materials and expertise available to compliance with various energy-efficiency standards, building codes, and comprehensive life-cycle costing analysis.

BUILDING ENVELOPE DESIGN

Building design, construction, and operations in a cold climate place a premium on the envelope. The reason is simple: the envelope is the primary system for retaining heat.

The building envelope is a passive system, working in conjunction with active systems within the building (HVAC, lighting, etc.) to provide thermal comfort and the other requirements of shelter. The design of an envelope system must inform the design of the heating system and vice versa. Specifically, the design and thermal performance of the envelope must be tied to the design and installation of the heating systems.

Integrating the design of the envelope with that of the heating system provides a framework within which thermal comfort and building efficiency are optimized. No other integration is likely to have as profound effect on building operations, durability, and user satisfaction.

Following is a list of common envelope terms as used in this book and a plain-language definition:

- **Envelope.** The entire enclosure of a building, including roof, walls, and floors exposed to the elements (including those below grade). Envelopes are made up of one or more assemblies. It is also referred to in many code books, design standards, and guides, as the *environmental separator*, *building fabric*, or *enclosure*.
- **Assembly.** The arrangement of components that make up walls, roofs, and floors.

- **Component.** Individual products and materials that go into an assembly. Components may have specific roles to play in the performance of an envelope (e.g., air barrier, insulation, cladding) or may be included in the envelope for practical reasons (e.g., structural members, fenestration, decorative elements).
- **Fenestration.** Doors and windows, sometimes skylights.
- **Insulation.** The material in an assembly providing the primary resistance to thermal conduction. There may be multiple layers in cold climates and each ought to be as contiguous as possible.
- **Air Barrier.** The material in an assembly designed and installed to resist uncontrolled air flow through an assembly.
- **Vapor Barrier.** The material in an assembly designed and installed to resist uncontrolled water vapor migration through an assembly.
- **Moisture Barrier.** The material in an assembly designed and installed to resist uncontrolled liquid water movement through an assembly.

Passive System

Although a well-designed envelope may provide some heat gain through fenestration, passive solar, and other factors, in most designs it is the envelope's passive resistance to the transfer of heat, air, and moisture from the building to the outdoor environment that is critical. Because it is passive in nature, envelope performance offers two fundamental benefits to the designer and operator of a building: predictability and longevity. *Predictability* means that the performance expectations can be well planned and calculated as part of the integrated design strength. This predictability will be addressed later in the sections related to thermal, vapor, and air transfer. *Longevity* is addressed in the selection of appropriate materials; design and installation of the air, vapor, and thermal control layers; and in building operation and maintenance. This chapter also addresses the active components of the building envelope such as windows, doors, and solar design, in following sections.

Getting it Right the First Time

A thoughtful and well-constructed building envelope should last a very long time and reduce operating, maintenance, and (potentially) capital costs. Most envelope materials, when maintained in an appropriate design environment, will last for several decades. Even components whose performances naturally degrade over time, such as certain types of insulation that lose a portion of their R-value as they age, do so in a controlled and predictable way that is usually easily accounted for. In addition, the envelope assemblies themselves are not typically user dependent. With the exception of a careless individual leaving a door or window open, most envelope components do their job day in and day out without anyone giving them second thought. Therefore, the performance of a building envelope should remain relatively constant throughout its life, provided the design and construction are appropriate to the application.

The need for predictability and longevity means that it is imperative to get the envelope right, preferably the first time. Although this applies to all buildings, it is particularly important to do so in cold climates. As noted previously, the envelope of a cold-climate building is optimized when it is integrated with other building systems. This integration means that changes in envelope performance or characteristics affect whole-building performance. In addition, the remoteness and severity of cold climates means that an envelope problem may quickly grow severe. These problems can include the following:

- Cost of failure
- Construction challenges
- Collateral effects on related systems
- Operating and replacement costs

Cost of Failure

Envelope failures in cold climates have costs beyond simple envelope repair. Although envelope failures common to buildings in all climates—such as moisture ingress, structural failures, and wind damage—are present, the effects on building occupants are often more severe in cold climates. An envelope failure that might prove a hindrance in a temperate climate may threaten health and life safety in a cold climate where the building heating system might simply be unable to cope with the increased demand and thus make the building (or portion of it) untenable. Whereas a carefully considered envelope in a temperate or even tropical climate may be fully functional when open to the outdoors, there is seldom any such provision in cold climates.

Cold climates, with corresponding long drying times and high thermal gradients within envelope assemblies, also exacerbate envelope failures through prolonged deposition of moisture. This has the potential of trapping large amounts of water in an assembly over the lengthy heating season. This will be explored in greater depth in later sections dealing with moisture control and mitigation in assemblies.

Construction Challenges

The geographic remoteness of many cold-climate sites and cold-construction temperatures means that repairs to an envelope failure are often delayed or impossible with materials on hand.

Collateral Effects

Because envelope failures (visible or not) are likely to degrade thermal resistivity of the building, they are likely to place additional strain on the active heating systems in the building. This has a negative effect on both heating system serviceability and operating costs.

Life Cycle

Given the far-reaching effects of envelope failures, building envelopes should be designed and constructed to last a very long time. Components such as air, moisture, and thermal control layers should remain in place and functional for the life of the building. This allows life-cycle costing to extend over 20, 30, and even 50 years for many envelope components. And, although it is outside the scope of this book to provide instruction on appropriate life-cycle costing models, envelope design components tend to have consistent performance and sufficient longevity to respond well to most of the approved models, whether a simple cost payback, amortized, or life-cycle assessment (in which the environmental effects are weighted in the decision) are used. There are many methods of doing life-cycle costing, and at least as many opinions on which of these is best or most applicable to a project.

Cost Savings

Because cold-climate design is often heavily dependent on mechanical systems for heat and ventilation, envelope and mechanical systems may often be traded off against each other, which lowers both capital and operating costs.

A well-performing envelope will always reduce operating costs. The amount of savings this represents, whether measured in energy units, dollars, or maintenance reduction, can vary substantially. It is recommended that both the envelope and the associated mechanical systems be put through a thorough life-cycle costing analysis. The inclusion of both systems ensures that cost trade-offs and dependencies are identified.

The latter point is particularly germane to costing in cold-climate systems. Clearly, improving the performance of the envelope may have a corresponding reduction in the mechanical requirements. These improvement gains are usually subject to the law of diminishing returns. However, what is sometimes not as well understood is that when sufficiently high performance is attained within a specific component, associated systems may be eliminated entirely.

A familiar example occurs in window selection. Improving window performance reduces the requirement of perimeter heating. However, in a building with, for example, a hydronic baseboard system to avoid condensation and drafts necessary in cold climates with standard windows, upgrading to triple or quadruple pane sealed units in a thermally insulating frame may remove the need for perimeter heating entirely. If the cost of the window upgrade is less than that of the perimeter heating system, it is a double win for the designer and the building owner: lower capital costs, lower operating costs, and improved thermal comfort. Known as *cost tunneling*, these types of cost trade-offs sometimes result in large gains in building performance with no capital cost increase (Figure 8.1).

Although beyond the scope of this book to delve deep into costing options, some good rules of thumb to identify cost tunneling potentials are as follows:

- Look at system dependencies that are obvious (e.g., envelope components and mechanical systems) and cost analyze these components together.

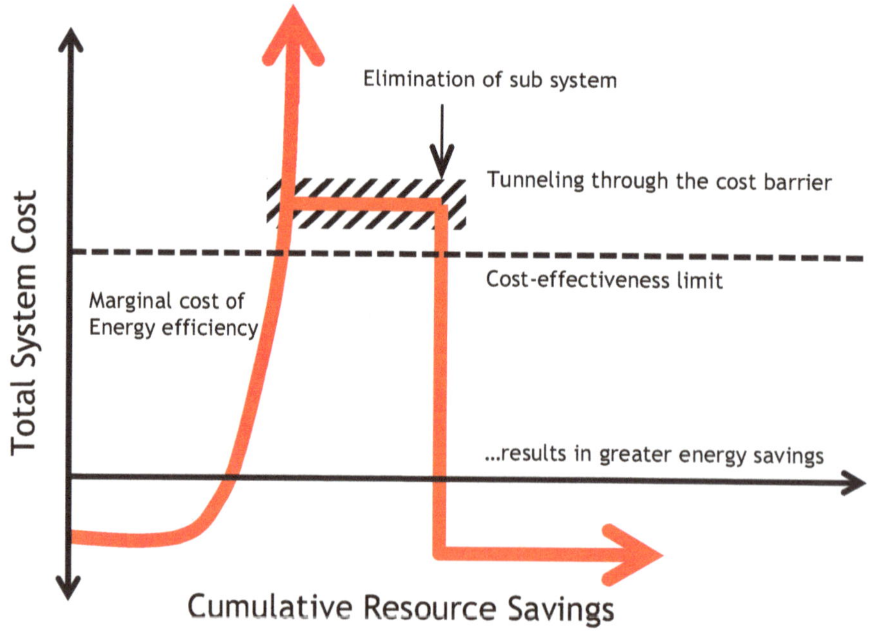

Figure 8.1 Cost tunneling: the performance gains of one system allow for significant reductions in or elimination of another.

Courtesy Four Great Walls

- The lowest-performance component in an envelope assembly is often the one most critically linked to cost tunneling—start the analysis with this one.
- Look to components that are not easily replaced for performance gains, such as insulation. Hard-to-replace translates into long service lives for these components. This makes long payback periods and smaller operational gains more attractive.
- Pay attention to details. If a system such as perimeter heating is eliminated, there is little room for errors in installation. Make sure that the installation will meet the performance objectives, because the eliminated system means that there is no ready backup plan.
- Do not hesitate to model alternatives and provide safety factors. Failure is costly.

Climate Considerations

Effective envelope design in cold climates must account for both the expected regional conditions as well as local microclimate conditions at the site.

Design Temperatures. Envelope design and component selection begins with understanding the expected outdoor temperature conditions, the desired indoor comfort levels, and the difference between them. In many cold climates, this difference (ΔT) may be more than 108°F (60°C), from 68°F (20°C) indoors to –40°F (–40°C) outdoor design temperature. This is a significantly greater difference in temperature than that encountered in hot climates (68°F [20°C] interior indoor temperatures, outdoor temperatures up to 113°F [45°C], ΔT of 45°F [25°C]), and even more severe than would be experienced in many flash freezer installations.

Whereas exterior design temperatures are tabulated (ANSI/ASHRAE/IES Standard 90.1 [ASHRAE 2019b]), as well as most building codes will furnish relevant data), interior design conditions are established by the design team in consultation with the occupants and owner. Spaces not requiring human comfort levels (e.g., equipment bays, some storage facilities, many mechanical rooms) may be designed to allow for greater swings in temperature and lower ΔT values. This effectively lowers the design temperature difference and potentially saves energy, costs, or both. Care must be taken, however, as modifying interior conditions may also affect condensation buildup in assemblies and drying potential.

Precipitation. Rain tends to flow away from buildings a little better than snow. Although temperate and tropical regions tend to experience greater volumes of precipitation, accumulation of snow over long periods and its tendency to drift against structures represent unique design challenges. See the section on Precipitation Control in this chapter for discussion on mitigating the severity of these issues.

Humidity. Increasing the amount of moisture in the air has a twofold concern for cold-climate design. First, water vapor does not increase the thermal conductivity of air, but it does typically adversely affect human comfort levels. This means occupants feel colder. Second, the increase in airborne moisture may also lead to condensation issues (particularly if the building mechanical system is not designed in accordance with the envelope) and a reduction in drying potential for condensation that does collect in envelope assemblies. The prevention and mitigation of condensation in assemblies is addressed in the section on Keeping Moisture In.

Duration. The length of time a building is exposed to cold weather must also be factored in. Although duration in itself does not typically affect envelope performance explicitly, it does tend to significantly exacerbate negative climate effects. For example, a wall assembly with a given thermal performance and water vapor diffusion rate may perform adequately in a climate where nighttime temperatures dip significantly below freezing (e.g., –22°F [–30°C]) if the daytime sun routinely heats the skin and cladding to above freezing, giving the wall time to dry out and warm the structural elements. However, if this same wall is exposed to the same nighttime temperatures for months on end in an arctic winter, it is likely to experience deleterious effects such as ice build-up, dust shadowing, condensation issues, reduction in thermal performance, rot, and resultant poor IAQ. In short, the longer an envelope is expected to be exposed to extreme temperatures, the more care and consideration it requires to maintain acceptable performance.

BUILDING ENVELOPE PERFORMANCE

Key Points

In cold climates, the building envelope is required to fulfill many performance objectives, including the following:

- Thermal control—the resistance to heat flow outward from conditioned space
- Water vapor control—controlling the movement and deposition of water vapor in and through the building envelope
- Precipitation control—snow drifting, moisture ingress, and ice damming in the building envelope
- Durability, serviceability—the ability of the envelope to maintain integrity over time
- Human habitability and comfort—maintaining favorable indoor conditions without excessive energy demands while maintaining appropriate daylighting and exterior views

The first three points relate directly to envelope performance and are readily quantifiable with standard metrics and measurements. The latter two are more qualitative in nature but contribute to the overall success of a building design.

Performance Objectives

Some of the challenges experienced in cold-climate design are universal to buildings in other climates. However, the geographic and climatic conditions experienced in cold climates heighten the effect of some of these challenges, as well as posing unique ones such as interior vapor migration and freeze/thaw cycles. This section highlights some of these concerns, so that professionals, owners, and contractors will have background information to more effectively analyze design options for envelopes and related building systems.

A properly designed and integrated building envelope should address the following objectives at all times:

- Thermal control
- Water vapor control
- Precipitation control
- Durability and serviceability
- Human habitability and comfort

To be successful, an envelope system must address these significant and disparate requirements in a site-constructed (potentially in harsh conditions) assembly, often in the absence of skilled labor and supervision. This is, at best, a daunting task, at worst, a disaster in slow motion. The task is made somewhat easier by having many components within the wall, roof, and floors to address one or more of these items specifically, and the assembly of these components can be intelligently arranged to help meet others. These objectives are discussed in following sections in more detail.

THERMAL CONTROL

A good cold-climate envelope design ensures that the energy produced within a building stays there. Controlling heat loss is really about understanding the physics of heat transfer.

The quantity of heat is dependent on several factors, including temperature, mass, and material. Thus, to increase the amount of heat in a system, it must be hotter, bigger, or have higher heat storage capacity. Reducing the heat loss from a building requires reducing the temperature difference between the heated volume and adjacent surfaces, reducing the amount of warm material flowing out of the building, and/or reducing the ability to conduct heat outward. This can be quite simple.

Remember that heat only flows in one direction—from a warmer object to a colder one. Regardless of the heat loss mode, the direction of heat flow will always be the same—hot to cold. However, in a building, there are leaks and drafts, convective currents, and differing materials that block or allow the passage of thermal energy, all of which make the flow of energy out of a building a little more involved.

To address all the heat loss paths in a building, identify the common paths of heat transfer. To effectively design an envelope, the following must be realized:

- Limit the transfer of thermal energy through the skin (heat loss)
- Limit the flow of material that carries energy out of the building (mass loss)

Effective control of these losses is key to sound cold-climate envelope design.

How Buildings Lose Thermal Energy

Heat loss through the envelope occurs via many paths, all of which are interrelated to some extent:

- Conduction
- Convection
- Radiation
- Advection

The relative importance of each of these depends on several factors, including the materials used, the quality with which the materials are installed, and the temperature difference of the respective surfaces.

Keeping Heat In: Modes of Heat Loss

Envelope assemblies must address all modes of heat transfer. This means paying attention to details and connections, as well as placement, material selection, and installation quality to ensure continuity of insulation and control layers. This section should not be read as an exhaustive explanation of the modes of heat loss, but as a primer to outline the principles and vocabulary so that designers may better understand how to reduce them.

Conduction. Conduction is what most people think of when envisioning heat loss. It is, in fact, what is measured by insulation manufacturers when indicating the performance of products. In conduction, heat is transferred through direct contact with a solid. To reduce conductive heat loss, either

- reduce the temperature difference between surfaces or
- reduce the thermal conductivity of the material.

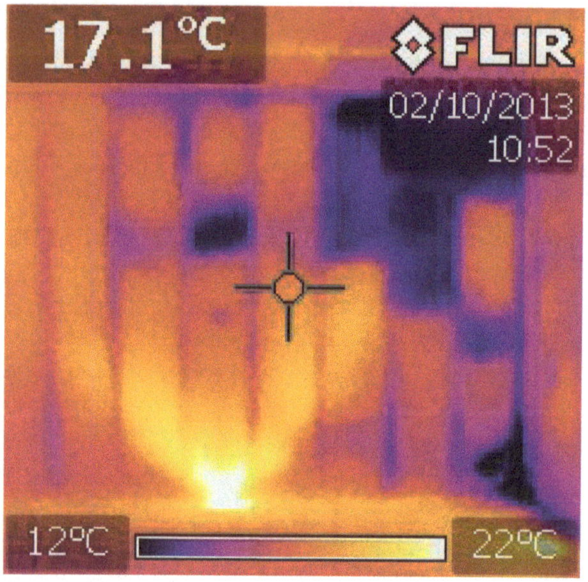

Figure 8.2 This thermographic image shows thermal loss through a wall. The darker colors indicate colder surfaces. Thermal bridging is apparent at the studs and a significant amount of heat is being lost in the top right-hand corner of the wall where insulation is missing.
Courtesy Four Great Walls

In temperate climates, outdoor conditions can work on the former. In cold climates, insulation does the latter (Figure 8.2).

Convection. Convection is the movement of heat through a fluid by imparted motion. In buildings, this fluid is air in the form of drafts. Moving and mixing air takes energy, which is lost as the air moves across colder surfaces.

Convection is sometimes confused with conduction, perhaps because the latter often leads to the former; a cold surface cools the fluid adjacent to it via conduction and the fluid moves because of convection, increasing the rate of heat loss. Convection will be limited in this section to air movement within the building and building envelope, but not leaking to the outdoor environment, which is a separate issue and separate mechanism of heat loss in buildings, and will be dealt with in the section on Advection.

Convection can be controlled most effectively by either restricting fluid flow or by reducing the temperature gradient across a surface. The former is used in multipane windows where the distance between individual glazing sheets limited to prevent the formation of convective currents (Figure 8.3), while the latter can be achieved by eliminating of thermal bridging in wall assemblies and reducing the formation of convection driven by surface temperature gradients.

Radiation. *Radiation* is defined by ASHRAE as thermal energy passing through space in the form of electromagnetic energy such as light or ultraviolet or infrared radiation (ASHRAE 2021b) (Figure 8.4).

At the absolute temperatures of buildings, the emission of thermal radiation is low. This makes radiation the smallest proportion of typical building energy loss. Heat loss from buildings through radiation is best addressed by lowering surface emissivity. Practically, this means installing a material with low emissivity, almost always a metal foil or derivative, somewhere in the envelope assembly, and always in combination with an air space. This material effectively reflects the thermal radiation back to the emitter, in this case, the interior of the building.

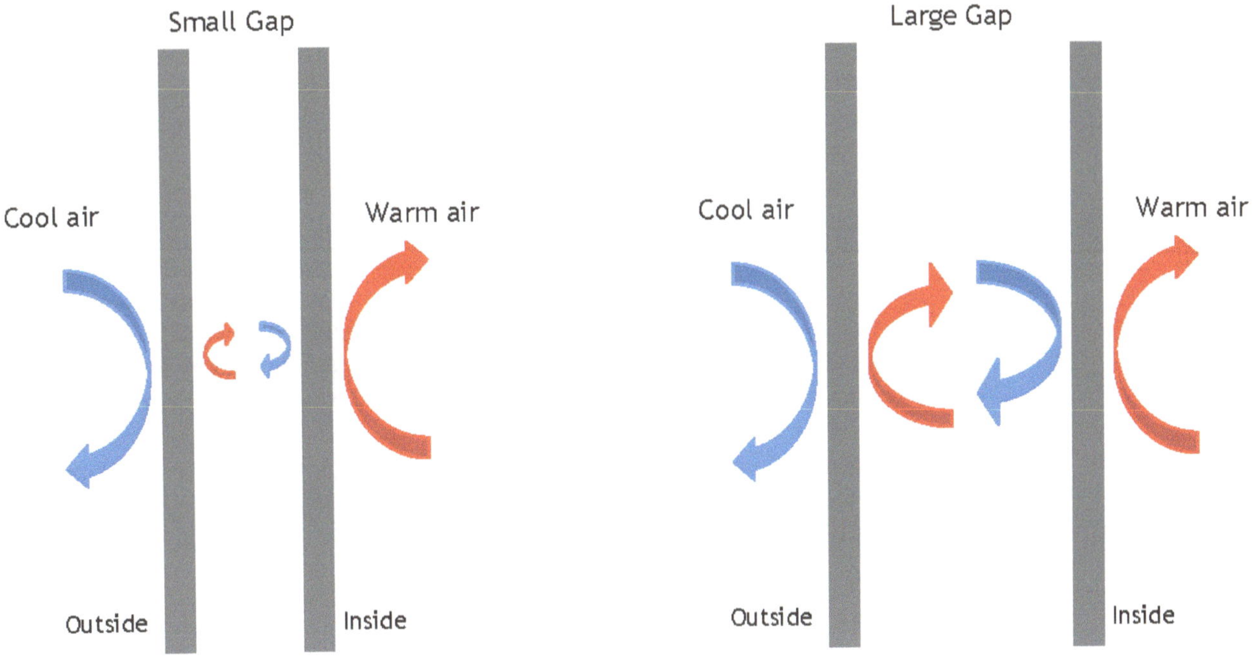

Figure 8.3 Sealed glazing units reduce convective currents by restricting fluid flow. The distance between the sheets of glass in a sealed unit is critical to its performance. With panes of glass relatively close together (left), boundary effects and fluid shear impedes the flow of air within the sealed unit, greatly reducing the creation of convective currents. If the distance is too large (right), convective loops can form within the unit.

Courtesy Four Great Walls

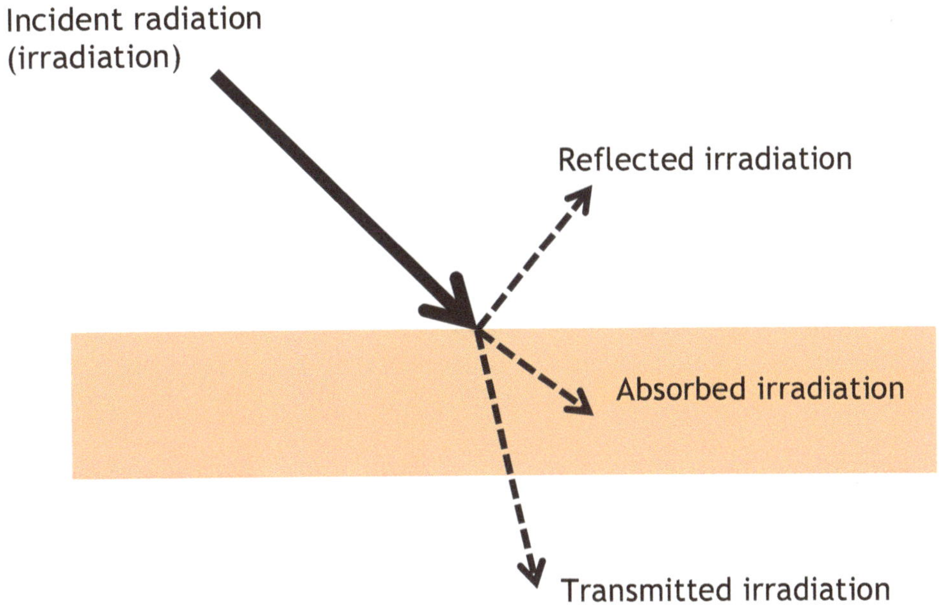

Figure 8.4 Thermal radiation—the infrared portion of the electromagnetic spectrum—has a complex propagation path through medium changes, undergoing reflection, refraction, and absorption. This complexity must be understood in order to control its transfer through a building envelope effectively.

Courtesy Four Great Walls

The complexity of radiation transfer means it is usually the least commonly understood mechanism of thermal energy loss, and probably subject to more exaggerated insulation product performance claims. Despite this, it does play a role—various films put on windows alter the emissivity and reflectivity of the glass surfaces and, when used correctly, can greatly improve the energy performance of a glazing unit. The thermal performance of a window is largely determined by the location and quality of the emissivity coating(s) to reduce radiative heat loss.

Advection. Advection is the transfer of thermal energy through moving a mass across a barrier. The more mass moved, the greater the heat loss. Buildings leak; warm air seeps out through cracks, voids, and gaps all the time. This warm air leaking out carries with it energy. Reducing the amount of air moving in and out of a building reduces the energy loss.

Controlling this air leakage is another matter. It is rare that a single large opening can be identified and easily remedied. Joints, differential movement, construction tolerances, and material porosity all contribute to the millions of little holes air can seep through. Some estimates place air leakage as 30% or more of the energy loss in modern buildings (Emmerich et al. 2005).

Controlling Heat Loss

This section addresses the principles behind reducing thermal energy loss by mechanism and recommends appropriate building envelope components to address each heat loss mode.

Conductive Heat Loss. Conductive heat loss in cold climates is best addressed through insulation. Most building products, for durability or structural reasons, are made from hard, dense materials. Materials with these properties usually also have high thermal conductivity. Properly installed insulation provides a continuous layer of high thermal resistance material to prevent heat conduction through these materials to the outdoor environment. To effectively reduce conductive heat loss, the insulation must be continuous. Heat, like all forms of energy, follows the path of least resistance. Design measures to reduce conductive heat loss include the following:

- Increase the thickness of insulation
- Increase the thermal resistance of the materials
- Eliminate (or minimize) thermal breaks

Each of these improvements may be done alone or in combination. In combination, these are the factors that will determine the effective thermal resistance of the envelope assembly under consideration.

Effective is stressed here as it takes into account the thermal bridging in the envelope. This is different from the nominal value, which is only determined by the first two points in the preceding list. Calculating effective thermal resistance for various assembly types can be complex. *ASHRAE Handbook—Fundamentals* (ASHRAE 2021a) is a good resource for the thermal conductivity of many building materials and the calculation methodologies to determine effective thermal resistance of assemblies. Effective thermal resistance reflects as-installed performance. Guides such as ANSI/ASHRAE/IES Standard 90.1 (ASHRAE 2019b) or most national or regional energy codes provide acceptable minimum effective thermal resistance values in floors, walls and roofs.

However, designers should understand that these are minimums. Increasing thermal resistance may be warranted for buildings seeking higher performance, or those that must maintain stringent interior conditions. In these cases the effective thermal resistance of the assembly might be well above these prescribed minimums.

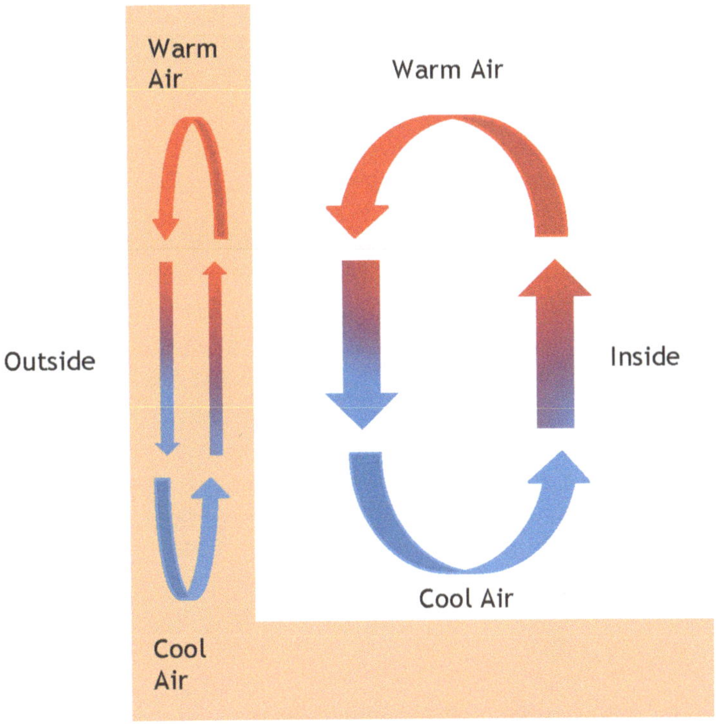

Figure 8.5 Typical convective current patterns in buildings. Larger arrows at right illustrate convective patterns that form within the occupied space of a building and are perceived as drafts by building occupants. Note that drafts can be caused by infiltration of cold exterior air into the building or by convective currents within the building as indicated. A smoke pencil or pressurization test can help determine the source of the drafts. Smaller arrows at left illustrate convective currents that form within the envelope assembly. These form when air within the assembly is allowed to move freely within a cavity. Note also how the formation and direction of flow is constant; cold, denser air moves down across exterior surface to be replaced by warmer, less dense air at the top to complete the cycle. This movement of air will continue as long as there is a temperature differential to drive it.
Courtesy Four Great Walls

Convective Heat Loss. Convective heat loss is remedied most effectively through the control of other heat loss modes. This is true for the two primary locations that convection currents form: on the interior surface of exterior walls where a portion of the envelope has a substantively lower thermal performance than adjacent surfaces (such as windows in opaque walls) or within cavities inside envelope assemblies if (1) the cavities are large enough or (2) the materials inside the wall are porous enough (Figure 8.5).

To address convective currents that form within the building, reduce the temperature differences that drive convective current formation. This is best accomplished through reducing conductive heat loss in assemblies and eliminating thermal bridges in the envelope.

Where convective currents form inside wall cavities, increasing outboard (that is, on the exterior side of the cavity) insulation to reduce conductive heat loss may or may not be an option depending on the envelope design. In cases where this is not possible, convection loops within the cavity may be controlled by ensuring the cavity is completely filled with insulation and specifying insulation types that have a high resistance to airflow. Both of these actions will prevent the conveyor of cold air from short-circuiting the insulation within the cavity.

Radiative Heat Loss. Radiative heat loss is the most difficult to control in building envelope design, because it does not require a medium to pass through (Figure 8.6). It is only effectively countered by including assembly designs featuring the following:

- A layer of material with high reflectivity/low emissivity
- An air space adjacent to the reflective layer

Both design features are critical to preventing radiative heat loss. Installing a highly reflective layer (typically, a shiny metallic foil product) alone does not address radiative heat loss, because this simply transfers the heat flow to the conductive path. Likewise, the inclusion of an air space without a reflective layer adjacent to it might be helpful from a serviceability point of view and to create thermal breaks within the envelope assembly, but it will not substantially address radiative heat loss. Radiative heat loss can only be addressed through including both design elements with the added provision that the reflective layer must be kept reasonably clean and dust free, and the adjacent air space should be continuous and at least 0.25 in. (6 mm) thick to reduce conductive and convective losses.

Advective Heat Loss. Controlling advective heat loss in a cold-climate envelope is dependent on reducing the rate of airflow through the envelope. This is effectively done by controlling either

- total area cross section of the opening(s) allowing airflow or
- flow rate through the opening(s).

As noted previously, the openings in an assembly are typically small in size but large in number, and often a direct function of workmanship of the air-barrier system. Carefully

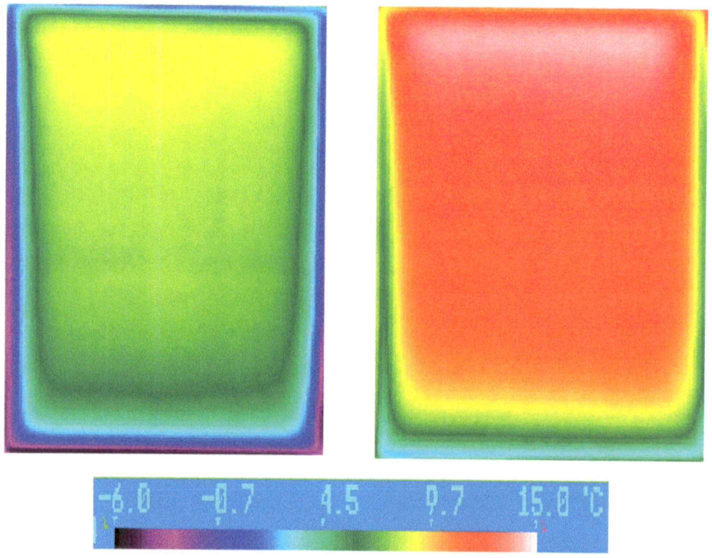

Figure 8.6 Low-emissivity (low-e) coatings. At left is a regular double-glazed sealed unit. At right is a double-glazed sealed unit with low-e coating and an insulated spacer bar. The thermograph is pictured from the inside with the cold exterior beyond the window. The much higher surface temperature of the window on the right represents tremendous energy savings, a reduced risk of condensation damage, and a more comfortable room with reduced drafts.
Courtesy Lawrence Berkeley National Laboratory

sealing penetrations, gaps, and transitions is critical. An envelope may have several layers of materials (often separated by a sizeable distance) combining to form an overall air-barrier system. Because of the interdependence of these related systems and the connections in plan, section, and across transitions, an effective air control strategy depends on ensuring that gaps in one material are compensated for in another.

To reduce the flow rate through any given opening in the building envelope, reduce the pressure differential driving the airflow. This may be done through very complex mechanical systems, but simple steps such as pressure-equalized rain screen cladding systems can also help reduce the pressure differentials across the envelope and lower the overall volume of air leakage in an envelope substantially.

Keeping Air In

Air Leakage. Sealing a building envelope against unwanted air infiltration and exfiltration is critical in cold climates. In a modern envelope design with reasonable insulation, it is estimated that uncontrolled air leakage through the envelope is responsible for 30% to 40% of the energy loss. Furthermore, air leakage is the most significant source of moisture deposition in envelope assemblies.

Although air control in the envelope may be contributed to by several layers and components in an envelope assembly, best practice stipulates that a single, continuous layer of material be identified and installed as the primary air control boundary, otherwise known as the *air-barrier system* (Figure 8.7). This ensures that tie-ins, penetrations, and gap sealing are more easily verified. The most important feature of an air-barrier system performance is continuity.

The list of requirements for air-barrier materials is found in Chapter 26 of *ASHRAE Handbook—Fundamentals* (ASHRAE 2021a). Using these materials is only part of designing an air-barrier system, however. Care must be taken in installation at all joints, penetrations, fenestration components, and corners. Installing materials that each act as an air barrier, but not transitioning between them appropriately or effectively, means there is no air barrier: the air simply flows through all the gaps between materials. ANSI/

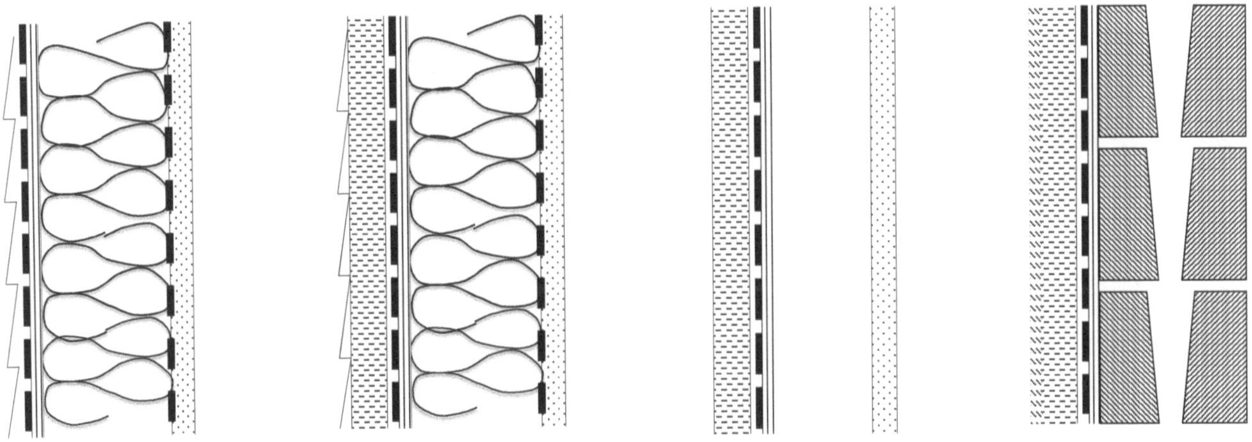

Figure 8.7 Wall sections illustrating common locations for air barrier components—indicated as heavy dashed lines—in building envelope assemblies. Note how the air barrier materials typically require suitable substrates for mechanical support and encapsulate porous insulations to reduce the formation of convective currents within the assembly.

Courtesy Four Great Walls

ASHRAE/IES Standard 90.1 (ASHRAE 2019b) requires an assembly to have an air leakage rate of not more than 0.04 cfm/ft^2 at 0.3 in. w.g. (2.0 L/[s·m^2] at 75 Pa) when tested to ASTM E2357 and CAN/ULC S742 (ASTM 2018, SCC 2011), reflecting the need to view the air leakage of the envelope assembly as a whole when considering energy efficiency (Figure 8.8). Some rules of thumb for calculating envelope air leakage are provided in Chapter 5 of this book.

Keeping Air Out

Air in buildings that is lost through leakage must be made up somehow. This usually means cold air infiltrating the envelope to replace the warm air that is leaking out. This helps explain why advection can account for so much of a building's heat loss; it is not just losing air at 72°F (22°C), it is taking untreated air at 20, 30, or 40 degrees below freezing and having to heat this directly as it is drawn in through the envelope. It also illustrates one of the fundamental differences between advection and other heat loss modes: it is a bidirectional flow. In a cold climate, the exterior temperature, humidity, and incident radiation are rarely higher than interior conditions; therefore, the flow of thermal energy through conduction is always in one direction, inside to outside. Air, on the other hand, can enter or exit a building depending on pressure loads (Figure 8.9). This carries with it particular concerns that are addressed in the Building Envelope Design section.

Keeping Moisture Out

Water accumulation in envelope assemblies degrades envelope performance, serviceability, and durability. This is particularly important in cold climates because the lower temperatures often result in long drying times for accumulated water and the potential for greater volume of moisture accumulation from freezing deposition. In addition to water entering from the exterior as precipitation, moisture may be deposited in the envelope through condensation of warm, humid air entering the wall assembly from the interior. This warm, humid interior air may enter the wall assembly through diffusion driven by vapor pressure or physical movement of air through the assembly via air leakage.

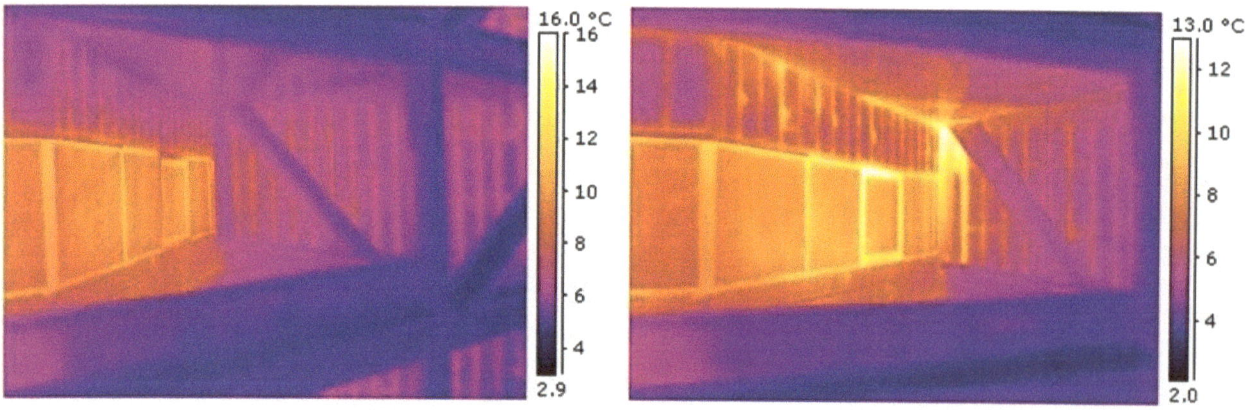

Figure 8.8 What air leakage looks like from an energy perspective. The thermograph on the left illustrates heat loss through an unpressurized building. The image on the right illustrates the significant increase in thermal transfer for the same window after pressurizing the building. The difference in thermal loss is due to advection.

Courtesy RDH Building Engineering Ltd.

Keeping Moisture In

In cold climates, large amounts of liquid water may be deposited inside an envelope as water vapor carried by warm interior air condenses into liquid water because of the temperature drop within the assembly toward the exterior. This deposition occurs in three ways:

- Water vapor condensing on cold interior surfaces and subsequently absorbed into assemblies as liquid water
- Water vapor migrating into and condensing within the envelope assembly
- Air leaking into or through the envelope assembly and depositing liquid water as it cools to the condensation point

Ice Buildup. Unique aspects of cold-climate precipitation control relate to accumulation of frozen material. As noted previously, rain doesn't drift. Wind-driven snow on roofs and around the base of buildings routinely reaches several feet in depth, and may be melted by heat transfer through the envelope. This results in having bulk water or ice accumulation in areas not usually exposed to precipitation. Care must be taken as a result to ensure that envelopes are sealed against water ingress even well above grade and below

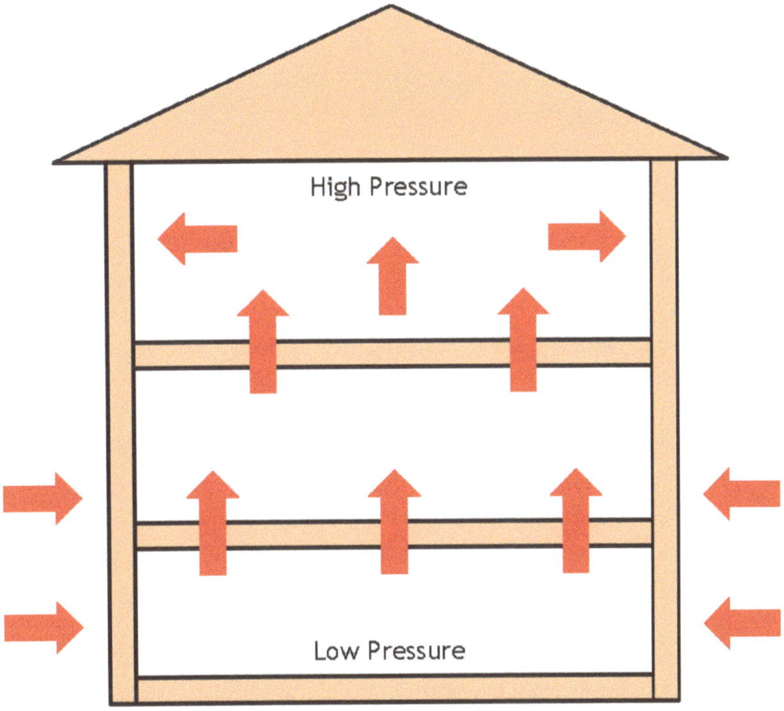

Figure 8.9 Simplified schematic of airflow in a building because of the stack effect. As warm air rises, it creates a high-pressure zone in the upper story, leading to exfiltration of air outward. The rising air creates a corresponding low-pressure area on the lower floor, leading to infiltration of cold exterior air at this level. Note that, although this illustration focuses on pressure differentials caused by stack effect, similar inward and outward (infiltration and exfiltration) movement of air can be driven my mechanical systems, wind, or changes to ambient air pressure. Unless compensated for by the mechanical system, exfiltration at one location in the building envelope will lead to infiltration in another.

Courtesy Four Great Walls

overhangs. Additionally, water that accumulates in envelope openings may freeze, creating larger openings by literally wedging apart the materials through repeated freeze-thaw cycles.

Likewise, any liquid water that accumulates in a cold climate has potential to freeze. This may result in large volumes of water being trapped within an assembly as ice. Upon melting, this sudden volume of liquid water may lead to severe damage to the building envelope.

Surface Condensation. The first area of the building envelope likely to experience condensation is cold interior surfaces. These typically include the following:

- Windows
- Window frames
- Structural elements that penetrate through insulation
- Locations with substantial thermal bridging

At such locations, heat loss through the envelope may be sufficient to lower the surface temperature below the dew point. This results in liquid water on the interior surface, which may cause the formation of mold, mildew, and degradation of materials, both physically and in appearance. Liquid water from surface condensation may also leak or be absorbed into the envelope assembly, causing additional damage. Surface condensation is avoided by ensuring surface temperatures remain above the dew-point temperature for the expected humidity levels. In addition to using the mechanical system to lower interior relative humidity levels, condensation may be controlled by

- providing additional insulation value to the assembly or components or
- designing mechanical systems to deliver additional heat to these areas, thereby raising the interior surface temperature.

For reasons of overall building efficiency, providing additional insulation value is strongly encouraged because simply adding heat via the mechanical system carries an obvious energy consumption penalty.

Diffusion. Water vapor also enters the building envelope through diffusion. *Diffusion* is the migration of water vapor through materials. It is a slow process, and controlled in an assembly by placing a vapor control barrier in the envelope assembly. The location of the vapor control barrier is critical to its function. To be effective, the vapor control barrier must be located where it will separate higher-humidity conditions from temperatures lower than the dew-point temperature of the interior air. In cold climates, where indoor humidity levels are normally higher than exterior humidity levels, this means locating the vapor control barrier inboard—the warm side—of the condensation zone temperature of the wall.

Air Leakage. By far the most damaging mechanism of moisture deposition in walls is through air leakage; some estimates put the amount of moisture deposited in envelope assemblies by air leakage at one thousand times the amount of water deposited by vapor migration. As noted previously, vapor diffusion in most buildings is a relatively slow process. This is not the case with air leakage, where high flow rates can be achieved if pressure differentials driving them are high enough. Air leakage has the potential to carry significantly more moisture into an envelope assembly and will deposit most of it in the vicinity of the air leak location, exacerbating the problem.

Air leakage is controlled by the air barrier. This may be the same material, construction, and installation as the air barrier installed to control advective heat loss. The salient features are also the same: the air barrier must be continuous, capable of withstanding the required pressure loads and structural movement of the design, and have sufficient protection from physical damage.

Envelope Drying Potential. Even the best-sealed and carefully constructed buildings will have air and moisture leaks. Therefore, consideration must also be given to ensuring that water that accumulates within envelope assemblies has some means to escape to the outdoors again. The ability to allow water within an envelope assembly to evaporate to the air is called *drying potential.* Several factors affect the drying potential of an envelope assembly, including the following:

- Weather
- Indoor humidity values
- Vapor permeance of the interior materials
- Ventilation rate of the cladding and cavity systems
- Insulation types
- Temperature of the wetted elements
- Materials
- Workmanship

Although software does exist to model drying potential for envelope systems, it is difficult to provide any real calculations to quantify drying potential. It is generally best practice to ensure that moisture and air leakage source control is used as the primary mechanism to address moisture ingress into envelope assemblies. Drying potential should be viewed as a form of insurance, to ensure longevity and performance from the envelope system.

There are practices to ensure envelope assemblies shed water that may collect within them, including the following:

- Provide only a single vapor barrier material in an assembly. Having two or more vapor barriers will prevent adequate drying between them. If the envelope assembly is required to have more than one vapor barrier layer, the detrimental effects can be mitigated by separating them with a well-ventilated cavity.
- Various construction materials have a wide tolerance to liquid water. Specify materials that are resistant to prolonged exposure to liquid water, such as concrete and masonry, in cases where the wall-drying potential is low.
- Air movement within a cavity removes water very quickly. Encourage airflow through ventilated cladding systems and allow air movement through conditioned service cavities to ensure that any moisture that might collect is carried away quickly.

Moisture buffering can help deal with liquid water entrapped in an envelope assembly. Solid wood materials, including log, heavy timber, and plywood sheathing, have the ability to absorb and retain a limited amount of moisture without significant long-term damage. This is only a viable option in cases where the seasonal variation in moisture results in extended periods of drying time to ensure the assembly can dry adequately.

Keeping Water Out—Precipitation Control

Preventing bulk water ingress from precipitation in cold climates is not very different from practice elsewhere, with two exceptions:

- Snow drifts
- Freeze accumulation

In short, the basic principles and best practices of precipitation control apply to cold-climate buildings: ensure flashings are in place and sloped correctly, provide gaps at material changes to eliminate capillary action, lap materials and architectural features to shed water to the outside of the building, slope sills and ledges away from the building, and other commonsense design principles. *The ASHRAE Guide for Buildings in Hot and Humid Climates* (ASHRAE 2009) provides further information on precipitation control measures.

Durability, Serviceability, and Human Comfort

Appropriate moisture control in envelope assemblies has a direct effect on occupant health and comfort, and building durability. The accumulation of liquid water fosters growth of mold and mildew. This has a direct negative effect on human health. It also degrades building appearance and longevity, and increases maintenance requirements. An energy-inefficient building might not meet performance objectives with respect to operating costs and environmental goals, but a building designed without sufficient care and diligence to moisture control in the envelope will likely have significantly more deleterious effects for occupants, including long-term health, repair, and tenability issues.

BUILDING A BUILDING

Design of an envelope system in a cold climate must reflect the need to conserve heat, maximize heat gain from solar sources, and meet user needs. This means it has two distinct design elements: (1) the envelope as building form (how the assemblies, fenestration, and components come together to form a cohesive building that meets user and performance objectives) and (2) the design of the individual assemblies to meet specific performance criteria. A successful cold-climate enclosure requires the coordination of the architect, engineers, building users, and, when present, the envelope consultant, to meet the necessary design briefs.

Orientation, Size, and Shape

Given the building design objective to conserve heat in cold-climate situations, there are general principles and considerations that make meeting performance objectives easier. Because cold-climate buildings are heating dominant, we will focus these design guidelines to address conservation of heat as a common design goal.

Orient the Building for Advantage

Building orientation is critical in cold-climate design. Orientation should capitalize on local climatic conditions including wind, temperature, humidity, solar access, and sun path. Key points to consider include the following:

- Exposing a large face of the building to the sun increases available passive heating through fenestration, increases building surface skin temperature to reduce heat loss, and furnishes space for supplementary energy gains such as photovoltaic panels.
- In extreme climates or where topography permits, reducing the exposure of the building to wind can have a notable positive effect on heating rates. Locate buildings low to the ground, in the wind shadow of vegetation, or in spots that are blocked from prevailing winds.
- Capitalize on other buildings by placing storage buildings, unheated parking garages, and other structures on the windward side of a building, which can help to dissipate wind and accompanying drifting snow. This can have a small but

noticeable effect on building performance and make a sizeable contribution to occupant comfort for those outside the building.

Having a building design tested in a drift tunnel helps identify problem areas, and can be used to help verify that entrances, fenestration, and ventilation openings are located to minimize snow buildup and remain clear and functioning.

KISS—Keep it Small in Size

The most economical space to heat is a space that is never built. Efficient planning and pruning of excess program, resulting in lower total floor area and decreased building volume, will reduce heating demand and energy consumption. Similarly, a smaller physical space means internal heat gains (electric lights, equipment and unregulated plug loads, body heat, and other sources) will meet a more sizeable proportion of the heat losses. The more efficient the plan, the easier it is to heat.

Surface Area Minimized. Closely related to the "KISS" principle, the shape of the building must also be considered. Forms with low surface-area-to-volume ratios (e.g., cubes) are inherently more efficient in retaining heat than those with high surface-area-to-volume ratios (e.g., fingers, long rectangles, or anything with a lot of wings or projections).

Heat loss is linearly dependent on exposure area; double the surface area, double the heat loss. Given this, it is good practice in cold climates to ensure that building form, layout, and projections minimize surface area while still meeting requirements for exterior views and lighting. This applies to both building massing and individual components such as balconies, floor slabs, and other bump-outs on the building exterior.

Plan for Heat Loss

Regardless of how insulated a building is, it will lose some heat through the envelope. To minimize this, some simple planning moves can help reduce this heat loss. Many of these are virtually free and can improve the performance of an envelope system substantially. These include the following:

- **Limit windows.** No window will ever perform as well as the adjacent opaque wall assembly from a heat loss perspective.Therefore, limiting window openings prevents excessive heat loss. Fenestration-to-wall ratio is an often-cited way to measure if a building is likely to lose excessive heat. For moderately cold climates, a 30% fenestration-to-wall ratio (i.e., 30% of the total wall area is window) is a reasonable limit. In severe cold climates, this number should be substantially less. Energy codes and energy efficiency standards such as ANSI/ASHRAE/IES Standard 90.1 (ASHRAE 2019b) and the National Energy Code of Canada for Buildings (NRCAN 2010) typically provide good guidelines for maximum allowable fenestration-to-wall ratios, often varying them with climate zone to reflect a reasonable energy balance.
- **Locate windows for advantage.** Implicit to limiting the window area is locating them so they do the most good. Windows on the dominant sun exposure side of a building (the south facade in the northern hemisphere, north face of the building in the southern hemisphere) allows both maximum daylight penetration and passive solar heating.
- **Create your own microclimates.** Certain spaces within a building (such as mechanical spaces or industrial process rooms) do not require typical indoor conditions and may have lower operating temperature ranges or less rigorous controls and heating duration requirements. Placing these on the perimeter of a

building provides a useful buffer to reduce heat loss from spaces that have more strenuous temperature and humidity requirements. Likewise, gathering up areas that have very rigid temperature and humidity requirements to the central locations is an effective strategy.

- **Vestibules and self-sheltering buildings help.** A good portion of the heat loss occurs at openings such as windows and doors. Much like the building massing benefits from placing buffering spaces on the outside vestibules, recessed windows, and even accumulated snow drifts can all meaningfully contribute to heat loss prevention in a building.

In short, the design team operating in cold climates should be aware of heat loss from the very start of massing and siting a building. As the climate becomes extreme, planning to reduce heat loss may be the primary design motivation of the project.

Building Envelope Design

The construction of all envelope assemblies—be it wall, roof, or underground assemblies—follow the same logic and design objectives. As stated in *The ASHRAE Guide for Buildings in Hot and Humid Climates* (ASHRAE 2009), all assemblies are similar in composition and requirements—only the orientation changes.

This section describes the process of designing an envelope assembly, highlighting the goals and various components of a vertically oriented wall assembly as an example. Although space and scope prevent undertaking a full explanation here, it is hoped that this discussion will give the reader insight into the process and considerations.

Objectives. All envelope assemblies must meet certain physical objectives. In all envelope assemblies, the migration (flow) of the following elements must be controlled by the building envelope:

- Thermal energy
- Air movement
- Vapor migration (water in the vapor state)
- Liquid water ingress

In cold climates, the direction of flow is always the same: thermal energy and water vapor flow from the interior to the exterior and water ingresses from the exterior inward. Airflow, though capable of flowing either direction (from interior to exterior or vice versa, depending on relative pressure differences) is represented as flowing from the interior out, because this is a source of water deposition in the envelope assembly.

To counter this natural movement of thermal energy, water, and air, specific materials and components should be installed into the envelope assembly as control barriers (Figure 8.10). These work to inhibit flow by increasing resistance to transfer. Given this, cold-climate envelope design is simply the systematic addition of components to provide appropriate resistance to these natural flows. A perfect envelope would eliminate all uncontrolled movement of energy, air, moisture, and water. In practical terms, it is more appropriate to limit them to acceptable levels. Intelligent envelope assembly design then is installing appropriate components in the right location with respect to the outdoors and each other.

Step 1: Insulation Installation—Addressing Conductive and Convective Heat Loss. By far the greatest loss of thermal energy through envelope assemblies is via conduction (Figure 8.11). Conductive heat loss also forms convective currents within the building by creating cold interior surfaces, leading to additional heat loss. Both these heat loss mechanisms are effectively addressed through inclusion of an appropriate insulation layer.

The insulation layer in the assembly is established first for several reasons. Insulation has large physical bulk (typically 4 to 12 in. [101.6 to 304.8 mm] thick in a cold climate, depending on type, design R-value, and climatic conditions), making it difficult to insert into an assembly as an afterthought. More importantly, the placement of the insulation layer(s) determines the thermal gradient through the wall assembly. This becomes critical as the other control layers in the assembly are located with respect to the thermal gradient and the resultant water vapor condensation point in the envelope assembly. The importance of this is clear as more components are added.

The following are points to consider in insulation design for the assembly:

- **Vapor Condensation/Thermal Gradient.** Because the insulation layer almost exclusively determines the thermal profile of the assembly, it will also establish

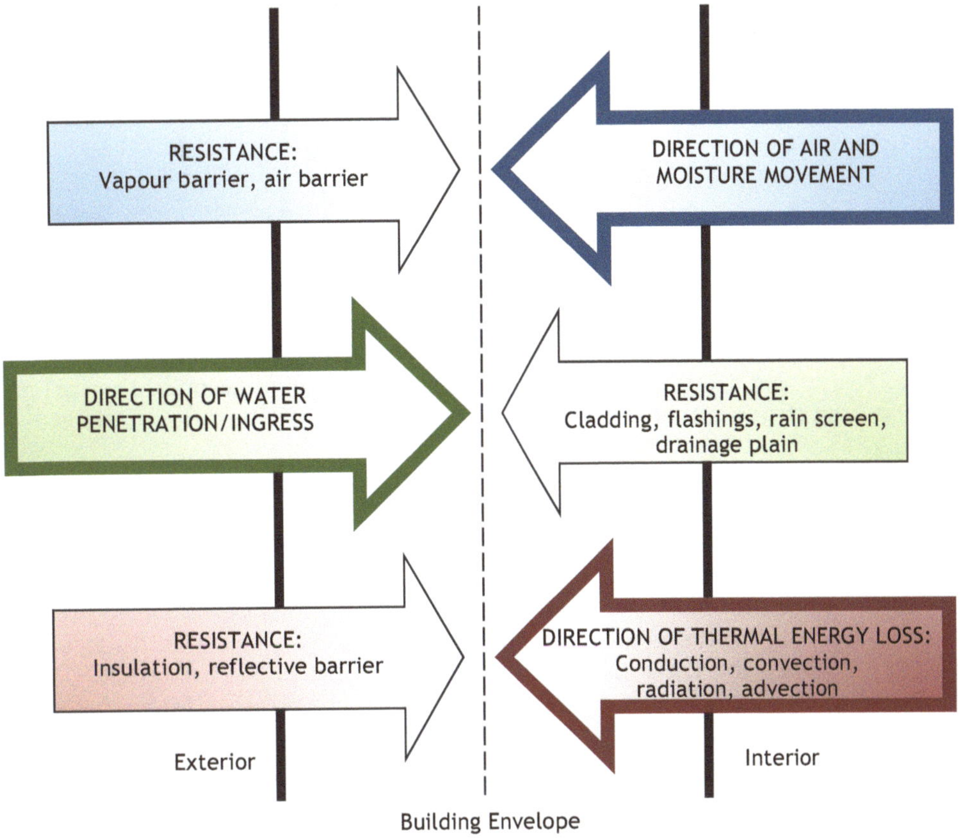

Figure 8.10 Graphical representation of envelope requirements. Large arrows indicate typical movement patterns of thermal energy, air, moisture, and water ingress through an assembly in a cold climate. Thermal energy, air, and vapor movement flow from interior to the exterior. The direction of water ingress is from the exterior to the interior. The direction of flow is dependent on simple physical laws: heat flows from higher temperatures to lower, moisture is driven by relative vapor partial pressure, and water ingress is driven by precipitation source. Although the magnitude of these drivers changes with region, the direction of movement is common to all cold climates. The smaller arrows indicate the application of resistance materials and building materials components to counter this natural movement. These resistance materials are called *control barriers*.

Courtesy Four Great Walls

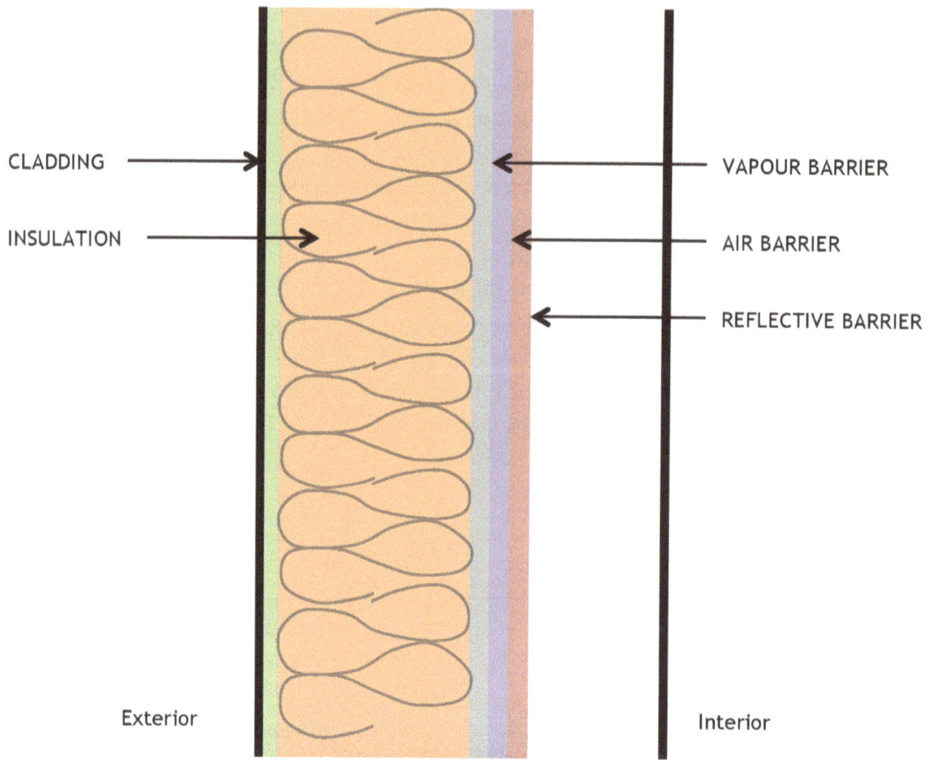

Figure 8.11 Control barriers' function and position in the building envelope.
Thermal control barrier: Insulation reduces conductive and convective heat loss. This barrier is optimally located toward the exterior to encapsulate structural elements and service systems in the conditioned space. Positioning it closer to the exterior also makes elimination of thermal bridges much easier, as well as limiting structural movement due to building temperature fluctuations and increasing durability.
Vapor control barrier: Water vapor movement is driven by vapor partial pressure. In cold climates, this almost always results in the net movement of water vapor outward through the envelope assembly to the exterior. To counter this, designers should create an envelope with the greatest resistance to vapor migration (lowest vapor permeability) near the interior surface. Materials of very low vapor permeability must always be placed on the interior side of the assembly dew point to prevent the water vapor from reaching the point of the wall where condensation is likely to occur. A successful and practical approach is to place the vapor barrier on the interior side of the insulation. Good practice stipulates that materials in cold climates should increase in vapor permeability as they move through the assembly toward the exterior.
Air barrier: An air barrier controls airflow through the assembly. Air barriers can be installed nearly anywhere within an assembly, depending on their vapor permeability. In this example, it is located on the inside of the primary insulation layer and, as such, is assured to be on the warm side of the condensation zone. A vapor-impermeable air barrier is preferred.
Cladding: Cladding requirements for cold climates are largely the same as for other climates. Rain screen installations are recommended. Cladding fastening can create significant thermal bridges through any external insulation type, so care should be taken to minimize this while still meeting the durability and structural requirements of the cladding system.
Reflective barrier: Controlling heat loss via radiation requires installing a low-emissivity (low-e) layer adjacent to an air space. A low-e coating can be added to windowpanes and within wall cavities. Depending on climatic conditions, this can have substantial benefit in reducing energy consumption.
Courtesy Four Great Walls

where any water vapor in the exfiltration air will condense in the assembly. Understanding where condensation can occur is critical to avoid the buildup of liquid water in an assembly, with corresponding degradation in performance and durability. This is particularly true if there are multiple layers of insulation in an assembly, or a dedicated vapor barrier where the thermal gradient of the assembly and vapor migration pattern is more complex.

- **Vapor Permeability.** Insulation may be either vapor permeable (e.g., fiberglass batt, mineral wool, or most organic fibers such as cellulose or cotton) or vapor impermeable (e.g., foil-faced insulations; extruded polystyrene; or most high-density urethane products, either spray-in-situ or sheet format). Knowing the vapor permeability of the insulation is critical to avoid trapping moisture within the envelope assembly. This is a particular concern in walls with two or more insulation layers or types, because they are likely to be physically separated. Being aware of the relationship between vapor permeability of materials, location of condensation zone in the assembly for the given building design parameters, and the drying potential of an assembly all factor into locating and specifying appropriate insulation layers within a cold-climate envelope assembly.
- **Insulate Outboard.** Placing the insulation layer outside of major building structural elements not only reduces thermal bridging, it promotes building durability by preventing the structural elements, fasteners, and various vapor, water, and air control membranes unnecessary stress and fatigue caused by movement resulting from thermal cycling.
- **Thermal Bridging.** Care should be taken when designing the insulation layer to ensure that it is continuous as possible in both plan (horizontal) and section (vertical). This includes placing insulation in the various assemblies so that it can continuously 'wrap' where assemblies meet (e.g., where walls intersect roofs).Designers must ensure that the installed thermal performance of the wall is understood and communicated to the entire consulting team. Thermal bridges, gaps, and structural elements that penetrate the envelope degrade the listed performance of the insulation. Even small, repetitive components such as Z-girts and cladding supports can have substantial effects on envelope performance.
- **Filled versus Empty Assembly Cavities.** At first glance, filling any available space within an assembly to increase thermal resistance would seem an excellent idea. However, because many of the cavities in question are on the interior side of the vapor and air-control barriers, this additional insulation may render the vapor barrier ineffective by shifting the condensation point toward the interior. On the other hand, filling voids reduces convective currents within a wall and thus improves thermal performance. A well-designed envelope will respect these competing aspects. A thermal gradient calculation is a must for any complex envelope assembly. Many cold-climate jurisdiction's building codes contain tables outlining the minimum ratio of vapor-impermeable outboard insulation to cavity insulation to ensure the cavity insulation remains above the condensation zone for the given climate zone. As with all building codes, this represents minimum performance values. Consultants working in extreme climates or seeking higher-performance values should still undertake a more sophisticated energy and condensation analysis, including computer simulation.

Step 2: Vapor Barrier Installation—Controlling Moisture. An envelope assembly must bridge interior space at ±36°F (±20°C) to exterior air at perhaps (–20°F, –40°F, or –60°F (–30°C, –40°C, or –50°C). Thus, the envelope assembly has a substantial thermal gradient within it. This thermal gradient will lead to condensation of any water vapor that enters the assembly.

As noted previously, water vapor enters an envelope assembly by either migrating through the materials or by air leakage through gaps, cracks, or discontinuities. Each of these is addressed separately.

Installing a vapor barrier helps control vapor migration into assemblies through materials (Figure 8.11). Most construction materials can transmit water vapor (i.e., are permeable) to some extent. Certainly, any joints and gaps between them allow water vapor to pass. Therefore, it is necessary in cold climates to install a component in any assembly to control the outward flow of water vapor from the interior to the exterior.

In cold climates, migration of water vapor to the outside of the envelope assembly deposits liquid water as it reaches the condensation zone. A successful envelope design will limit the outward flow of water vapor while simultaneously allowing any water vapor within the wall to migrate out to the environment. As a result, the vapor permeability of materials and construction layers outside of the vapor barrier (i.e., closer to the exterior of the envelope) should be higher to allow water vapor to pass through them more easily than those of the vapor barrier and of materials and construction layers closer to the interior of the envelope. Thus, water vapor—or liquid water—that finds its way from the warm/humid interior of the building into the building envelope assembly can continue its migration to the outdoors.

The following are points to consider in vapor barrier design for an assembly:

- **Modeling.** As rule of thumb, a vapor barrier is beneficial when interior water vapor partial pressure is higher than outdoors and there is potential for water vapor migration to the exterior environment. In most cold climates, where cold winter conditions result in dry exterior air, this is true. However, as exterior humidity increases, such as in cold, coastal environments, this may not always be true. In such cases, it is recommended that a computer simulation be done on the assembly design to determine water vapor deposition, condensation, and migration rates to determine if a vapor barrier is beneficial. Good practice stipulates that if a vapor barrier is deemed to not be required, it is good practice to design the assembly to have a high drying potential to allow for vapor to migrate through the assembly completely. Several software programs exist that will model vapor accumulation in assemblies.
- **Materials.** Industry standard states a material to be vapor impermeable below 1.0 perm (60 ng/[Pa·s·m^2]). There are also materials known as *semipermeable*, and classification of air barrier types based on material vapor permeance. *ASHRAE Handbook—Fundamentals* (ASHRAE 2021a) cites three categories of vapor barrier from ICC (2007):
  - Class I: 0.1 perm or less (0–5.7 ng/[Pa·s·m^2])
  - Class II: more than 0.1 perm but less than or equal to 1.0 perm (>5.7 ng/[Pa·s·m^2] to < 57 ng/[Pa·s·m^2])
  - Class III: more than 1.0 perm but less than or equal to 10 perm (57 ng/[Pa·s·m^2] or more)
- **Dry to the interior?** In those rare cold climates with very high exterior humidity levels, it may in fact be detrimental to install a vapor barrier on the inside. If there is no possibility for water vapor to migrate outwards, it may benefit assembly longevity and performance by allowing drying to the interior. This is will be

extremely uncommon in cold climates and should always be verified via computer simulation prior to finalizing assembly design.

- **Semipermeable or impermeable?** For most applications, the differentiation between impermeable and semipermeable is not critical. However, in cases of extremely cold climates or high interior humidity conditions (such as swimming pools or restaurants), it may be necessary to specify vapor barrier materials that are impermeable (i.e., with a Class I vapor permeance rating).
- **Vapor barrier continuity?** Vapor barrier materials are designed to resist water vapor migration by diffusion, which is driven by vapor partial pressure. This means vapor flow through the wall is approximately as a direct (straight) line process. As a result, vapor barrier continuity is not as critical as thermal or air barriers, which can follow fairly complex paths through the assembly. That is why many professionals refer to the vapor barriers as *vapor retarders*. This reflects acceptable installation procedures and code requirements where vapor barriers are usually required to be continuous, but not necessarily sealed like an air barrier system must be. In general, a vapor barrier's continuity requirements must increase as the building's interior humidity levels increase; a swimming pool or steam room should have a higher-performing, more tightly sealed vapor barrier system than a residence or office space because the vapor partial pressure is much higher in the former. Again, if in doubt, a computer simulation may be in order. If the envelope design uses the same material to act as the vapor and air barrier(s), then it must be sealed at all joints, gaps, and penetrations.
- **Number of vapor barriers.** Can you have too many vapor barriers? Yes. Is more than one vapor barrier always a bad design? Not necessarily. Multiple vapor barriers may be successfully integrated into an assembly provided they are all located on the interior side of the expected dew point of the assembly. Although a single vapor barrier is preferred, there are occasions in which multiple vapor-impermeable layers are present in a wall. If several vapor barriers are included in the wall, it is good practice to ventilate between them if this does not create a thermal short circuit or have gaps in the second and subsequent vapor barrier materials to allow for some wall drying to occur.

Step 3: Air Barrier. Having installed insulation to control conductive and convective heat loss and a vapor barrier to prevent moisture migration into the assembly, the envelope now must become airtight. Air leakage is a significant cause of energy loss through advection, as well as the dominant source of moisture deposition in assemblies. Controlling airflow through assemblies is of utmost importance for building performance and durability.

Air barriers perform three roles in a cold climate envelope:

- Control movement of air from the interior of the building to the exterior
- Reduce the formation of convective currents within the envelope assemblies
- Prevent moisture-laden air from moving from the interior and condensing inside the building envelope

Air barriers can be located most anywhere in an assembly (Figure 8.11). In most industrial, institutional, and commercial buildings, the air barrier system is located well within the wall assembly (1) to prevent damage from ultraviolet or physical abrasion from the outside, and (2) to minimize the number of penetrations through it for plumbing, electrical, and structural elements.

One of the most important characteristics of air barriers is water vapor permeability. Those considered vapor-permeable (typically, materials with a vapor permeance greater than 1.0 perm (60 ng/[Pa·s·m^2]), will permit the transmission of water vapor through them via diffusion. As mentioned previously, the vapor permeability of air barriers may be a benefit or a challenge, depending on the location in the assembly with respect to the thermal gradient and condensation zone. In general, if the air barrier is inboard of the condensation zone of the assembly, it is likely to be vapor impermeable and also perform as the vapor barrier in the assembly. If it is exterior to the condensation zone, it should be vapor permeable to ensure any water vapor within the building envelope can migrate to the exterior environment.

The following are points to consider when designing an air barrier for an assembly:

- **Number of Air Barriers.** Including more than one air barrier in an assembly is not only acceptable, but may be beneficial, particularly if it encapsulates an air-permeable insulation and reduces convective currents within the insulation material or adjacent cavity. However, care must be taken by the designer to ensure that additional air-barrier layers do not adversely affect the vapor-permeance characteristics of the design. Additionally, the greater an envelope assembly's resistance to airflow, the lower the drying potential. Any water that enters the assembly may take a long time to evaporate out.
- **Protecting Insulation.** Some insulation, such as foamed plastics, high-density mineral wool, and most rigid board-, foil-, or paper-faced products, inherently resist airflow and may not require an air-barrier system to prevent convective currents from forming within the insulating material. Other insulation, such as fiberglass batt insulation, has relatively high air permeance and should be encapsulated within a cavity that limits airflow to ensure maximum effectiveness. A good rule of thumb is that if the insulation is soft to the touch, it is probably air permeable. A more scientific method is to consult the manufacturer's specifications and recommendations.
- **Products.** Many air-barrier systems consist of a membrane, either sheet or fluid, applied over an approved rigid substrate. Care must be taken to ensure the substrate conditions and materials are compatible with the specified air-barrier product to ensure performance criteria—during installation and long-term—are met.
- **Continuity.** Air barriers must be applied to all faces of the building, including roofs, walls, and assemblies below grade. They must be continuous and sealed. Air movement is driven by pressure differences, so air leakage can occur up, down, or sideways. For buildings with several air-barrier materials, it is important to ensure that any transitions in the primary air barrier, for instance, at window and door frames and mechanical penetrations, have appropriate transitions to ensure continuity. This is a matter of careful detailing and installation.
- **Allow for Servicing.** Including a wall cavity on the interior side of the air barrier where building services (e.g., electrical and plumbing) may be routed without having to penetrate the air-barrier layer is a good idea. This reduces the number of penetrations in the air barrier and many potential leakage sites. It also eases the application of maintenance and upgrades without damaging the air barrier.
- **Sealing the Air Barrier.** Where penetrations in the air barrier must occur, they must be sealed according to manufacturers specifications. Particular attention must be given to maintaining air-barrier continuity at areas where movement or joints occur, such as the following:

- Substrate transitions
- Joints in sheet applied membranes
- Penetrations, such as for electrical, mechanical, or structural elements
- Fenestration and mechanical openings
- Parapets and foundation-wall transition zones

The vast majority of air leakage in buildings occurs at these joints, junctions, and transitions. Attention must be paid to the design and execution of these details at both the design and construction phase.

Step 4: Cladding—Controlling Liquid Water. One obvious requirement for a building envelope is to keep precipitation out and protect building materials and systems from moisture and ultraviolet degradation. Envelope assemblies must ensure that water on the outside stays on the outside. In this respect, cold-climate envelope design has the same goals as any environmental separator—carefully flash windows and other penetrations to shed water out, avoid roof details that may drain water onto walls, and provide moisture planes within the assembly to drain any water that enters to the outside—all the normal best practices in shedding precipitation apply. They might require more careful and exhaustive application in cold climates because of drifting snow and freezing, but in principle, they are the same (Figure 8.11).

Specific cladding considerations unique to cold climates largely relate to durability and maintenance. Building materials become brittle at very cold temperatures. Sealants, cladding, substrates, and membrane materials all become susceptible to cracking and stress failure. Material contraction in the cold must also be factored in; unintended gaps allowing moisture penetration in the building may form as a result of materials contracting during cold periods.

Shingled, cladding, and rain screen applications that (1) do not rely on face sealing to accommodate movement in the materials, and (2) tolerate gaps and differential movement are highly recommended. Rain screen applications also offer additional benefits in promoting air movement within the cladding system and providing additional assembly drying potential.

The following are points to consider in cladding design for the assembly:

- **Materials.** Material selection is important in cold climates. Thermal expansion/contraction, resistance to moisture, cold-weather ductility, appearance, and weather resistance are just some of the considerations that should be factored into the choice of cladding material. In addition, because cold climates are also often characterized by remote communities, things like maintenance, resistance to physical damage, and availability of replacement components may be concerns.
- **Source locally.** Using local materials—for instance, by studying the traditional design vernacular—will likely furnish a good starting place for cladding materials. What has worked in the past has a good chance of working in the future. Locally sourced materials also allow for much easier replacement should damage occur.
- **Select colors.** Color selection on a cold-climate building is an aesthetic choice, but can also affect building performance. Darker colors absorb more solar radiation, but also radiate over a longer wavelength, thus cooling down faster at night. A color scheme that concentrates darker colors on the sun-exposed side and has lighter colors on façades not directly in sunlight will actually use a little less energy than the opposite configuration.

- **Select textures.** Buildings with large expanses of walls can benefit from cladding materials and configurations with rough textures. This creates a multitude of small air pockets around the building that facilitate energy conservation by increasing the still air layer around the building, similar to the function of fur on an animal.

Step 5: Radiation Reduction. The final element to consider is control of heat loss through radiation (Figure 8.11). This is by far the smallest component of energy loss in buildings. As noted earlier, radiation loss is difficult to address effectively in buildings, but as envelopes become more energy efficient, it can play a role, even in opaque building assemblies.

The following are points to consider in radiation heat control design for the assembly:

- **Materials.** As noted previously, controlling heat loss via radiation requires a material of low emissivity adjacent to airspace. The air space should be as airtight as possible—both to prevent convective currents and to prevent dust from accumulating on the low emissivity surface—both of which reduce the effectiveness of the material to control radiative heat loss by increasing surface emissivity.
- **Cost versus benefit.** Installing a reflective insulation layer expressly to control radiation heat loss may be valuable for high-performing or extreme-climate buildings. This is particularly true in very cold climates where the temperature difference between the indoors and outdoors is large, or in temperate climates or buildings with high internal heat gain where cooling loads may be reduced in the summer months. This can, in fact, be a case of cost tunneling, where an air conditioning system might be eliminated if solar heat gain can be limited.In most buildings, other heat loss modes (conductive, convective, and advective) are significantly larger in magnitude, meaning the performance gain for the opaque portions of the envelope may be small.For transparent portions of the envelope, such as windows, low-emissivity coatings are almost always cost effective.
- **Double up tasks.** A well-planned assembly can capitalize on materials with dual purposes to reduce radiative heat loss. For example, the foil layer of foil-faced insulation typically has air and moisture control properties in addition to providing a low-emissivity surface.

Assembly Guidelines

This section discusses common concerns for different envelope assemblies in cold climates. The envelope design team must take into consideration all local and professional guidelines to develop an appropriate, cost-effective envelope assembly while working in concert with the rest of the design team. This section discusses the base (foundation) through the building envelope, covering design considerations and common pitfalls for envelope design in cold climates. The following points are valuable, but cannot be considered an exhaustive checklist.

Foundations and Below-Grade Spaces. Foundation design in cold climates must balance heat loss yet still provide freeze protection for foundation support components. Some things to consider are as follows:

- Ensure that there is no water under or adjacent to the foundation if freezing is a possibility. Clean, washed gravel fill around foundations and footings helps ensure that water does not collect at these locations, as does appropriate drainage tile installation.

Figure 8.12 Building on piles.

Erich Binder

- In locations where the ground has permafrost, heat must be prevented from flowing into the ground and melting it at all costs. An effective means to counter heat transfer into the ground is to elevate the building on uninsulated piles. This allows the piles to dissipate heat to the air instead of transferring to and thawing the soil. (See Figure 8.12.)
- If the foundation design requires frost protection, installation of insulation may be vertical or horizontal. Both orientations are effective and have various advantages; the designer must consider which is best.
- Grade beams and other foundation designs using piles for support must allow for ground movement in frost conditions.
- Geotechnical reports are critical in cold climates to aid foundation design and should include ground conductivity, if possible, to help determine expected heat loss and water content of the soil.
- Insulating interior spaces adjacent to foundations must be carefully considered. Overinsulation can lead to ground freezing, excessive movement of the foundation, and premature failure.
- Even walls and floors below grade can have significant air leakage. It is not good practice to rely on soil or back fill to act as an air barrier in the system. A proposed air-control layer should be applied to all walls and floors, even those in contact with the ground.

Crawlspaces. Not all buildings have crawlspaces, but those that do must answer the following design questions:

- **Heated or unheated?** Both can work in cold climates. If heated, it must be adequately insulated (including preventing heat loss to the ground) while still allowing heat flow to prevent freezing of foundations. If the crawlspace is unheated, then the floor separating the crawlspace must be insulated from conditioned space above. If the crawlspace is unheated, the foundation design must assume freezing conditions and be designed accordingly. Human comfort may be affected if the crawlspace is unheated and floor structural members may develop cold bands on floors due to thermal bridging.
- **Ventilated or nonventilated?** If there is an unheated crawlspace, it should be ventilated. If heated, it cannot be ventilated. Do not mix these two approaches or do either partially.

Air leakage in heated crawlspaces can be a major source of energy loss and may lead to premature envelope failure because of condensation on adjacent structural members. If a crawlspace is heated, it should meet all the same envelope requirements as occupied space with respect to air leakage and moisture control.

In unheated crawlspaces, building services such as heating and ventilation ducts, plumbing lines, and other mechanical devices must be protected from cold as though exposed to exterior temperatures.

Floors. In many buildings, floor edges are either fully or partially exposed to the elements, so they must be accommodated in the envelope, with the following considerations:

- **Air Loss.** Given the structural penetrations typical of most floor assemblies, air sealing at floor edges is difficult. Several solutions exist, depending on the floor structure and air-sealing strategy for the building. It must, however, be a designed solution; deciding during construction how to air-seal floor ends is a recipe for failure.
- **Moisture Control.** Like air leakage, moisture control at floor edges is complicated by the multitude of structural elements intersecting each other, especially in light wood-frame buildings. Employing a spray-applied moisture retarder (such as site-applied, high-density urethane foam insulation) is often preferred to sheet-applied moisture barriers such as polyethylene in these locations.
- **Insulation Value.** In cold climates, exposed floor edges should be insulated to the same thermal resistance as adjacent walls.
- **Cantilevers.** In cold climates, cantilevered floors are a severe design challenge, and should be avoided if possible. Floors that cantilever out from the building MUST be insulated or be designed with a thermal break in the structural elements. Like balconies, these are effectively cooling fins on the envelope and should be insulated to the same standard or better than adjacent wall assemblies. Many regional energy codes indicate minimum values for cantilevered floor assemblies. Use of continuous insulation on the underside and ends is recommended. Also, as the structural portion of the cantilever floor is rarely heated directly, consideration must be given to air and moisture ingress at these locations. The prolonged and severe cold cantilevers experience can lead to condensation and rot. Appropriate consideration to effective air sealing, vapor barriers, and assembly-drying potential must be given to these areas.

Walls. Walls typically make up the largest areas of the envelope assembly and therefore must be well designed in cold climates. Wall performance standards have changed over time in response to better understanding of how thermal energy, air, and moisture move through wall assemblies (Figures 8.13, 8.14, 8.15, and 8.16). Wall design considerations include the following:

- Wall performance should be as consistent as possible throughout the space. Significant differences in thermal performance along a wall face, either caused by large thermal bridges or inconsistent insulation application, can lead to cold spots and interior condensation problems.
- Care must be taken with cladding materials that absorb moisture, such as masonry. As seasonal temperature fluctuations occur (prevalent in more temperate cold climates), they can release months of accumulated moisture on warmer days, causing reverse vapor diffusion through the wall by increasing the humidity levels within cavities if not properly vented. This is a particular concern with cavity-insulated walls using a vapor barrier on the innermost side, but may also be a concern in cold climates with high exterior humidity levels.
- Interfaces between wall assemblies and fenestration are traditionally weak points in the envelope with respect to air leakage and water ingress. Careful detailing is critical at these junctions, with air, vapor, precipitation, and insulation layers all maintaining continuity.
- Designing a wall assembly with interior service spaces (i.e., inside the thermal, air, or vapor barriers) increases serviceability and greatly reduces the number of penetrations within the wall system that can be sources of energy loss. Even simple furring works in this regard if spaced and sized appropriately.
- Like every assembly, air leakage through walls is a major cause of envelope failures and energy loss. In choosing an assembly design, it is often beneficial to choose a system that has been tested as an assembly. Testing standards such as ASTM E2357 (ASTM 2018) and CAN/ULC S742 (SPC 2011) describe how to measure the air leakage performance of an assembly (including openings, penetrations, and joints). These provide not only quantitative performance measures for the assembly but also manufacturer's details that may be site-verified to help ensure installed performance expectations are met.

Fenestration. One of the biggest design opportunities (and sources of failure) is fenestration design. The relationship between fenestration, heat loss and gains, and mechanical system design is complex, and it is difficult to provide general rules. The design team should be aware of this and make the most of the integrated process, making the layout, specifications, and integration of fenestration a top priority. The following are design considerations for fenestration:

- Window glazing properties make a huge difference. Pay particular attention to the glass specification in cold climates, maximizing for heat gain on southern exposures and minimizing heat loss on the others. Specify high visible-light transmittance and high solar heat gain coefficients (in the range of 0.6 and above for very cold climates). Lower visible light and solar heat gain coefficient values are a concern only in buildings with extreme internal heat loads, or those in milder cold climates where buildings are also air conditioned.

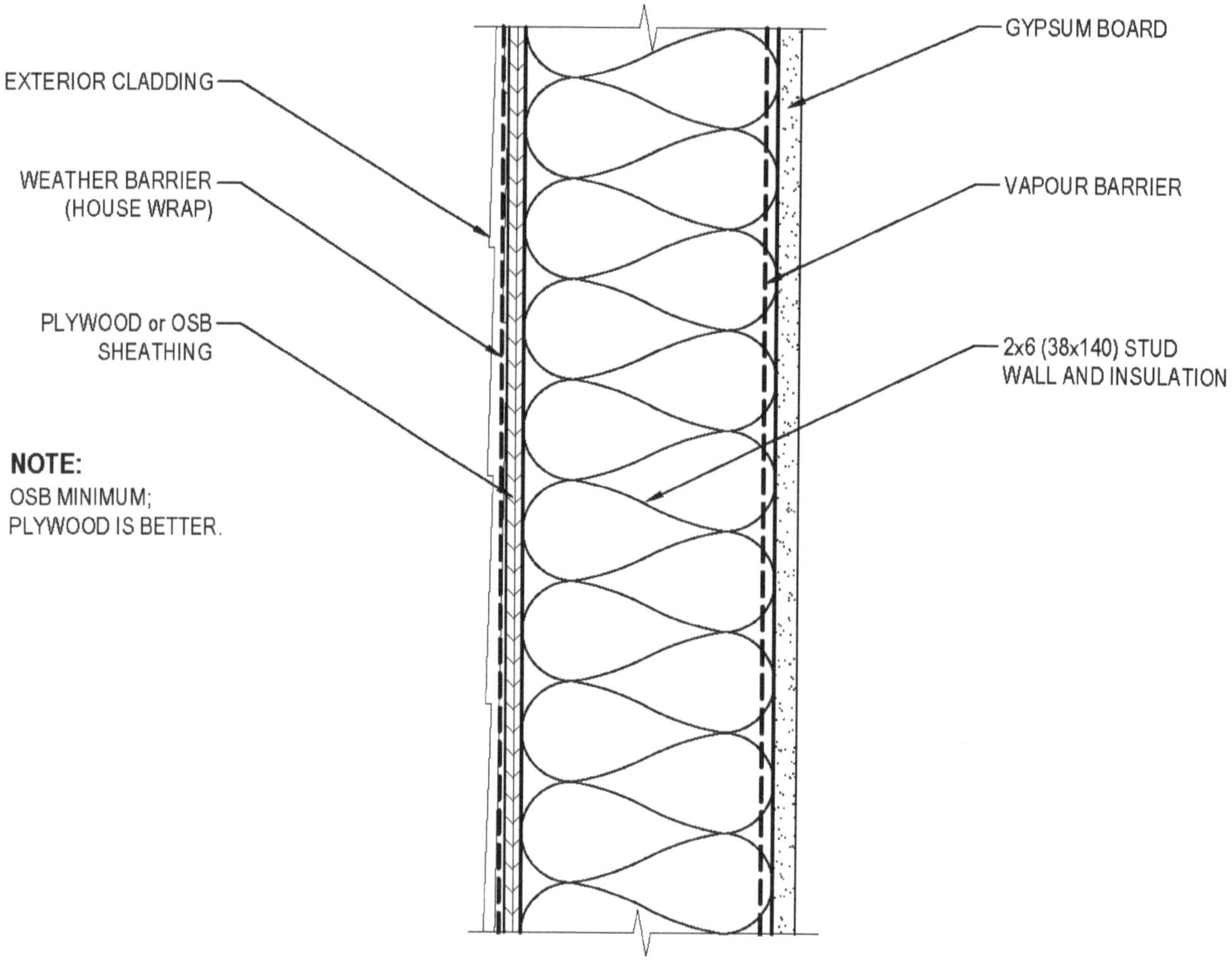

Figure 8.13 Obsolete wall section. This 2x6 framed wall is now obsolete in cold climates because of its many issues.

1. There is thermal bridging at the studs, top plates, bottom plate, and at door and window framing.
RESULT: Heat loss

2. The vapor barrier is compromised by electrical outlets, switches, and gypsum board screws and is not continuous between floors.
RESULT: Moisture may enter the wall cavity and condense causing mold growth and building material damage.

3. There is no drainage gap behind the cladding to allow drainage of water that penetrates past the cladding.

4. Penetrations in the combined air/vapor barrier membrane lead to positive air leakage and subsequent water deposition in the wall cavity.

Courtesy Scott Bundziak, Axiom Architecture Inc.

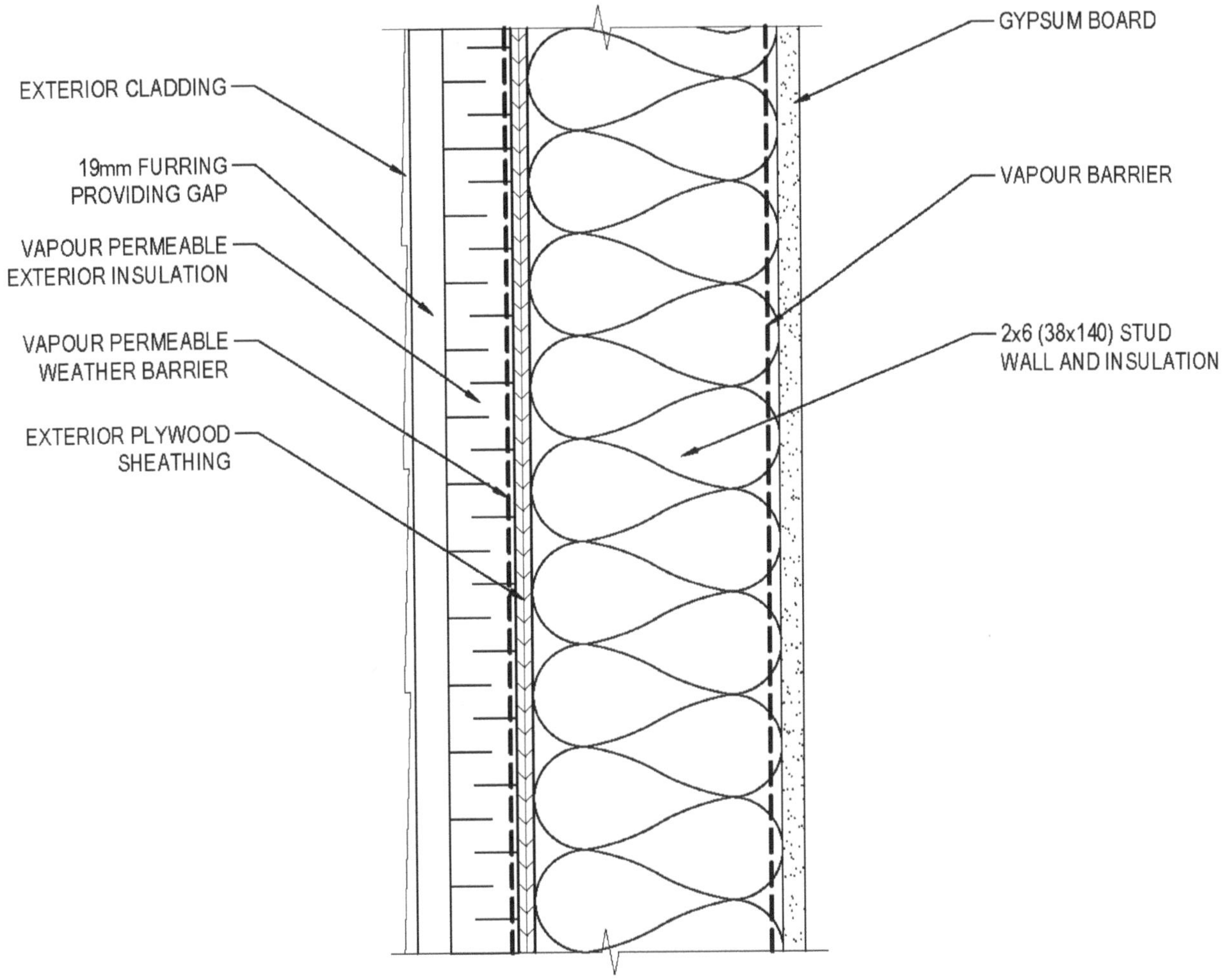

Figure 8.14 Performance minimum wall section. This wall is the minimum building code standard in many cold climate regions. It has many advantages over the obsolete wall section.
1. The addition of the exterior insulation provides a thermal break at the wall-framing members.
2. The exterior insulation keeps the wall cavity warmer.
RESULT: Less risk of condensation inside the wall cavity.
3. The cladding, drainage gap and weather (water resistant) barrier provide an effective rain screen.
CAUTION: Care must be taken in specifying the type of exterior insulation as both installed R-value and vapor permeance may lead to moisture condensation occurring within the wall cavity and subsequent water buildup inside the wall cavity.

Courtesy Scott Bundziak, Axiom Architecture Inc.

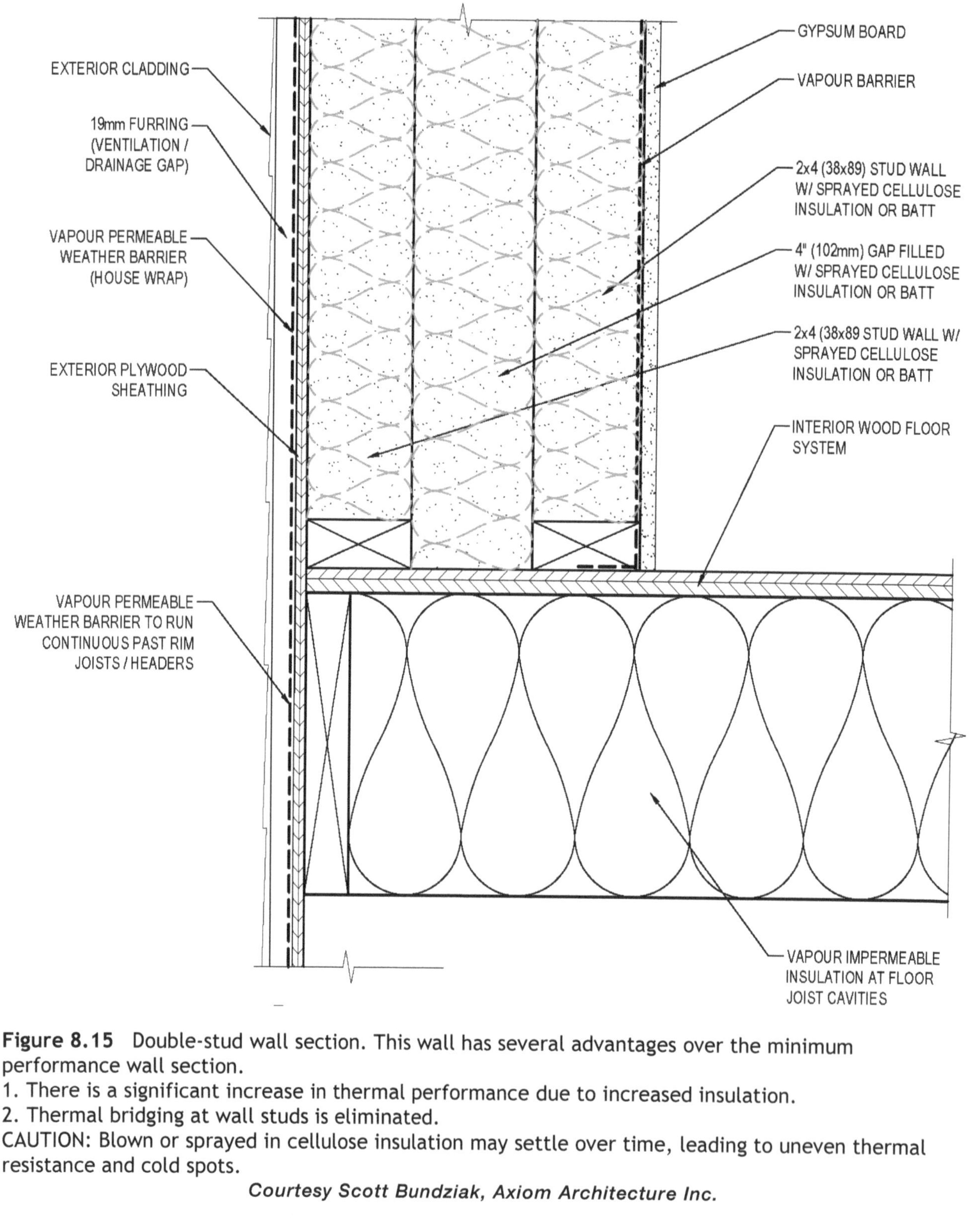

Figure 8.15 Double-stud wall section. This wall has several advantages over the minimum performance wall section.
1. There is a significant increase in thermal performance due to increased insulation.
2. Thermal bridging at wall studs is eliminated.
CAUTION: Blown or sprayed in cellulose insulation may settle over time, leading to uneven thermal resistance and cold spots.

Courtesy Scott Bundziak, Axiom Architecture Inc.

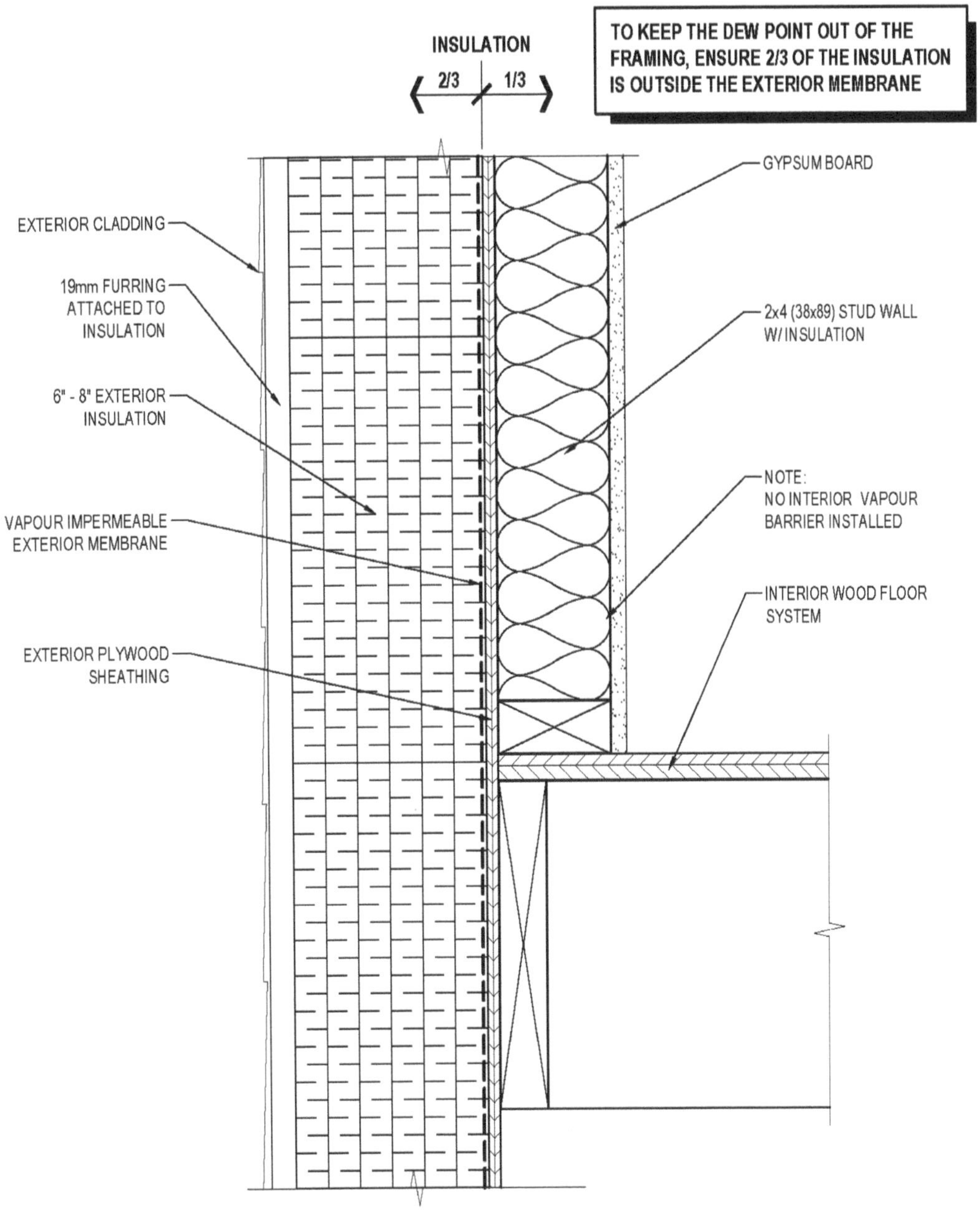

Figure 8.16 Exterior insulation wall section. This wall features primarily exterior insulation and has many more advantages than the other wall section types.

1. Insulation is continuous over framing.

RESULT: Thermal bridging at wall studs, top plates, bottom plate, door and window framing and even at floor connections is eliminated.

2. Super insulated equals energy efficient.
3. 2 × 4 exterior wall framing can be used instead of 2 × 6 framing.

RESULT: Less expense (subject to structural limitations—consult local codes)

4. The vapor is never compromised because it is located in a protected area between the exterior insulation and the exterior plywood sheathing.
5. The vapor barrier is continuous between floors.
6. The vapor impermeable exterior membrane is also a perfect air barrier, meaning no air leakage.
7. Eliminates the need for interior vapor barrier and floor joist cavity insulation.

Courtesy Scott Bundziak, Axiom Architecture Inc.

- Window frames are a major source of heat loss for fenestration components. Fiberglass, vinyl, and wood are all suitable framing materials for cold climates, having significantly better thermal resistance than comparable aluminum frames. Regardless of materials, all frames in cold climates should be thermally broken to limit heat loss.
- In cold climates, low-emissivity coatings are typically applied to the No. 3 face of double-glazed sealed units, or the No. 5 face of triple-glazed. These coatings, which are almost always written into the glass specification in of the project, are highly beneficial in controlling radiative heat loss through windows.
- Multipane sealed glazed units can be filled with inert gases (argon and krypton are common) or partially evacuated to increase their R-value. This is a useful performance gain, but does have the potential to degrade over time as the seals of the glazing unit become more porous over time. Designing the mechanical system for the aged thermal conductivity value of the glazing can help avoid future condensation problems.
- In buildings with severe space restrictions and requiring thinner wall sections, it can be difficult to accommodate the thickness of triple or quadruple pane glazed sealing units from a constructability point of view. In such cases, some window manufacturers offer multipane sealed units with reduced spacers between the sheets of glass. Be aware that performance of such units decreases when the distance between elements of a sealed unit strays from the optimum distance between panes (about 0.5 in. [12 mm]).
- Doors should always be insulation filled in thermally broken frames. Storm doors or vestibules are again a helpful addition. Pay careful attention to weather stripping and ensure that the space between the door frame and the wall assembly is well sealed and insulated, preferably with an expanding air-barrier-type material such as spray polyurethane foam.
- The thermal insulation of a window is typically aligned with the sealed glazing unit, or the frame has offsets built into it to enable the sealed unit to maintain thermal continuity with surrounding materials. It is important to keep these offsets and alignments in mind. Significant thermal bridges can be created simply by installing a window unit out of alignment.
- Similar to wall assemblies, fenestration components are also rated for air leakage, wind resistance, and water ingress (static and pressure-driven). Ensure that windows and doors selected for the project match expected climatic conditions.
- Most manufacturers of fenestration components provide installation details. These should be followed explicitly on site. If no manufacturer's installation materials exist, fenestration standards such as CSA-A440 (CSA 2014) provide a range of installation and sealing options to follow.

Roofs. Roofs are treated the same as any other assembly, with the major differences coming from the horizontal orientation of the surface. This places a particularly high demand on precipitation control. However, all other control layers (thermal, air, and vapor) must also be incorporated for the roof to meet its performance and durability expectations.The following are considerations for roof design:

- Just because a roof is watertight, do not assume that it is airtight. Roofs are designed to shed water and, even if designed to withstand ponding, only do so under a very limited range of water depths and pressures. Air leakage, driven by pressure differentials, is much more difficult to contain than surface water. Roof areas such as parapet caps, service penetrations, curbing, and roof-to-wall transitions are only designed to shed water from above and may leak significant amounts of air. Proper air sealing—not just precipitation control—must be incorporated into the roof and parapet design and at all transitions from horizontal to vertical assemblies.

- Roofs are generally designed with lower thermal conductivity than other assemblies, partially to combat the stack effect forcing warmer air to the underside, and partially to compensate for still air layers on horizontal surfaces being less effective in upward heat-flow situations. Practically speaking, it is generally easier to increase the insulation thickness of roofs compared with walls because of their horizontal construction. This is of practical use to the design team when considering the overall thermal loss of a building, where a trade-off in wall performance may be compensated for by increasing roof insulation.
- Like crawlspaces, both ventilated (cold attic space) and unvented (generally, flat roofs or heated attic situations) roofs can be made to work in a cold climate. If a ventilated roof is chosen, particular care must be taken to seal the ceiling below the cold attic against air and vapor leakage. The stack effect and lower pressure zones created by the movement of exterior air or attic vents can put additional pressure on the air and vapor barriers at this location. This can lead to large amounts of moisture being deposited on (cold) roof structural members, leading to rot and mold problems. In unvented roof systems, particular care should be given to continuity of insulation and the control of vapor ingress into the assembly, as there is little opportunity for condensed vapor to escape outward. Careful selection of materials with high vapor permeance may help in this regard. Of particular note in this discussion are peaked roofs (shingle roofing) insulated by applying insulation directly to the underside of the roof sheathing. This is typically urethane spray foam to act as combined thermal, air, and vapor barrier. In these circumstances, because of the many penetrations by service and structural elements, it is difficult to ensure proper air and vapor barrier continuity. It is recommended that either a supplementary air barrier be applied outside the roof structure or a layer of exterior insulation board be applied prior to shingle installation. Both actions seek to provide the same insurance against cold air leaking through the penetrations in the spray foam air barrier and condensing on the underside of cold roof sheathing.

HIGH-PERFORMANCE BUILDINGS

The most rigorous high-performance building standard is the Passive House Standard (Figure 8.17). This standard goes well beyond current building codes by specifying building envelopes that are extremely well insulated and airtight, solar energy use, high-performance windows, and heat recovery ventilation. In most cold-climate regions, a furnace is not required, resulting in reduced heating mechanical system requirements and obvious cost tunneling benefits. The insulation system is so extreme that heat loss through the building envelope's assembly via conduction, convention, advection, and radiation is drastically reduced. Great care is undertaken during construction to produce a continuous air barrier. Passive House is unique in that it addresses global issues of climate change and energy security by achieving a dramatic reduction of 80% to 90% in energy use (Passive House Canada n.d.).

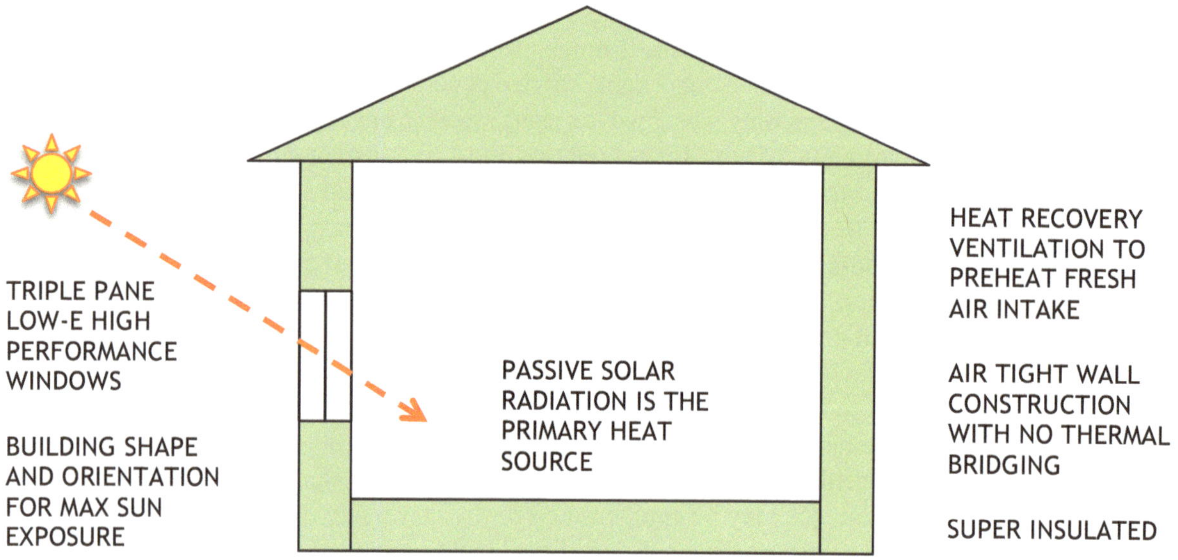

Figure 8.17 Passive House Standard is a revolutionary and scientific approach to building design and construction. It is the world's most rigorous and most advanced building performance standard. Passive House performance standard was developed in Germany in the early 1990s and is based on research conducted by the Saskatchewan Research Council in the 1970s. Design fundamentals include site orientation, building shape, super insulated, airtight construction details, thermal-bridge-free construction, high-performance windows, and heat recovery ventilation. Benefits of Passive House technology include extraordinary energy efficiency, superior IAQ, increased thermal comfort, sound insulation, durability, low maintenance, and sustainability.

Courtesy Four Great Walls

SUMMARY

Cold climates place additional challenges on the building envelope. But with careful consideration to the design and installation of building envelope components, heat loss, moisture movement, and airflow are controlled and the envelope will remain functional over the lifetime of the building, even in the harshest environments.

Using the guidelines in this chapter, one will find many viable options for constructing a building envelope suitable for cold climates. Building designers today are mindful and aware of the challenges posed by climate change and are responding with innovative and inspiring new designs that are not only environmentally responsible, but affordable, energy efficient, healthy, and resilient. As the cost gap continues to narrow between building code minimum and advanced sustainable design, and where communities and cultures embrace the idea that construction can make the world a better place, we are limited only by our imagination and the political will to make it happen.

HOW COLD IS IT?

It is so cold ... we had to open the fridge to heat the house!

9 Design Considerations

GENERAL DESIGN CONSIDERATIONS

Based on the variations of deep and sometimes extreme temperature ranges, design considerations for HVAC systems in cold climates are different from those for other climate zones.

Severe weather events must be anticipated and planned for. For example, winter ice storms can cause utility failures and result in prolonged blackouts and disruptions of heating systems. The North American Ice Storm of 1998 is one such event that caused massive damage and untold costs and time to recover. For this reason, emergency power and heating must be considered for life-risk buildings such as hospitals and seniors' homes.

Cold startup of equipment must be considered during the design stage of a project. Supply and return fans and motors should be designed with low-temperature operating conditions in mind. It is recommended that low-temperature grease be used for fans and motors that will have bearings exposed to low temperatures for extended periods of time. Fluid coils must be protected from extended exposure to freezing temperatures.

Locating equipment in a mechanical room, penthouse, or on a rooftop is dictated by many parameters, including snow loading, wind conditions, insulation thickness, and R-values. Snow loading may dictate heavy construction to withstand heavy roof loads, which can be mitigated by installing the equipment on stilts or raised platforms rather than on a roof curb. In this manner, snow will then blow under the unit, eliminating the snow drifting problem (Figure 9.1).

The preferred locations for boilers and HVAC units are as follows:

- Internally installed (within the building), provided the units do not increase the size of the building.
- External installation (platform-mounted, wall-mounted), if access is provided for routine maintenance activities.
- Grade installation (on local raised platform above snow level).
- Roof-mounted with adequate maintenance access. Consider use of enclosed maintenance corridors for outdoor roof-mounted units.

Consideration should be made for anchors, intake, and exhaust opening orientation relative to prevailing winds.

If equipment is to be installed on a roof, attention should be given to maintenance access, fixed stairs (required for gas-fired equipment), ladders, code requirements for two

Figure 9.1 Roof-mounted condensing units.

points of egress, and winter service. A common practice is to provide a walk-in heated service corridor with lights. Care should be taken to ensure that access points are not restricted by ice and snow drifting.

For extreme low ambient temperatures, consider high R-value insulation in rooftop air-handling equipment. Exterior ductwork is not recommended in extreme low temperature conditions, but if provided, a high R-value, close-cell insulation with exterior metal jacketing is recommended. Thermal bridging of metal (e.g., supports) between the indoors and outdoors must be limited to control condensation in the ductwork and rooftop equipment.

Ventilation Openings

As noted in Chapter 5, outdoor air, exhaust, and relief air openings need to be located on a building with consideration to year-round wind directions and snow levels.

When louvers and hoods are located on the predominant wind direction side of a building, snow can easily be entrained in the wind and blown into the openings. This is why it is preferred to not have air intakes facing into the predominant wind direction. Snow can be very light or granular in extreme cold climates, making it even more susceptible to entrainment.

Intake openings located too low or close to a vertical surface next to an adjacent roof level allow snowdrifts to accumulate and either block air intakes or have snow entrained into the airstream. Snow can also be entrained inside of a building because of high winds or hoods designed with high-velocity airflow rates.

Lower air velocities at air intakes reduce the possibility of snow being pulled into the building. It is recommended to keep the free area intake face velocity below 500 fpm

Figure 9.2 Collapsed intake hood.

(2.54 m/s) with some designers preferring values below 300 fpm (1.52 m/s) in colder temperatures where the snow is more granular. This velocity is recommended for both louver and hood sizing. Note that for louver applications, the velocity is based on the free area of the louver, not the velocity across the entire face. The recommended velocities are for peak airflow conditions during winter operation. Velocities at air intakes can be higher during summer economizer mode when snow is not present, unless wind-blown rain is also a design consideration.

When snow enters a mechanical intake, it can clog filters or intake plenums, resulting in reduced airflow and IAQ issues. Systems completely clogged by snow and ice can create significant negative pressures in an HVAC system. Figure 9.2 shows an instance where an exterior hood collapsed as a result of exterior snow and ice obstruction.

Melting snow inside of a building can cause building damage. Designs can anticipate the potential of snow and rain getting through the intakes by providing sloped floors or watertight ducts and low-point drains that will drain under expected operating pressures. Baffles and plenums can also be used to slow the airflow, allowing the snow to "drop out" of the air stream in controlled locations. Heat tracing of damper operators and floor drain piping is recommended.

Access doors with view ports should be provided for intake and exhaust ducts to allow visual inspection for snow. For larger intake plenums, provide a large access door and duct support capable of allowing a person to physically remove the snow, including walking inside the duct. Be cautious of relying on sloping ducts to drain to the exterior as the water will freeze upon exiting the warm envelope, resulting in plugged drains, ice dams, or icicles.

Exhaust and building relief air discharge points should also be located away from the predominant wind direction. Backpressure from the wind can make these systems not operate correctly and result in building overpressurization.

Openings in structures and air-handling units (AHUs) such as intakes and exhaust must be provided with adequate means of closing during non-operating hours. Insulated and thermally broken control dampers with edge seals rated for cold temperatures are rec-

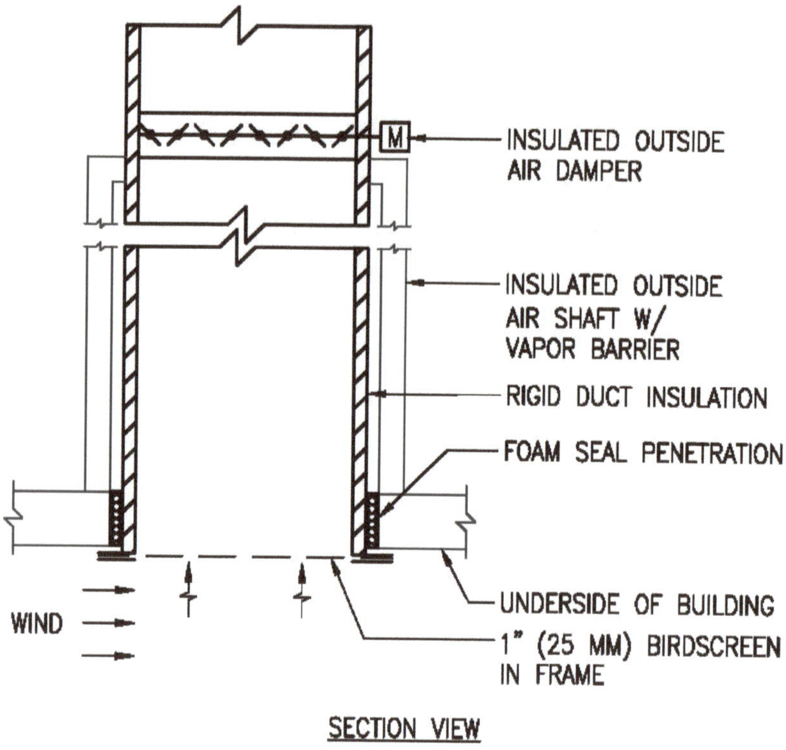

Figure 9.3 Outdoor air intake below elevated structure.
Courtesy DLB Associates

ommended. It is also recommended that control dampers be located tight against the exterior termination point rather than at the equipment. This will reduce the possibility of snow intake during non-operating hours as well as reduce heat loss. For exhaust and relief air, backdraft dampers should be located on the warm side of control dampers to keep them from freezing shut.

For elevated facilities, such as those installed on pilings, consider using the space under the building for mechanical openings. The opening should be flush with the underside of the building to limit negative pressure pockets and a large size bird screen should be provided to keep animals out. The intake shaft should be treated the same as an exterior wall complete with weatherproof lining, insulation, and vapor barrier. Figure 9.3 shows a diagram of this type of installation.

Unique to cold climates is the formation of larger grains of frost called *hoarfrost* (Figures 9.4 and 9.5), the direct condensation of water vapor to ice at temperatures below freezing occurring when air is brought to its frost point by cooling. It is analogous the process by which dew is formed, except that, in the case of dew, the saturation point of the air mass is above freezing. The occurrence of temperatures below 32°F (0°C) is not enough to guarantee the formation of hoarfrost; the air must be initially damp enough so that, when cooled, it reaches saturation and any additional cooling will cause condensation to occur.

Louvers and inlet hoods must be designed with consideration for hoarfrost and ice buildup. Clogged or blocked air-intake hoods can cause ductwork to collapse and equipment to fail.

Providing hinged bird screens (Figure 9.6) on inlet hoods is a common practice to prevent hoarfrost from blocking the airflow. As the frost builds up on the screen, the negative

Figure 9.4 Hoarfrost buildup on inlet screen.

Figure 9.5 Hoarfrost can form on outdoor air intake hoods.

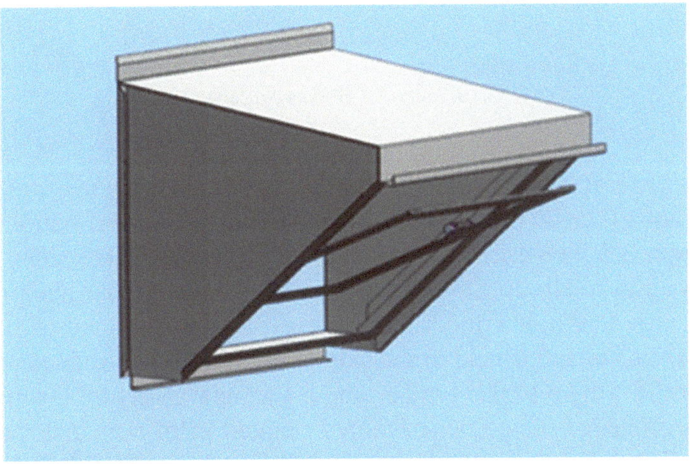

Figure 9.6 Hinged bird screen on inlet hood.
Courtesy Scott Springfield Manufacturing

pressure will cause the screen to lift and relieve the pressure before it can cause damage. Another frost prevention strategy is bathing the bird screen with infrared radiation.

Hood Designs

When aesthetically possible, hoods provide superior protection from snow and wind compared to louvers. A hood can either be built into a building façade or, more commonly, be made out of sheet metal. There are several different design options for hoods depending on the climactic conditions.

Figure 9.7 shows a traditional, downturned hood. The hood is designed with heavy-gage sheet metal to deflect snow that may slide off the roof and is held to the building with steel angle-iron so that it is not susceptible to high winds damaging the building envelope.

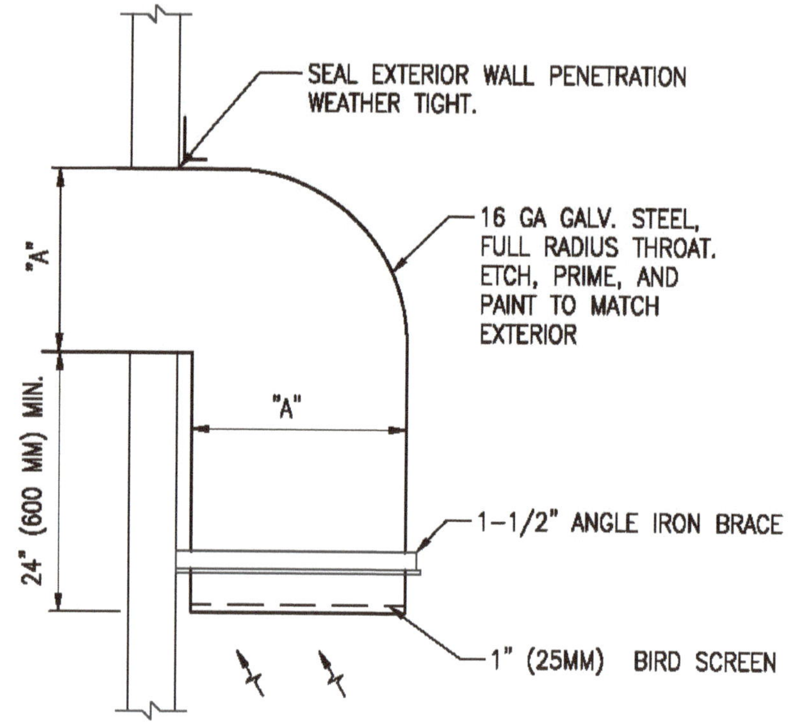

Figure 9.7 Traditional downturned hood.

Courtesy DLB Associates

One type of hood design is an *arctic tee*. There are numerous variations of this design in use, but the basic principle is that the hood has a solid bottom that keeps wind-blown snow from being pushed up into hoods. Air intakes on both sides of the hood are intended to allow air to still be able to enter and leave the hood in case one side is plugged with snow.

Like the traditional hood, the arctic tee is also designed with heavy-gage sheet metal to deflect snow, and is held to the side of the building to reduce wind damage. Because of the size and weight of this type of hood, secondary structural support is required.

The arctic tee sometimes becomes clogged with snow in extreme wind and snow conditions. The hood design used in these extreme conditions, including in Antarctica, is a modified downturned hood that captures many of the design features of the previously described hoods. This design is shown in Figure 9.8.

It incorporates a 6 in. (150 mm) offset from the building to allow windblown snow to go around the hood. The 45° angle on the intake both reduces the velocity at the face of the intake as well as creates a pressure zone behind the hood to help keep snow out. As noted in the arctic tee design, this hood also has an angle-iron support for both structural and wind damage protection.

The hood intake is sized at the recommended lower velocities noted previously. For outdoor air intakes, a secondary louver can be installed at the face of the building. The louver-free area is selected to maintain the low intake velocity. This provides a secondary barrier for snow but also significantly reduces the velocity of air, allowing snow to drop out of the airstream. For comparison, an arctic tee is shown in Figure 9.9.

Filter Design

Where filters are prone to snow or ice clogging, it is recommended that filter design accommodate for a winter and a summer location for the prefilter section. The summer

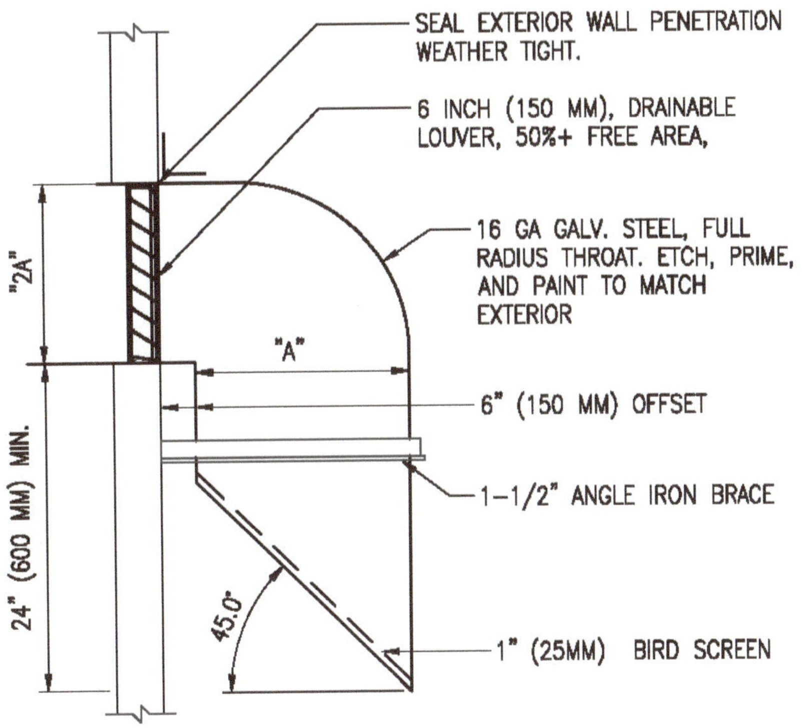

Figure 9.8 Extreme low temperature hood design.

Courtesy DLB Associates

Figure 9.9 Arctic tee.

Craig Fredeen

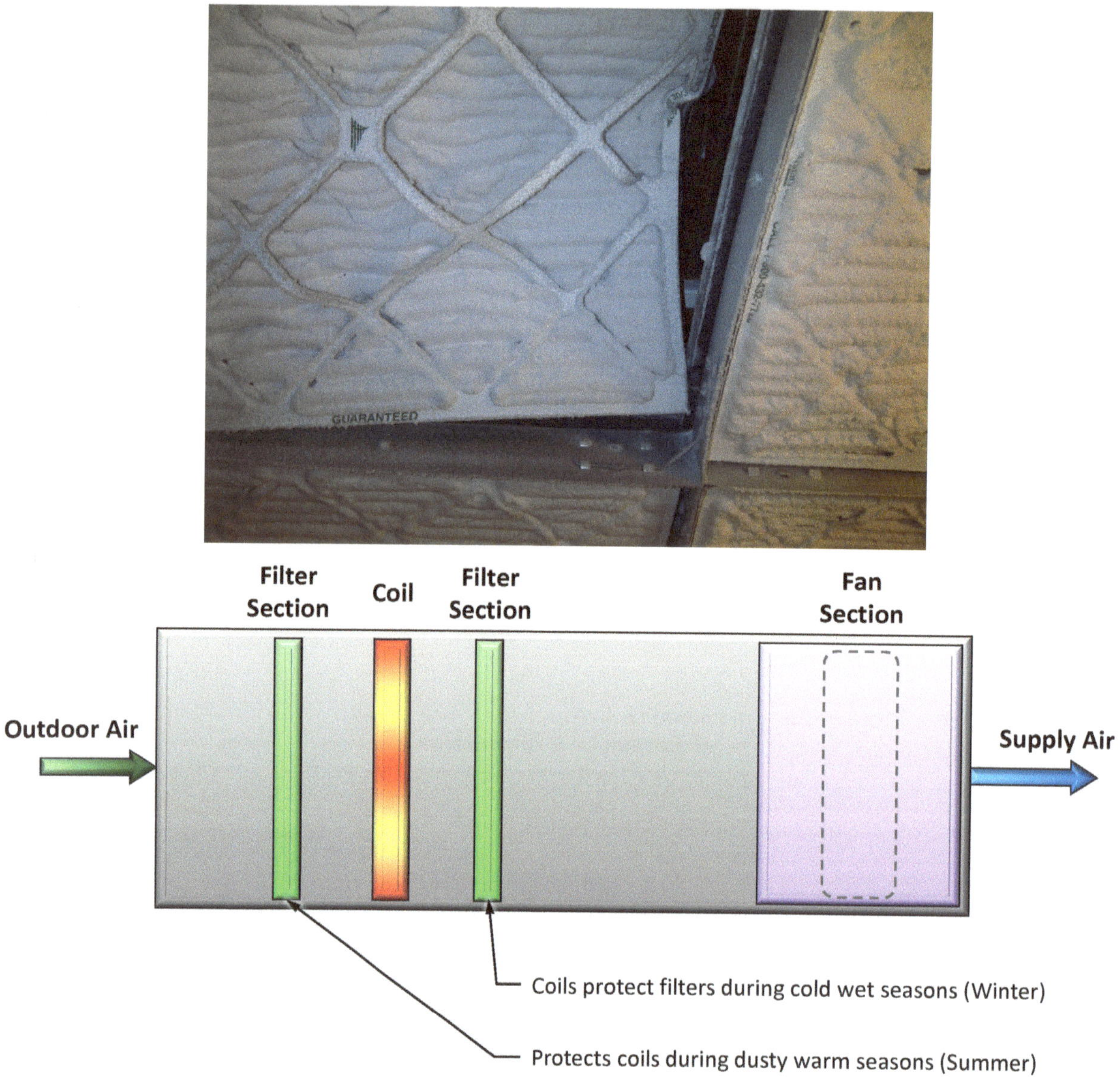

Figure 9.10 Summer/winter filter.

Courtesy DLB Associates

location is after the inlet damper and the winter location is downstream of the preheat coil or other heat source in the AHU. If the unit has a mixing box with return air mixing, this may not be required. Clogged filters may have an effect on the building pressurization and must be monitored accordingly (Figure 9.10).

Is very important to ensure pleated filters are installed vertically to allow moisture and heavy particles to fall down in the filter frame.

Piping Circuits

There are a number of ways to create piping circuits within building HVAC systems. Chapter 13 of *ASHRAE Handbook—HVAC Systems and Equipment* (ASHRAE 2020a) describes in detail the advantages and disadvantages for the reverse-return and direct-return systems. In general, reverse-return systems are used most often to facilitate system balancing.

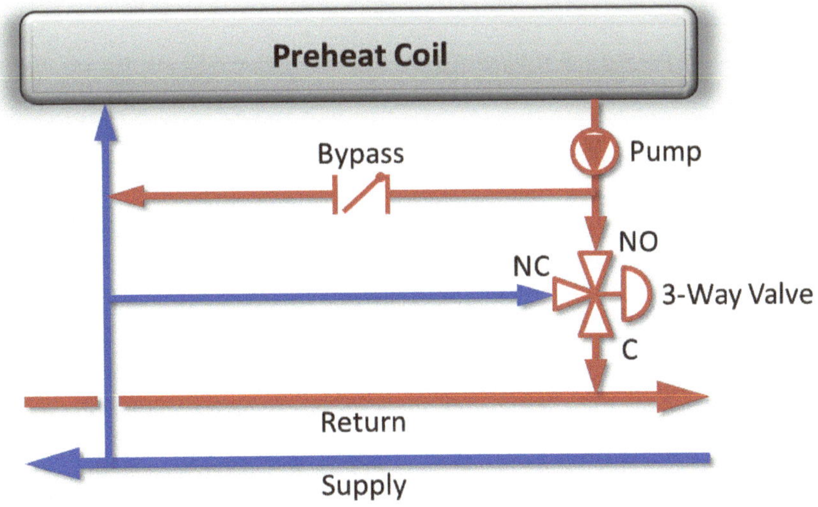

Figure 9.11 Preheat coil piping arrangement.
Courtesy DLB Associates

To reduce circulated flow within the building piping systems with resulting reduced pump energy usage, higher temperature drops within the piping circuit are used more frequently. This can be achieved through terminal unit and coil selections and/or cascading of terminal systems. The higher temperature drops fit well with the use of condensing boilers, which operate at higher efficiencies with lower return water temperatures.

In colder climates, two-pipe distribution systems for heating and cooling systems can be used rather than dedicated four-pipe systems, because there is typically a larger dead band zone between the heating and cooling system activation, particularly when economizer systems are used.

Freeze Protection

When hydronic fluids are exposed to outdoor air, such as in AHU coils, antifreeze solutions such as glycol mixtures should be used to prevent freeze-ups. Hydronic coils in the AHU should use glycol mixtures of sufficient strength relative to the climate zone to prevent the fluid from freezing. Since the viscosity of the pumped solutions rise with the increase in glycol mixture percentages, a mixed-hydronic system can be used for building heating. Hot water can be circulated for the hydronic radiation terminal equipment, while heating glycol can be generated via a heat exchanger to be circulated to the AHU coils or other equipment exposed to outdoor air. In mild climates or where mixed air temperatures reduce the chance of freezing temperatures, providing a pump to initiate constant flow of water through the coil may prevent freezing (Figure 9.11).

Chilled-water coils used for cooling purposes, which use water and not glycol solutions, must be provided with isolation valves, drains, and vents to allow flushing and drainage of the coils prior to cold conditions.

For water-based coils, control safeties such as freezestats can be provided downstream of the coil to identify freezing conditions and shut down the unit to protect the coil. These devices have a capillary tube that is stretched across the downstream face of the coil. When any one foot of the capillary tube senses temperatures below the set point, a switch is activated and the unit shut down. These safeties can be provided in standalone control or as part of a BAS.

Air Stratification

When cold outdoor ambient air is mixed with warm return air, the mixture remains stratified. The cold air, being heavier than the hot air, does not mix very well through a standard damper section without additional methods being used. A stratified airstream will freeze the bottom of water coils as well as create nuisance freeze stat alarms on glycol coils.

Strategies to encourage mixing, such as locating the outdoor air damper on the top of the unit and return air on the backside or bottom of the unit, along with positioning opposed- or parallel-blade dampers, assist in directing airstreams together, but stratification will still occur. Components such as an air blender or air mixer aid in reducing stratification. Temperature sensors located downstream of a blender are less exposed to tempered stratified airstreams and provide more consistent and accurate temperature readings. The use of these devices requires a notable amount of upstream and downstream distance that must be taken into consideration early during the layout of air-handler spaces.

COMBUSTION AIR AND FLUE TERMINATIONS

Most local codes and manufacturers' installation manuals provide guidelines on how to size and construct the combustion air and exhaust flue systems for various types of equipment. Verify with the local authority which code requirements (and local amendments to these codes) are in place. Because many of the national code recommendations for these systems are not recommended for cold climates, verify if engineered solutions are acceptable.

It is important that exhaust flue systems remain within the warm building envelope as much as possible to minimize flue gas spillage during appliance start-up. This spillage would be a result of the appliance fan not being able to overcome the cold air stack within the flue should the gases cool down sufficiently.

For residential and small commercial applications, indirect-fired, high-efficiency furnaces (Figure 9.12) should be considered in the design of the building. The design of high-efficiency furnaces requires condensing of the flue gases to achieve the high ratings. Condensation is a concern in cold-climate conditions.

Direct-Vent Systems

The design of high-efficiency boilers, water heaters, and furnaces requires condensing of the flue gases in order to achieve the high energy-efficiency ratings. Combustion air is often a *closed-loop* or *direct-vent* system, meaning the combustion air is drawn from the outdoors, drawn through the heat exchanger/burner, and exhausted to the outdoors (Figure 9.13). The high-moisture-content exhaust gases condense when exposed to the cold ambient temperature.

Uncontrolled dispersion of condensate in cold climates can cause damage to the building structure, roof, and surrounding areas, thus requiring care in management and control of the exhaust systems. One solution is to have collected and treated condensate from equipment flow through insulated, heat-traced drainage pipes to approved drains.

The sealed combustion air connection to the device also must be considered. The designers should contact the manufacturer to verify the minimum air temperature that can be directly connected to an appliance.

Some equipment is not tested or listed under extreme cold temperatures. Hoarfrost and wind-blown snow can also obstruct the air intake, resulting in tripping of the equipment flow sensor alarms and shutting equipment off.

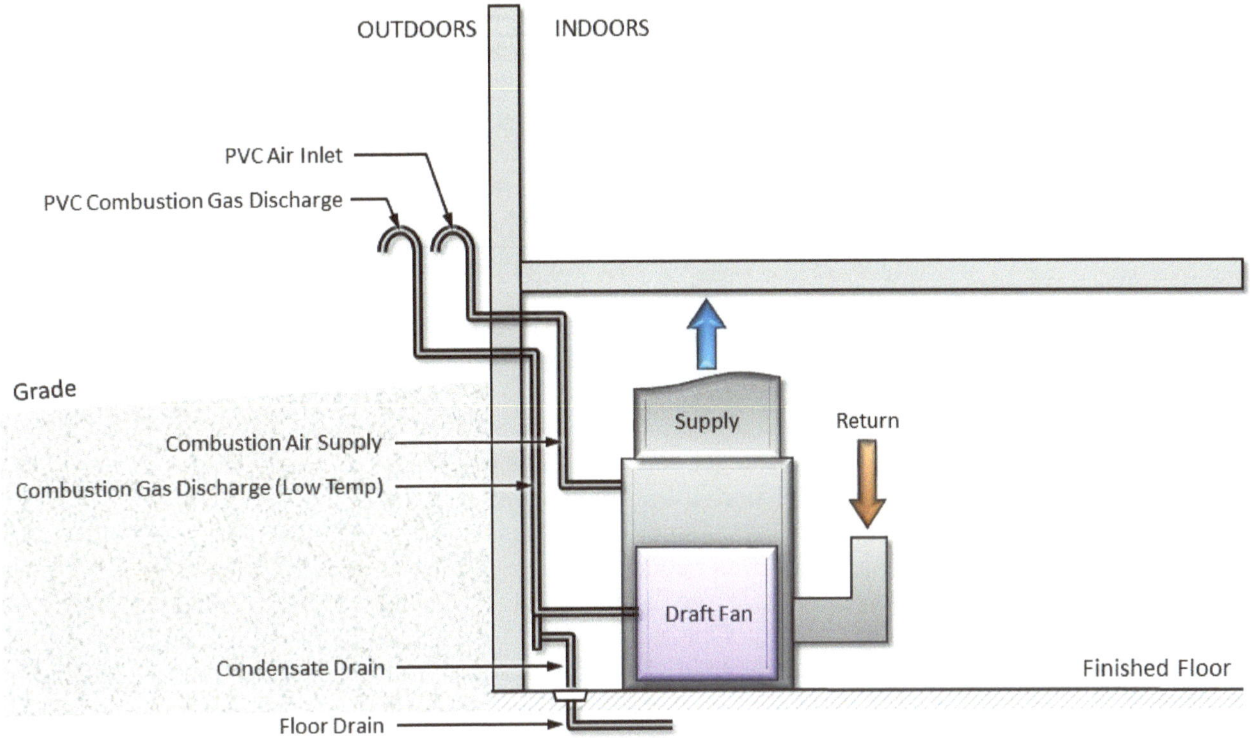

Figure 9.12 High-efficiency furnace installation.
Courtesy DLB Associates

Figure 9.13 Uncontrolled condensate buildup.
Stan Yackel

Piping materials must be considered in colder climates; materials that become prone to damage or even shattering should not be used (e.g., PVC).

Care must be taken when piping the condensate from a high-efficiency furnace (heat exchanger). If the condensate pipe is run through an outdoor wall, there is a potential for freezeup of the line, causing a backup of condensate that will eventually flood the heat exchanger or cause it to shut off when it is required to operate. During combustion, products of nitrogen are produced that will combine with the condensate to produce nitric acid. For every 100,000 Btu of gas input, there is approximately 1 gal of condensate produced.

Building codes may not allow this condensate to be directly connected to a drain and require neutralizer tanks to neutralize the acid before it can be flushed down a drain.

Indirect Combustion

In some instances and for some equipment, it may be more beneficial to bring the combustion air into the mechanical room where it is then supplied to the heat exchanger and exhausted through the chimney flue. Flue condensate is dispersed as a vapor cloud in the cold ambient air and water condensate is piped to a proper drain location.

Combustion air brought into the building will be at cold ambient temperature and must be dealt with accordingly. One method is to have the combustion air delivered into the mechanical room from a high level and dispensed at floor level. To control the inundation of cold air along the floor, the combustion air duct drop may be installed in a bucket trap to minimize cold-air seepage when the furnace is not calling for combustion air (Figures 9.14 and 9.15). This trap does not prevent cold air from entering the mechanical room.

Oversized combustion air openings, sometimes mandated by codes, can lead to excess amounts of outdoor air brought into a space. This not only leads to the possibility of freezing conditions, but also leads some operators to cover up parts of the combustion air system. It is recommended to provide an engineered combustion air system that includes stoichiometric calculations on the required volume of combustion air as well as its associated distribution duct system. See Chapter 7 for more information.

For larger installations with many demands for combustion air from multiple boilers, furnaces, and hot-water heaters, design considerations for combustion air tempering may

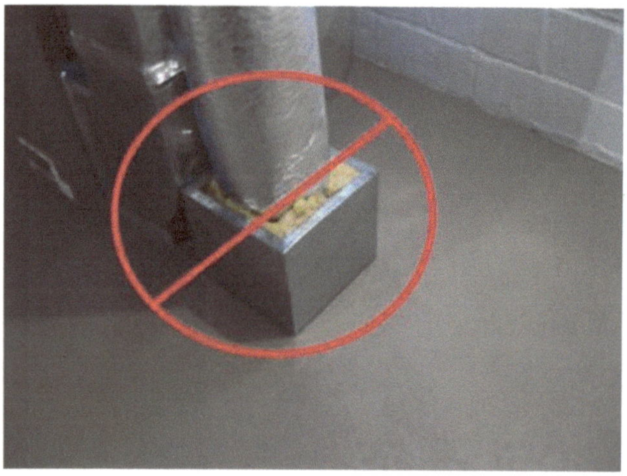

Figure 9.14 Improperly installed combustion air duct.

Stan Yackel

Figure 9.15 Properly installed combustion air duct with bucket trap.

Stan Yackel

be required. A common approach is to preheat the combustion air before dispersing in the mechanical room by providing a duct drop (beside the combustion air duct) that is provided with preheated air, mixing and heating the cold outdoor air as it enters the mechanical room (Figure 9.16).

In addition to combustion air, the designer must take into account mechanical room cooling. For rooms with fuel-fired equipment and pumps, the heat generation can be significant. Many electrical support devices in the mechanical space have recommended operating temperatures at 105°F (40°C) or lower that must be maintained. This is commonly done through economizer mode ventilation.

Another methodology of providing combustion air to the space as well as ventilation air for cooling is shown in Figure 9.17. The combustion air is provided through a single opening, extended to within 12 in. (300 mm) of the ceiling. Some jurisdictions require this termination distance. The upturned elbow provides a heat trap to prevent excess overcooling of the space. Without the elbow, heat will be able to freely leave the mechanical space. No damper is provided on the combustion air opening, so no interlocking with appliances is required.

The amount of combustion air required is calculated as explained in Chapter 7 and in *ASHRAE Handbook—Fundamentals*, Chapter 28 (ASHRAE 2021a). The sizing of the exterior hood/louver as well as the internal ductwork depends on the pressure drop allowed for the various types of fuel-fired appliances in the room. Traditionally, this is a

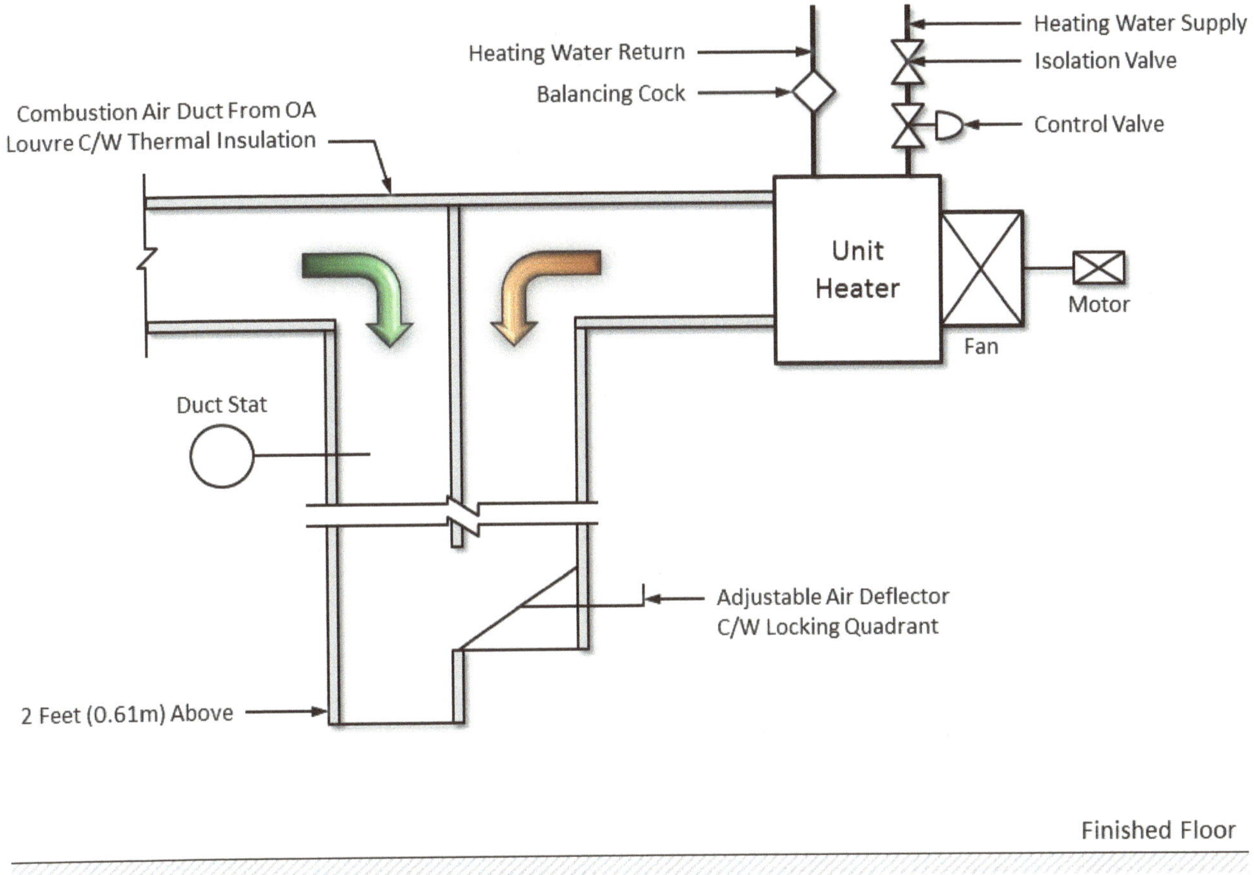

Figure 9.16 Combustion air unit heater control detail.
Courtesy DLB Associates

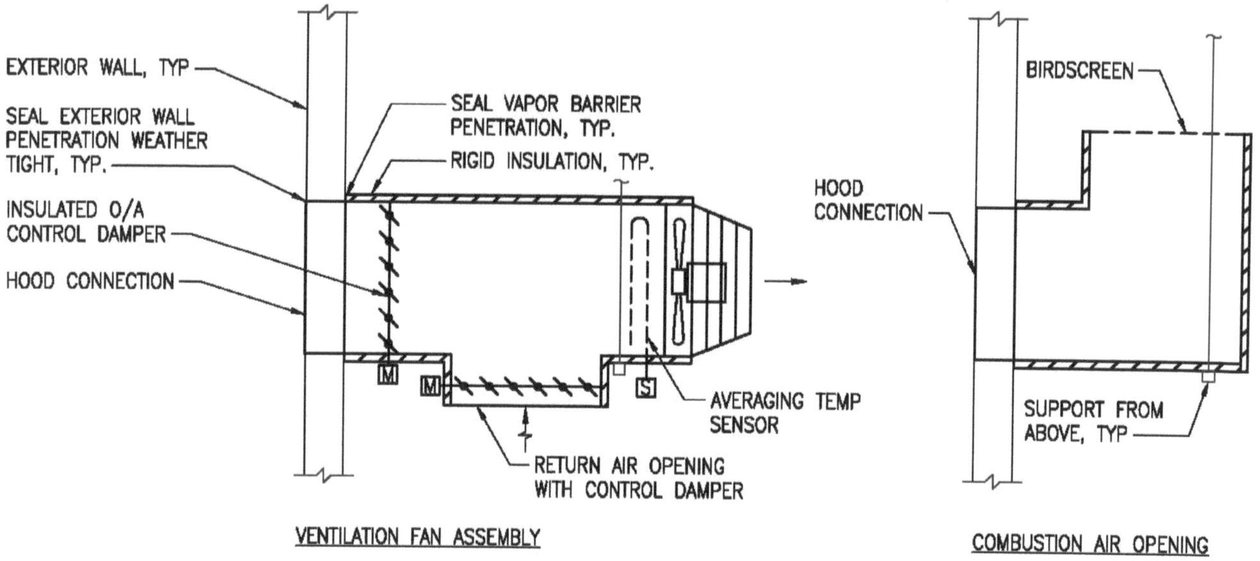

Figure 9.17 Combination combustion air and ventilation fan assembly.
Courtesy DLB Associates

maximum total pressure drop of 0.05 in. w.g. (12.5 Pa) for fan-assisted/forced-draft appliances and 0.02 in. w.g. (5 Pa) for natural draft/barometric damper appliances.

A separate ventilation fan assembly is provided to cool the space. The assembly mixes outdoor air and air from the mechanical space to provide a tempered cooling air temperature of 55°F (13°C). The mixed air temperature ensures that the cooling air will not be so cold as to freeze pipes or affect the performance of lights or smoke detectors.

The amount of cooling air that the fan is sized for depends on the heat generation in the space. Considerations for this calculation include the following:

- Pump operation (note not to include standby pumps)
- Standby losses from equipment jackets
- Appliance flue heat loss
- Heat gain from electrical equipment such as transformers
- Heat gain from other appliances such as air-conditioning unit condensers

The calculation should be completed for winter design conditions when the greatest number of appliances and pumps will be operational, cooling air at 55°F (13°C), as well as during summer conditions when less equipment is in operation but the supply air temperature is at ambient outdoor air temperatures.

Boiler room temperature set points should be lower than the maximum electrical equipment operating temperatures, usually at a more desirable maintenance and operational space temperature of 80°F to 85°F (26°C to 29°C). Higher temperatures than this may lead to complaints from facility staff.

When the ventilation fan is operated based on a call for cooling, the combustion air opening acts as the relief air opening. It is important to note that the fan is designed to supply air into the space and not exhaust it from the mechanical room, which may create a backdraft condition on atmospheric appliances. A direct-drive fan should be selected to reduce maintenance associated with belts, and should also be configured so that the motor is exposed to the space to facilitate easy replacement.

Combustion Air Installation Caution

Combustion air systems must not be blocked during cold conditions because it may starve the appliance and cause a buildup of carbon monoxide, which has led to many deaths during the winter season. All combustion air systems are prone to blockages. Installation of carbon monoxide detectors is highly recommended, and is a code requirement in many regions of the world.

There are other combustion air methodologies, allowed in many codes, that are common practice in warmer climates but that are not recommended for cold climates.

Indirect combustion air systems should not draw air from cold attics. Warm air from the mechanical room will rise into the cold attic, creating ice-damming issues on the roof as well as frozen condensation on the insulation from moisture in the mechanical space.

Two combustion air connections to the outdoors, one high and one low, are also not recommended. These systems are designed for warmer climates to provide passive cooling of mechanical spaces. In cold climates, this tends to overcool the mechanical space, leading to the potential freezing of water pipes. It is recommended to use a single opening at the top of the room, with warm-air traps to prevent excess heat leaving the space.

Caution should be exercised with fan-powered combustion air systems or control dampers in the combustion air duct. Most codes require these type of systems to prove airflow before the appliance can fire. Since these are mechanical devices, the fan and control damper will fail at some point and prevent the heating appliances from firing. This may, in turn, create a freezing condition in the building. Plus, this requirement may involve installing the air-proving relay in-line with the burner circuits, which may void the burner listing.

When an open combustion air system is used, care should be taken in locating water and sprinkler pipes around the combustion air opening. High-wind conditions outdoors may cause infiltration that could freeze these pipes.

Air-Volume Temperature Compensation

In a cold-climate makeup air application, cold- to very-cold-temperature outdoor air is heated, increasing the volume of air delivered from the heat exchanger. In some instances, this increase may be negligible and disregarded; in other cases, the increase can have several effects, such as an increase in duct velocity, leading to higher noise levels and inadvertent undersizing of the heat exchanger. For example, air heated from –36°F to 75°F (–32.8°C to 24°C), increases the volume of air delivered from the heat exchanger by approximately 26%. Adjusting the airflow through the fan for these conditions may be accomplished by adjusting dampers and/or using a variable-frequency drive and discharge airflow sensor to control the speed of the fan motor.

HEAT RECOVERY

Heat recovery designs may include heat pipes, glycol runaround loops, plate heat exchangers, heat pumps, and heat wheels. Heat is recovered from the building exhaust air stream, hot air captured from boiler breaching via a recovery coil or other source, sewer water, chiller/heat pump condenser water, or other waste heat source and transferred to the incoming cold outdoor air stream. Because of the temperature and moisture differences between the incoming outdoor air and the return/exhaust air, the potential exists for significant heat recovery. The design limitation for heat recovery is frosting of the exhaust side of the heat exchanger, discussed further in the following sections on Frost Prevention and Defrost Strategies.

Various types of heat recovery systems have different attributes to achieve energy savings. Designers should look at the overall energy used to determine the amount of

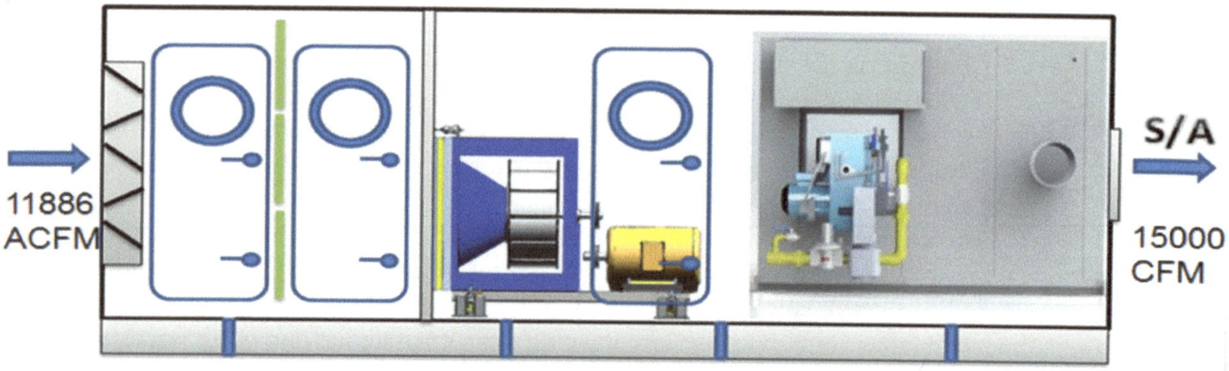

Figure 9.18 Indirect gas-fired makeup air unit—Air volume increase.
Courtesy Scott Springfield Manufacturing

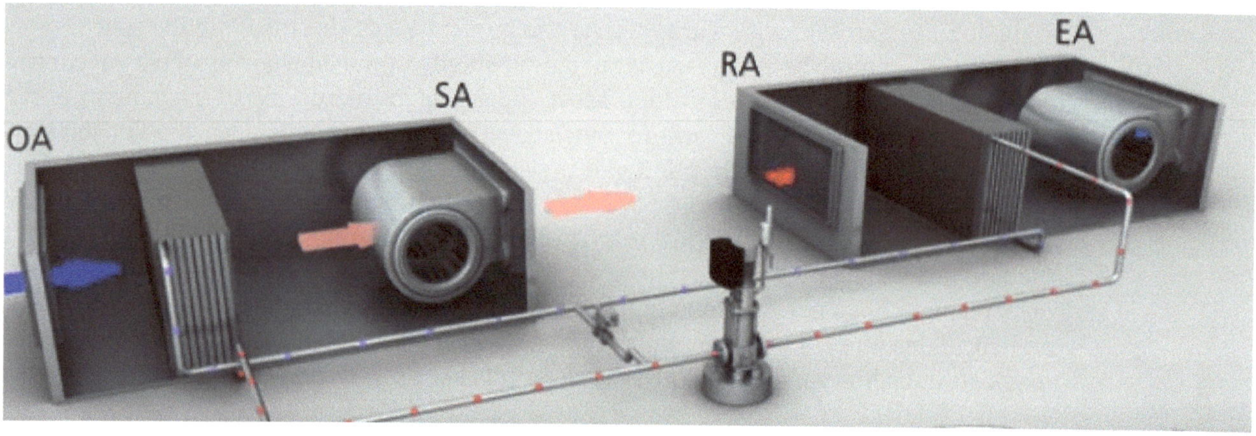

Figure 9.19 Heat recovery runaround loop.
Courtesy Konvekta AG

energy saved and what type of system is needed to achieve these savings. A total view of the design and its regional energy cost should be factored in the selection of the energy recovery system. When designing an energy recovery system, consider the lack of energy when a building is started cold or the waste heat source is off line.

Heat pumps are very versatile and can be used in many applications (air, water, ground, solar, and industrial process). Regional temperatures are a huge factor in what type of design is required to achieve the design criteria. Preheating the outdoor air may be required so that the exhaust air coil will not freeze when the exhaust air temperature drops below its dew point. The energy (power) used for the coils and compressors should be taken into account to consider the system net efficiency and the system's overall effectiveness. Coils and compressor power should be taken into account, considering the viability of the heat pump option.

A runaround loop is probably the oldest system used for energy recovery. Traditional system designs take the extreme outdoor supply air temperature in sizing the coil, taking into account the glycol water mixture at a fixed flow to achieve its effectiveness of 40 to 50% at design condition without considering the added coil pressure drop or pump energy. European runaround energy recovery systems adjust the fluid flow with variable-frequency drives, DOE-2 programs, mapping coils and pumps, and bypass valves to

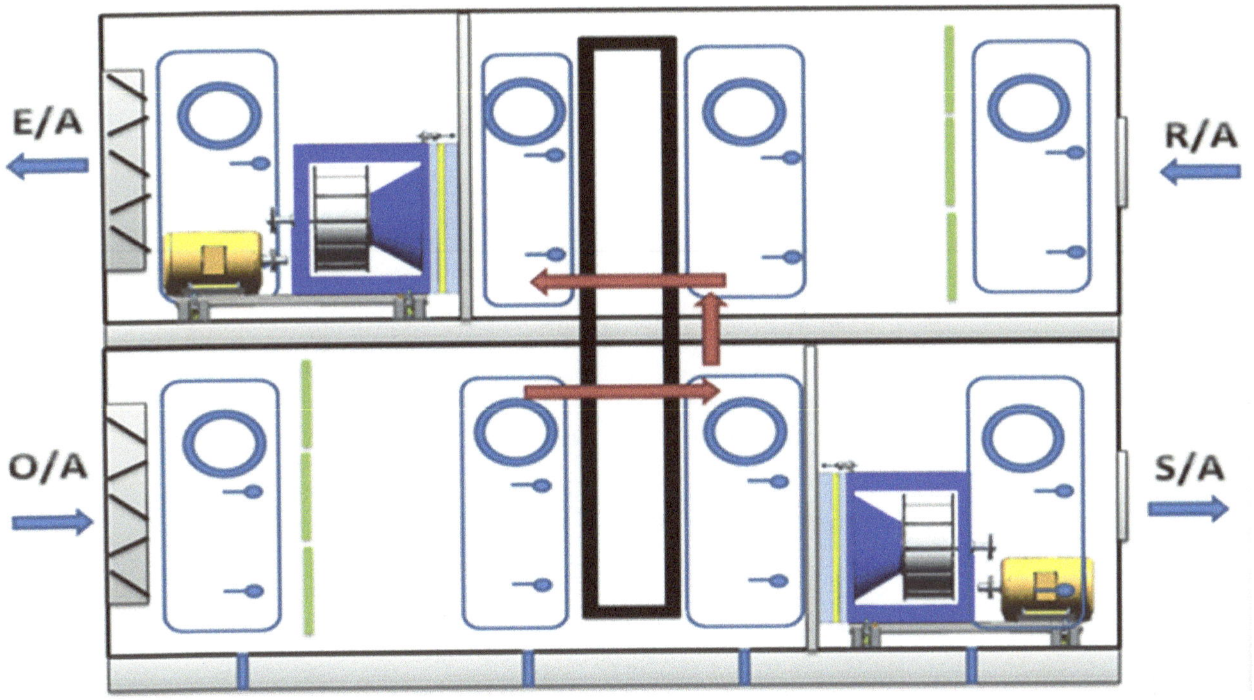

Figure 9.20 Heat recovery wheel inside HVAC unit (red arrows indicate purge air flow).
Courtesy Scott Springfield Manufacturing

achieve levels of energy recovery of 70 to 90%. These systems can be used without any preheats, because the sensors and bypass valves can redirect a portion of the heat to the exhaust coil preventing freezing. These systems can be retrofitted into existing systems. Always take into account the net values of energy savings by including the pump energy and added coil pressure drop.

Energy recovery wheels, also known as *enthalpy wheels* or *heat wheels*, have grown in popularity in the past decade (Figure 9.20). Energy wheels allow the transfer of latent energy (humidity/moisture) into the outdoor air stream, which can be beneficial in drier climates. They are effective, but must be designed into new facilities, because ductwork must be coupled into the same location to achieve the heat transfer. Preheat coils, bypass dampers, and variable speeds must be considered to prevent the freezing of the exhaust air below its dew point. The total cost of the system, added ductwork, preheated coils, and the pressure drop of the wheel itself will be needed to determine the net value of energy savings.

Another form of air-to-air heat recovery is the use of fixed-plate, cross-flow heat exchangers. These units use a nonmoving core, as shown in Figure 9.21, that exchanges heat from the return/exhaust air to the outdoor air. Because these systems have no moving parts, they require less routine maintenance. They also have no cross leakage, so they are preferred for laboratories and health care facilities. Energy wheel technology can have a cross leakage rate of 2% to 5%. Plate-type units traditionally could not transfer latent energy into the outdoor air stream, but newer membrane technology is developing where these types of units can also now transfer some latent energy.

The use of solar heating systems has been growing in Europe. Many countries mandate inclusion of these systems in new buildings. Depending on the system, in cold climates the heat transfer medium must be considered along with the snow loads on roof tops and the weight of the equipment for structural integrity.

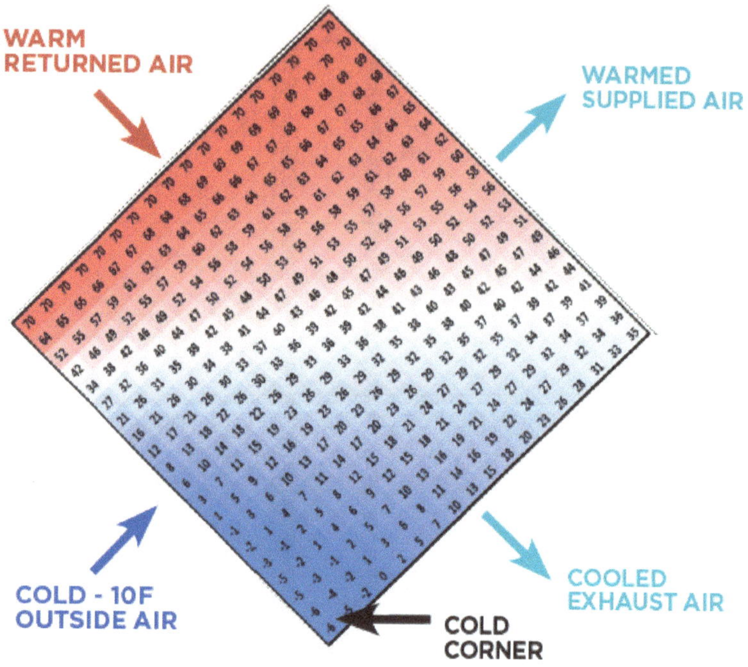

Figure 9.21 Plate heat exchanger.

Courtesy Innergy Tech

Frost Prevention

Several methods are available for mitigating frost and ice formation in the heat exchange core. Frost prevention strategies are common in commercial applications, and defrost strategies are common in residential equipment.

These strategies allow the exhaust air to be cooled down to the dew point when the cold side of the heat exchanger is below freezing. This restricts the amount of heat recovered, because the heat can only be recovered from the exhaust air above its dew point temperature. Other methods for consideration are face-and-bypass damper arrangements, reducing the speed of the heat wheel rotation, reducing the tilt angle of a heat pipe, or providing a preheat coil in the outdoor air upstream of the heat exchanger. Refer to Chapter 12 for details on control strategies.

The cross-flow nature of a plate exchanger leads to variation in its leaving air temperature. Using computer calculations, it's possible to show this variation (Figure 9.21). The coldest area of a plate exchanger, called the *cold corner*, is always at the junction of the entering outdoor air and leaving return air. The location of the temperature sensor is important to be in a good averaging mixed air temperature.

Defrost Strategies

Defrost strategies include control algorithms that allow condensation and freezing in the heat exchanger core, then periodically stop cold airflow through the core while continuing to pass warm air through the core to melt out the accumulated frost and ice. In cold climates, defrost strategies can facilitate much higher levels of heat recovery. The design must ensure that interruptions in fresh air during the defrost cycle do not degrade proper IAQ. If defrost-based system interruptions are not acceptable, then the use of a preheat coil can be used. The preheat coil is controlled based on maintaining the exhaust air discharge temperature above its dew point.

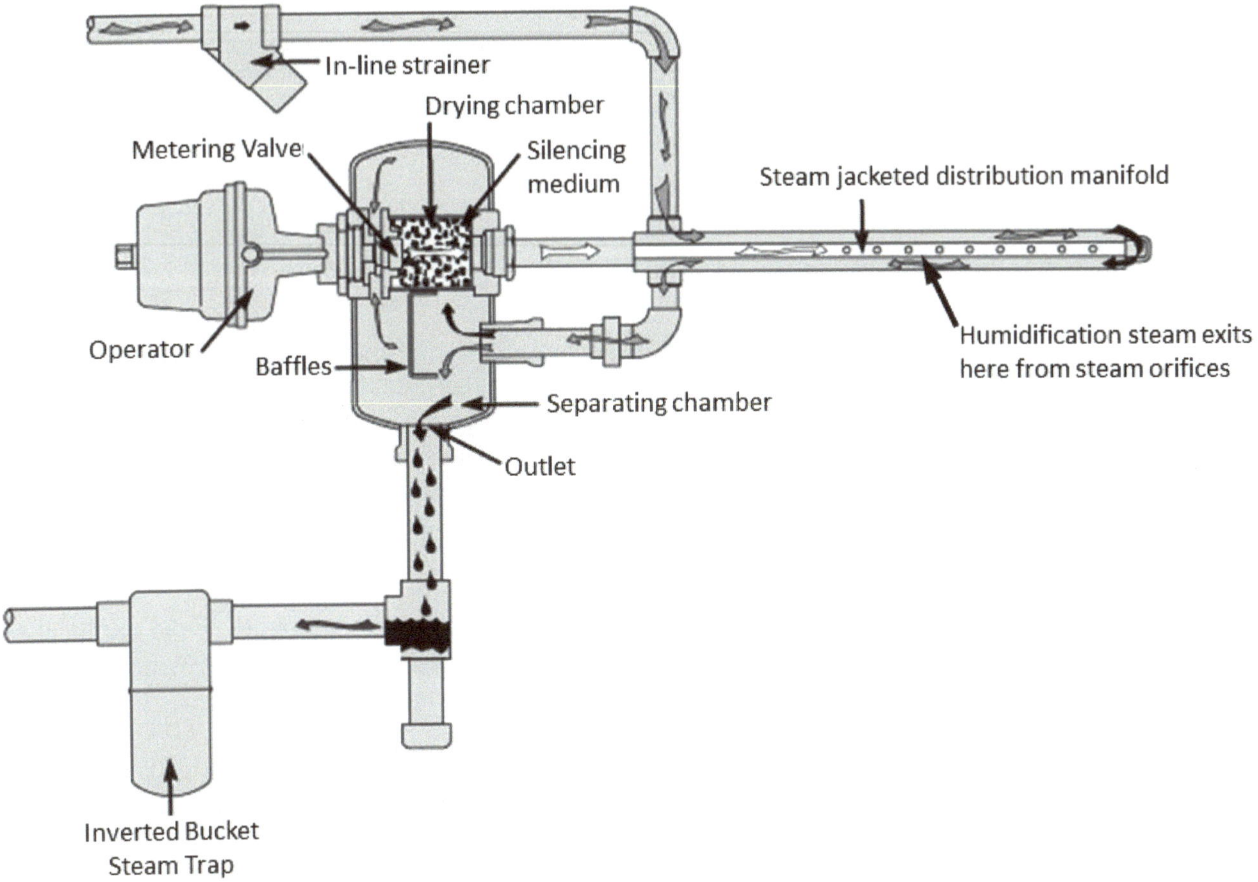

Figure 9.22 Direct-injection steam humidification.
Courtesy DLB Associates

When sizing the primary heat exchanger, the temptation to reduce the size/load based on heat recovered from the recovery device must be weighted with the potential upset conditions. The heat from the exhaust system may not always be available for several reasons, such as internal heat loads from occupied spaces not present, heat not available during defrost mode when the exhaust air bypasses the heat exchanger, failed heat recovery wheel operation, cold start-up, and quick recovery from prolonged winter storm power outages.

HUMIDIFICATION

At one time, humidification was only for human comfort; however, with more emphasis on IAQ for health reasons, humidification is now understood to be a key component to achieve healthy indoor environments and comfort. This is the same with air filters, as they were initially designed for coil protection but are used today as an indoor environmental quality device. Other applications where humidification is sometimes a requirement is for manufacturing processes and in health care facilities. Humidifiers can be installed as standalone devices or in packaged air-handling equipment. The humidifier grids are normally located in the supply air section. Humidifiers come in a variety of types. They are typically steam grids with gas/electric or boiler steam generators, but other methods such as humidifier media or spray nozzles are also common where permitted.

Table 9.1 Indoor Relative Humidity Levels that Keep Thermal Double-Pane Windows Condensation Free

Outdoor Air Temperature	Maximum Indoor Relative Humidity at 68°F (20°C)
−22°F (−30°C) or below	15%
−22°F to −11°F (−30°C to −24°C)	20%
−11°F to 0°F (−24°C to −18°C)	25%
0°F to 10°F (−18°C to −12°C)	35%
10°F to 32°F (−12°C to 0°C)	40%

During cold conditions, the relative humidity in the space must be controlled relative to the outdoor air ambient temperature to limit the humidity level below the dew point temperature of the most susceptible structural member in the building envelope, mainly the window glazing components. As the outdoor ambient temperature decreases, the relative humidity level maintained in the space must also decrease. Control of the AHU discharge humidity level is critical to limiting condensation downstream of the humidifier. High-limit humidity sensors should be strategically located to ensure high-humidity levels are controlled.

If providing humidification, the designer must work closely with the architect to ensure that the building envelope is carefully designed, including proper window performance for the anticipated relative humidity to protect against condensation. The required hygrothermal design calculations and analysis is described in *ASHRAE Handbook—Fundamentals*, Chapter 25 (ASHRAE 2021a). Modeling software and tools are also available.

Humidifiers require regular maintenance and typically higher energy usage. A higher relative humidity in the exhaust air stream can also increase the defrost cycle time in heat recovery units.

Extreme Design Considerations

Humidification in extreme cold climates and remote locations must be carefully considered, as building envelope condensation and cost can be major issues. Humidification should be a client-driven option. Table 9.1 can be used to determine the humidity levels in the given space based on outdoor air temperature conditions. Please note that the better the glass R-value, the less condensation is formed on a window; triple-glazed windows with inert gas provide better thermal insulation.

Where high relative humidity levels are required in extreme cold environments, hygrothermal calculations are recommended to ensure that condensation will not occur at window surfaces or within the building envelope. Building pressurization also needs to be closely controlled with alarm conditions provided under excessive positive pressure conditions. Driving moisture into the building envelope, or having condensation occur within the envelope, will result in costly envelope damage as well as encourage the growth of mold. Natatoriums in cold environments have a history of expensive lawsuits stemming from improperly designed envelopes and mechanical systems that do not properly control humidity and building pressure.

Humidification should also be restricted to only those rooms that require it. This can be accomplished through the use of vestibules/anterooms and installing vapor barriers around the humidified spaces. Residual humidity in the rest of the facility can create issues such as frosting at exterior door thresholds and seals. Plus, as noted, cold-temperature air is extremely dry, so allowing humidification to leak out of the designated location will result in higher operational costs in conditioning and generate more water vapor than is necessary.

COOLING SYSTEMS

Cooling Coils

Cooling coils may be required in cold climates during the summer months when the outdoor ambient temperatures rise above indoor design temperature set points. The cooling source for the coils may be chilled water, chilled glycol, or a refrigerant. Cooling coils that operate with water must be provided with drains and vents to prevent damage of the coil tubes during winter operation. Preventive measures, such as constant water flow through the coil, using the cooling coil as a heating coil, or flushing the chilled-water system with glycol while draining the system should be considered.

Drain pans are required to prevent the condensate from the coil from collecting in the AHU. Drain connections and sufficiently deep P-traps for removing the condensate from the unit while operating must be provided. If the cooling coil is only operational during the summer months, the condensate may be directed to the roof drainage system. If the cooling coil is operational during cold temperatures, the condensate connection and piping must be heat traced or located and routed to a drain through heated spaces.

Cooling coils that are required to operate during cold ambient conditions may be glycol coils, but usually the requirement dictates that the use of mechanical cooling systems using refrigerant systems be considered. Industrial applications may dictate that outdoor air quantities be restricted to a minimum, and therefore full outdoor air volumes are not available for economizer or free-cooling mode or operation.

Condensing Units

Condensing units that are required to operate during cold-climate conditions require design considerations for operating within a wide ambient temperature range. The condensing coil must be sized for the warmest ambient design temperature, normally 10 Fahrenheit degrees (6 Celsius degrees) above ambient to take into consideration roof temperature. The condensing unit coil and condenser fan arrangement will therefore be considerably oversized for operating during cold ambient temperatures, resulting in poor operation of the compressor system.

For the system to operate properly during cold ambient conditions, the active surface of the condenser coil is effectively reduced by flooding a portion of the coil with excess refrigerant. This flooding is achieved using a form of head pressure control.

The system must also accommodate this excess refrigerant during the summer operation; this requires a receiver to hold the refrigerant when it is not in the condensing coil.

Crankcase heaters for the compressor and receiver are a requirement. In extreme subzero ambient conditions, the compressor and receiver are located in an insulated and heated space (Figure 9.23).

Chillers

Air-cooled chillers must have a cold-weather package and a hot-gas bypass, They require extra refrigerant to back up in the condenser to give them a phantom load and stop them from tripping on low-suction pressure. Depending on the system—whether it is remote condensing or it is a packaged air cooled chiller—designers should consider the efficiency lost due to glycol. In this case, an oversized chiller can compensate for efficiency differences. The direct-expansion system will need a reservoir (cold-weather kit) and a hot-gas bypass.

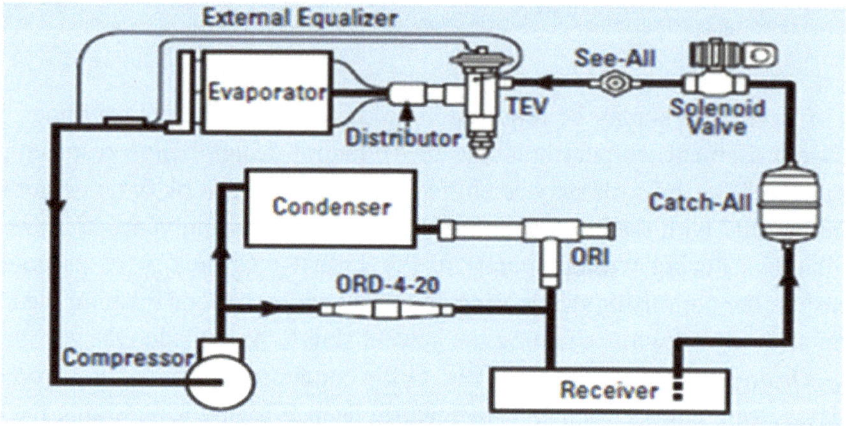

Figure 9.23 Low ambient cooling refrigerant schematic.
Courtesy Sporlan Division, Parker Hannifin Corp

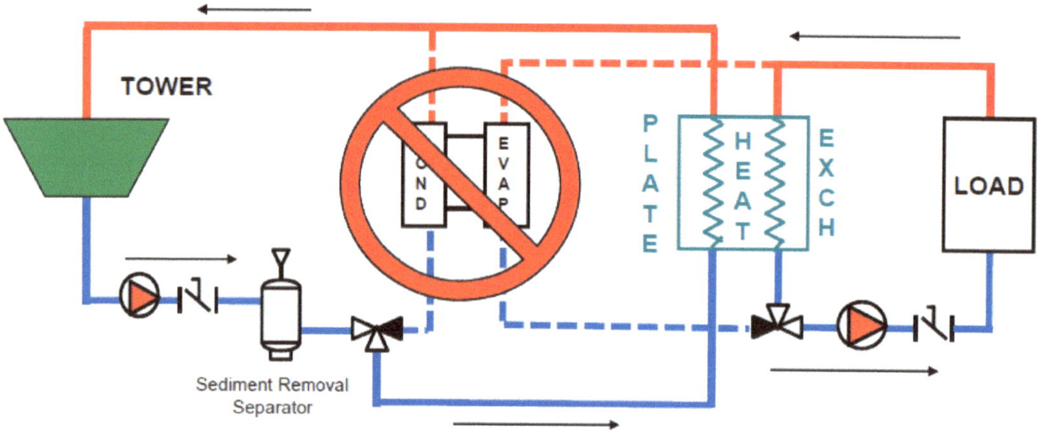

Figure 9.24 Cooling tower piping schematic showing a free cooling state, bypassing the chiller.
Courtesy Delta Cooling Towers

Cooling Towers

The following are ways to prevent the water in the cooling tower from freezing when the air temperature is below freezing:

- Bypass valves and fans for warm water to fall into the sump tank
- Sump tank heater
- Indoor sump tank
- Bypass so the water does not go over the top of the cooling tower

Dry Coolers

A dry cooler is similar to a condensing unit except it traditionally is a closed-circuit hydronic loop and uses glycol as the cooling medium. Dry coolers are piped similarly to the cooling tower diagram in Figure 9.24. During the summer, a dry cooler is an isolated loop on the condenser side of a chiller. During the winter, the chiller is bypassed and the dry cooler acts like a free cooling fan-coil unit directly connected to the terminal cooling equipment.

Figure 9.25 Snow-covered cooling tower.

NATURAL VENTILATION AND GRAVITY INTAKE SYSTEMS

Another means of ventilating a space is through operable windows/dampers, gravity air vent openings, and other natural ventilation design strategies. These types of systems are typically code compliant and are gaining attention in the sustainable design community. In temperate climates, these strategies could be implemented with careful selection of air discharge locations in respect to occupant comfort. In colder climates, these systems can create uncomfortable drafts or cold spots in a facility. Through-wall intakes can also create condensate situations where the ambient air temperature is lower than the indoor air wet-bulb temperature. The designer should take great care when designing these types of systems in cold climates.

HEATING SYSTEMS

When designing a heating system, the designer should first ask the following "big-picture" questions, including:

- What fuel sources are available?
- Will there be a central heating plant or distributed heat producing units?
- What are the driving factors for the design, including first cost, maintenance, comfort, and energy efficiency?

A more thorough discussion on fuel considerations is found in Chapter 4.

Heating System Types

A central heating plant normally consists of a single large or multiple heat-producing appliances and a distribution system that serves multiple spaces. A central-plant system typically has a lower long-term maintenance cost, because it reduces the number of pieces of equipment that require maintenance and places them in one location. Aside from central air-heating systems, a central heating plant is typically more expensive than a distributed heating system because of the cost of the distribution system.

A distributed heating system is one where multiple heat-producing devices are located throughout a facility. An example of this may be a condominium with each resident unit having a furnace or heat pump. The advantages and disadvantages of this system are the opposite of those of a central plant: it allows tenant billing of utilities consumed and provides a lower first cost at the expense of having to maintain multiple appliances.

Central Air Heating

The least expensive heating system is a central air-heating system. This typically is an air handler with an electric or fuel-fired heating coil (heat exchanger) that distributes both ventilation and heating through the air as well as cooling during the summer months. In addition to being low in cost, these types of systems also require minimal floor area inside of a facility, with equipment typically having been located either inside a fan room or installed on the roof.

The disadvantages of an air-heating system include that air is an exceptionally inefficient heat transfer medium compared to steam or hydronic heating. It also typically has limited zoning controls unless provided with area-based reheat coils.

Fuel-based heat exchangers are prone to failure. A failed heat exchanger can introduce carbon monoxide and other products of combustion into the airstream. Annual inspections are recommended on fuel-fired air handlers, such as furnaces, to ensure that the heat exchanger is in good condition.

When designing air-heating systems in cold climates, care should be taken in locating and selecting the air diffusers. The heat must reach the occupant. Heated air distributed at the ceiling level has to fight room stratification to get to the occupant level. This requires careful diffuser selection based on throw and positioning of the diffusers so that air will get to the floor. Consider also that, during cooling mode, this same throw pattern may also result in uncomfortable cold drafts. The industry has responded to this dilemma by having temperature-based diffusers that automatically change the air pattern based on discharge air temperature.

Another solution is varying the amount of airflow based on heating and cooling conditions through the use of control dampers or Variable Air Volume (VAV) boxes.

Distribution ducts, whether used for mechanical cooling or for heating, should be insulated. Energy guidelines such as ANSI/ASHRAE/IES Standard 90.1 (ASHRAE 2019b) provide recommendations for the insulation thickness.

Steam Systems

Steam Central Plant

A central steam plant normally consists of two or more steam-producing boilers, a steam and condensate return distribution system, and steam terminal units with condensate traps. The boilers can be electric, though they are usually fuel-fired, either with natural gas, fuel oil, or coal.

To provide redundancy, optimal efficiency, and turndown capability during part-load periods, multiple boilers are provided sized at various percentages of total anticipated heat load. Redundancy is normally included in these percentages to ensure that facilities will stay above freezing if one or more boilers fail or are off-line for maintenance. Common percentages include two boilers at 66% or 75% peak load, three boilers at 50% peak load, and four boilers at 33% peak load. Some clients require 100% redundancy in their systems, which will change these values.

Central steam plants (Figure 9.26) used to be very common and can still be found in central-utility/campus-type heating plants. Distributed steam systems can transport energy great distances with minimal energy loss as compared to central hydronic plants. Steam also has a significant capacity to carry energy compared to air and hydronic systems.

Figure 9.26 Central steam plant.
Courtesy David Lima, Simoneau

Steam systems, with their chemical treatment and distributed condensate traps, can require significant maintenance compared to other systems. For cold climates, freeze protection for the condensate systems must be designed into the system. Steam plant operators are also becoming harder to find as these systems are slowly being replaced with distributed hydronic systems, and technical training is not as commonly available. Operational costs for a central steam plant must be considered as part of the overall system selection.

If the distribution system includes underground pipe routing outside of the facility, either direct buried or through utilidors, the impact of heat loss into the surrounding ground must be taken into account. If the heat melts surrounding permafrost, the pipe or utilidor system may fail because of ground subsiding. Ground heaving from freeze/thaw cycles around the pipe may also cause failure in the distribution system. A civil engineer experienced in arctic design should be consulted for these types of distribution systems in arctic conditions.

Steam Coils

If steam is used as the main heat source for a building or AHU, steam-to-glycol heat exchangers should be used if possible versus directly installing the steam heating coil in the airstream; coils, traps and piping can be damaged by frozen condensate. Where the steam coils must be installed by necessity, designers should consult with manufactures offering nonfreeze strategies. This includes various construction types such as tube within a tube, vertical tube, single- and double-row vertical feed coils, and integrated face-and-bypass damper assemblies, among others.

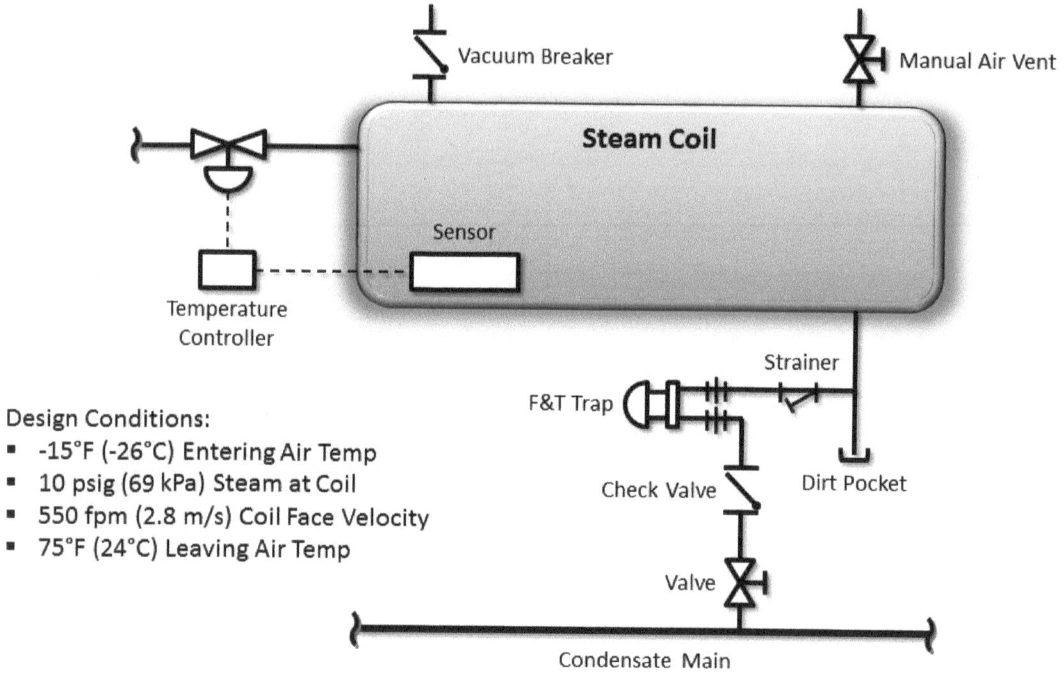

Figure 9.27 Steam coil piping arrangement.
Courtesy DLB Associates

To prevent flooding of condensate within the coil, positive steam pressure must be maintained. This requires the control valve to provide constant steam flow through the coil even though the supply air temperature is satisfied. Integrated face-and-bypass coils are a common practice. Full face-and-bypass dampers for temperature control can be used in conjunction with an antiwipe damper located downstream of the coil to prevent overheating.

Hydronic Systems

In cold climates, hydronic heating systems are often used for space heating in commercial buildings. One significant benefit is that hydronic systems can transport more energy with less space than an all-air system. There are different methods of distributing the heat to the conditioned spaces as follows:

- Radiant—low temperature
- Terminal units—reheat, fan coils, wall fins
- AHU coils

Hydronic Central Plant

Hydronic heating plants normally consist of a single large or multiple smaller hydronic boilers. These boilers can either be electric, fuel-fired, or biomass-based units. The distribution system for a hydronic system consists of distribution pumps, supply and return distribution piping, and terminal hydronic heaters. System accessories normally include an expansion tank, air separator, and either a water or glycol makeup system.

Similar to the steam boiler plant, hydronic heating plants are also recommended to be composed of multiple boilers to provide redundancy and maximum efficiency during part-load conditions.

Fuel-fired hydronic boilers that condense the products of combustion can have combustion efficiencies above 90%. Standard-efficiency boilers are noncondensing and usually are below 88% efficient.

Energy can be very expensive in colder regions because of transportation costs. Therefore, higher-efficiency equipment can provide significant energy and utility cost savings. However, condensing equipment is typically more complicated than noncondensing equipment, and it can require more maintenance and more sophisticated maintenance personnel.

Similarly to steam systems, designers of underground distribution systems outside of a facility must be cognizant of thawing and freezing cycles of the surrounding ground.

Terminal Hydronic Heaters

The types of terminal hydronic heaters and their application are varied. The units typically have a hydronic control valve. In cold climates, it is desirable to have these valves fail to the full-heat position unless otherwise noted.

Baseboard/Finned Tube. Perimeter hydronic finned tubes provide the highest economical heat output when coupled with a high temperature 180°F (82°C) supply hydronic heat plant, but at the cost of lower efficiency. Boiler efficiency is a function of return water temperatures. Derating factors should be applied to calculate baseboard output when operating with a larger ΔT and/or lower supply temperatures. Multiple tiers within an enclosure or greater lengths of baseboard maybe required to deliver design performance at reduced temperatures.

Radiant Ceiling Panels. Hydronic or electric radiant ceiling panels have the benefit of not taking up floor space and are a more sanitary installation for health care spaces and restrooms. In exterior office spaces, the space under a perimeter desk will not be "seen" by the radiant ceiling panels and therefore will be a lower temperature. This may cause comfort issues if the exterior envelope is weak, such as in spandrel glass buildings. Designers are advised to refer to ANSI/ASHRAE Standard 55 or ISO 7730 (ASHRAE 2020b; ISO 2005) for evaluating fluid temperatures and resulting radiant asymmetry to prevent discomfort from excessive surface temperatures. The heat output of radiant ceiling panels steeply derates as hydronic supply temperatures are lowered. Care should also be taken to limit the total flow through each panel circuit so as not to generate noise.

Cabinet Unit Heaters. Hydronic fan-coil cabinet unit heaters can either be installed on a wall surface or at the ceiling. Because of their high heat output and high sound levels, these units are best applied to entry vestibules and at exterior doors. For cold climates, the ideal installation is to install a cabinet unit heater low on a wall with inverted airflow that takes return air from the top of the unit and discharges it low to the floor. Using an inverted airflow pattern distributes heated air across the floor of a vestibule/entry, which assists in the melting and evaporation of snow and rain. This reduces long-term maintenance associated with molding walk-off mats, or worse yet, safety issues related to icing floor surfaces. Ceiling-mounted units can also be installed in vestibules where wall surfaces are not available to the designer, but a cabinet unit heater that is installed too high above a floor or in too large of a vestibule will not be able to throw the heated air to the floor to sufficiently melt and evaporate the surface. Cabinet unit heaters that are used with water-based heating systems and located in areas prone to freezing, such as an entry vestibule, may have their control valves removed and "run wild" to ensure heat is in the coil at all times. In these situations, only the fan is activated on a call for heating. Overheating of the vestibules during summer months is a possibility if the heating system is kept on.

Hydronic Unit Heaters. Hydronic fan-coil unit heaters are available in either horizontal or vertical discharge configurations. These workhorses of cold-climate heating systems are durable and provide an exceptional amount of heat. With the directional diffuser on the discharge of the units, these units can also easily throw heated air at the floor level. Because of aesthetic issues and the units' exceptionally high sound levels, they are typically used in unoccupied spaces such as mechanical or storage rooms, or in high bay applications such as vehicle maintenance facilities.

Alternative/Renewable Energy Heat Sources

Alternative heating sources including biomass and solar/photovoltaic systems are covered in Chapter 2.

Radiant Floor Systems

Radiant floor systems can be heated with either hydronic heat or electric heat. Radiant floor systems are excellent in cold climates because they meet established radiant requirements and floor conditions identified in ANSI/ASHRAE Standard 55 (ASHRAE 2020b) and ISO 7730 (ISO 2005). These systems can provide user comfort in office towers, schools, and homes but are also very popular in maintenance bay applications because they assist in evaporating moisture from melted snow as well as thawing the undersides of vehicles.

The lower temperature of a radiant floor distribution system makes it an ideal companion with low-temperature, high-efficiency condensing equipment as well as alternative fuel heating sources such as ground-source heat pumps and biomass equipment.

Because floor temperatures are regulated by comfort standards, their output is controlled to these conditions. The upper limit for offices and homes is approximately 30 Btuh/ft^2 (95 W/m^2) and shops/warehouses approximately 50 Btuh/ft^2 (158 W/m^2). In extreme cold climates, this may not be enough to provide heating output and may need to be supplemented with other radiant surfaces (walls/ceiling) or heated air during peak heating conditions.

Depending on the installation, a high-mass radiant floor system can also be slow to adapt to changes in room temperature, such as late-afternoon solar gain or large, transient occupant loads. The reciprocal event is also possible, in which sudden heat loss events occur, such as an overhead door in a maintenance facility being opened. Secondary pickup load equipment, such as hydronic unit heaters, can assist in taking care of these sporadic events.

Snowmelt Systems

Snowmelt systems are very similar to radiant floor systems except they are installed outside of the facility. Snowmelt systems are used in cold climates on sidewalks and concrete pads to ensure a safe, dry walking route or to reduce the seasonal use of salt and gravel, which can shorten the life of concrete surfaces. Though rare, snowmelt systems have been used on roofs and other structures to reduce the weight of accumulated snow to levels that an existing structure can handle.

Snowmelt systems require slightly higher hydronic temperatures than radiant floor systems. They are sized based on wind conditions and what the lowest outdoor air temperature is when there is the potential for snow. The systems are designed not to operate when it is either warmer or colder than the regionally expected snow conditions.

Snowmelt systems can be operated based on manual control, automatic activation based on a temperature and liquid sensor, or run continuously to maintain a slab temperature above freezing. Continuous operation, or also known as *idling*, is rare and used only in critical applications such as a hospital drop-off area.

For hydronic systems, the heating medium is glycol at a percentage that protects it from slushing at ambient temperatures. The use of a heat exchanger between the snowmelt system and the boiler is recommended to both protect the boiler from low-temperature shocking but also protect the building heating system from draining in case of exterior pipe damage that may occur from snow plows or frozen ground shifting.

The use of closed-cell foam insulation, both below and on the sides of snowmelt areas, are necessary to reduce energy usage and increase the reaction time of the snowmelt system. Details should be provided to protect the tubing at concrete expansion joints.

Boilers

There are many types, operating conditions, and fuels used to produce hot water or steam to provide heat in cold climates. Operating boilers, temperatures, and pressures vary depending on the type of boiler used. In cold climates, heating boilers (hot-water and low- and high-pressure steam) typically use natural gas, oil, or propane for fuel. Boilers are available in noncondensing (>80% efficiency) and condensing (>90% efficiency) designs. Condensing boiler installations have been increasing over the last few years to improve the overall building energy efficiencies and help meet the requirements of Leadership in Energy and Environmental Design® (LEED®) Green Building Rating System or *International Green Construction Code*® (ICC 2018).

One advantage of the condensing boiler versus a conventional hydronic boiler used in the past is that the concern of maintaining a minimum hot-water return temperature through a technique such as a three-way valve system is no longer required.

Condensing boiler heat exchanger materials must be resistant to the corrosive effect from the acids formed during flue gas condensation. The materials vary between boiler manufacturers and can include stainless steel, aluminum, copper, and cast iron.

The efficiencies of most condensing boilers improve with a lower hot-water return temperature. This helps the designer to take advantage of distribution systems with lower-temperature, high-delta temperature loops. This not only improves the boiler efficiencies, but also reduces the pump energy required, because of lower-flow requirements.

When using condensing boiler technology, special venting requirements must be considered. The venting materials must be resistant to acid resulting from condensation formation from the flue gases. These materials have included AL29-4C stainless steel and polypropylene. The venting materials must be selected and sized in accordance with the boiler manufacturer requirements.

Formation of ice at the venting terminus is also a concern, and a typical rain cap is not allowed in many jurisdictions. Reducing cones have been used to increase the velocity at the termination point, thus reducing the potential for icing.

In cold climates it is advisable to preheat the combustion air from the flue exhaust to increase boiler efficiency on noncondensing boilers.

In standby noncondensing boilers, care must be taken to prevent downdrafting, which can cause severe damage to boiler tubes and drums. Pressurizing the mechanical room or installing a control damper in the flue stack (if permitted by local code authorities) are two options to prevent downdrafting.

Another concern with standby noncondensing boilers is the potential for thermal shock because of low return-water temperatures when the boiler is energized. This risk of thermal shock can be alleviated by continuous flow of supply hot water back to the boiler return header (although inefficient) or using a three-way valve system controlled to maintain a minimum boiler return temperature.

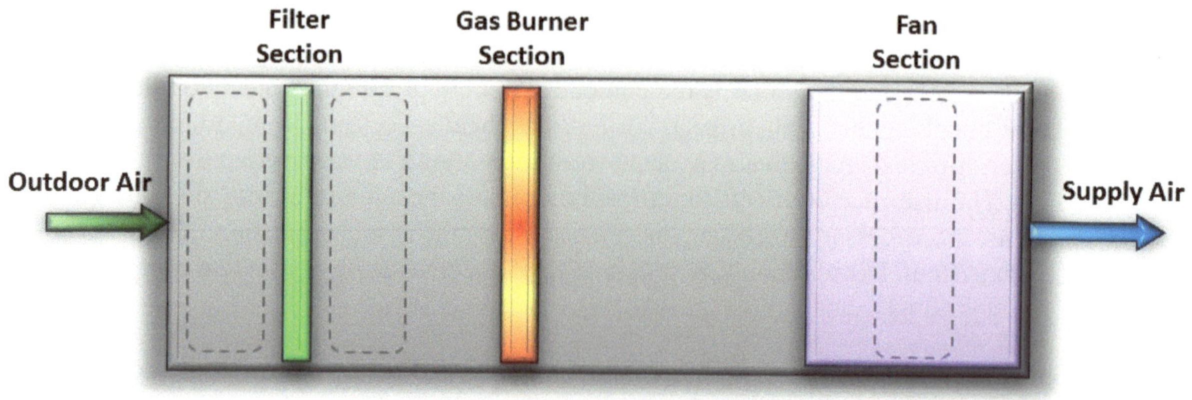

Figure 9.28 Direct-fired makeup air unit.
Courtesy DLB Associates

Gas-Fired Heat Exchangers

Gas-fired heat exchangers may be either direct fired or indirect fired. The heat source for an AHU may be a heat exchanger (combustion chamber) using natural gas, propane gas, or fuel oil. Indirect-fired burners are recommended for use where human occupancy and building codes require that the products of combustion are not in the airstream. Direct-fired burners are most often used for makeup air applications interlocked with an equal amount of exhaust air and are typically found in places such as restaurants, industrial applications, and vehicle repair shops.

When cold outdoor air is heated, the volume of air increases. If airflow quantity is critical, then an airflow measuring station located downstream of the heating section with variable-frequency drive control of the air fan is recommended.

Indirect-fired heat exchangers require the fan to be located upstream of the heat exchanger; the fan is therefore located in the cold-air stream. The need for arctic-grade motors needs to be considered in this application.

Gas-Fired Radiant Heaters

Where natural gas or propane is available, gas-fired radiant heaters can be a very efficient way to provide heat to occupants. These are particularly useful in high-bay applications such as warehouses, ice rinks, and maintenance garages. The use of radiant heat has been shown to keep occupants comfortable at lower space temperatures, which lowers energy costs.

Radiant energy also reduces the stratification issues associated with ceiling-based air heat, which is common in open bay spaces.

The designer should reference the section on design considerations for natural gas and propane systems in cold climates (Chapter 4) as well as combustion air and flue requirements for the equipment noted in this chapter.

Fuel-Fired Space Heaters

Solid-fuel (wood) fireplaces and oil-pot stoves are proven heating technologies. These systems typically do not require electricity to operate, which can be important in locations prone to power outages. They are simple, proven, and reliable systems with ready fuel sources. A properly designed system will radiate heat into the surrounding structure, prolonging the heating effect. Fuel-fired space heaters make excellent backup heating sources for open-/single-zone spaces or spaces that open into a common, central zone.

The designer should remember that these units can have higher surface temperatures that may burn occupants or create fire concerns if installed too closely to combustible materials. Combustion air also must be taken into account with these units. Fireplaces and oil-pot stoves can still be a source of deadly carbon monoxide.

Newer technology has developed fuel-fired space heaters that are simple, efficient units that operate using oil, propane, or natural gas. These units commonly come with stand-alone controls and concentric, through-the-wall combustion air and exhaust air flue kits that reduce installation costs. Fuel-oil-fired space heaters are workhorses in rural cold-climate communities.

GEOTHERMAL SYSTEMS

To improve building energy efficiencies, geothermal systems are being used more often. The following are some design considerations for geothermal systems:

- To make the geothermal system cost-effective based on life-cycling costs, it is very important to rightsize the geothermal loop according to the building heating and/or cooling loads. This rightsizing typically provides a smaller ground-heat exchanger than if rule-of-thumb loads are being used.
- One design technique used for the geothermal HVAC system is to size the ground-heat exchanger for the smaller load (heating or cooling) and provide supplemental boiler or fluid cooler to top up the capacity. This is typically called a *hybrid system*.
- Consideration should be given to providing backup boiler(s), gas or electric, and/or fluid coolers for the geothermal systems.
- One of the challenges in a cold-climate location is the seasonal heat balance for the geothermal loop. Because the building may be heating-dominant throughout the majority of the year, the geothermal system may extract more heat out of the ground than it has the capability of returning during the cooling season. This effect can be exacerbated by the operational schedules of buildings such as schools, which are typically closed during the summer seasons. Some of the possible solutions to maintaining the seasonal balance are solar heating or exhaust air heat recovery.

PLUMBING SYSTEMS

The plumbing design considerations are similar to those of the previously mentioned heating sources. No piping is to be routed outside of the vapor barrier, or it will likely freeze.

Plumbing Walls

In general, plumbing fixtures are not located on exterior walls unless a furred wall or cavity is provided between the fixture and the exterior vapor barrier where piping can be routed within the conditioned space. Even under these conditions, infiltration can freeze pipes. It is not uncommon to have plumbing chases be provided with a zone of bare, hydronic finned tube inside the plumbing wall to keep the space from freezing.

Plumbing walls, or *wet walls*, can also experience freezing conditions when located perpendicular to an exterior wall and even inside of a building. The cause of this is poor insulating and vapor barrier installation that allows cold air to enter into the plumbing chase from the exterior walls or even down from cold roofs.

Indirect-Fired Water Heaters

Where hydronic boilers are present, it may be advantageous to use hydronic, indirect-fired water heaters for domestic hot water. These units use boiler water circulated through a heat exchanger to transfer heat to the domestic hot water system.

Because of the heat capacity of the boiler, indirect-fired water heaters can typically provide a significant amount of hot water compared to standalone electric- or gas-fired units. When fuel oil is the predominant fuel source, the use of indirect-fired water heaters is preferred over having additional fuel-oil-fired burners and their associated high maintenance.

The disadvantage of indirect-fired units is that the boilers cannot be turned off during the summer, and the boiler supply temperature cannot be lowered below 150°F to 140°F (66°C to 60°C) to create water at a discharge temperature of 120°F (49°C). This can limit the energy-efficiency capabilities of low-temperature condensing boilers. Most condensing boiler manufacturers have piping arrangements for indirect fired water heaters with a "hot-water demand" control function that temporarily increases the boiler supply temperature to produce hot water and then reverts back to the lower supply temperature when the domestic hot-water load is satisfied.

Plumbing Vents

Because plumbing vents contain warm, moist air from the sanitary sewer and other waste systems, condensing and freezing of the condensate needs to be taken into account in cold environments.

In general, the plumbing venting system should be routed on the warm side of the building envelope until it is ready to exit the building. Vents should be insulated, as they pass through unconditioned spaces such as attics and cold roofs. It is common to extend insulation on plumbing vents a minimum of 3 ft (900 mm) from the point of leaving the conditioned space to reduce condensation on the surface of the piping. This insulation is extended through the envelope opening up to the flashing.

In colder environments, frost can build up and completely cover a plumbing vent termination. A frosted-over vent will affect fixture drainage. Good design practice is to increase the size of the vent termination by one pipe size to help offset this condition. The next level of protection is to extend the vent insulation through the roof all the way to the termination point. The outside of the insulation is protected with sheet metal, including a metal cap.

In extreme cold conditions, the insulation solution is further enhanced by the addition of electric heat trace wrapped around the vent pipe from inside the building to the vent termination.

Storm Drains

Storm drain systems can be open to ambient temperatures from the roof drain and overflow drain to the outlet location. Because of this, all storm drainage piping should be insulated, with a continuous vapor barrier, throughout the building. Condensation that forms on the outside of uninsulated pipes can create mold conditions inside concealed spaces.

Storm drain systems are active throughout the winter and spring to remove melting snow. Snow can melt from direct sun exposure while the ambient temperature is still below freezing. This can result in freezing of roof drains and scuppers. Where roof drains are located above unconditioned spaces, such as in a building overhang, heat trace may be installed inside the pipe from the roof drain all the way to the warm conditioned space. In extreme cold conditions, the roof drain bodies and scuppers are heat traced in all locations. Locating the heat trace on the inside of the pipe provides a path for water to travel.

Where overflow spouts are provided, heat trace may be desired to ensure they remain free of ice and are there to protect the structure if need be. It is good practice in cold climates to provide an overflow spout where the storm water system leaves the perimeter of the building in case the storm water discharge system freezes closed. Where there are open outfalls, those too are heat traced, with the heat trace running inside the pipe from the outfall to the interior of the building.

The discharge location of storm drains and overflows should be coordinated with the architect to ensure that spring melting does not create unsafe icing conditions on walkways and surrounding building surfaces.

ELECTRICAL SYSTEMS

Most mechanical systems require electrical power as well as interfaces with the signal and fire detection systems within a facility. This section covers some of the items that a mechanical designer should coordinate with the electrical designer for projects.

Power Conditioning

Chapter 4 covers design considerations for electrical utilities in regards to power quality. Sensitive electronic equipment can be susceptible to power fluctuations. The use of over/under voltage protectors at the piece of equipment can protect against brown-outs that might damage electronics. Variable-frequency drives also provide this protection through how they inherently create a modulating power signal.

Redundancy

As noted in Chapter 5, mechanical systems in cold climates need to have redundancy built into them in order to maintain occupant comfort and usability of the facility upon the event of equipment failure.

When redundant pieces of equipment are provided, such as a 100% redundant pump system piped and controlled in a primary/backup configuration, it is important to provide separate power circuits to each piece of equipment to ensure that a single breaker or disconnect malfunction does not inadvertently create a freezing condition.

Generators

The use of standby power generators are common for public and larger cold-climate facilities. The standby power system typically covers equipment and systems critical to keeping the facility occupied and safe from freezing, but not necessarily maintain occupant comfort under design conditions. This does not typically include ventilation systems. The mechanical designer should coordinate with the electrical designer to ensure that all parts of the heating plant, distribution system, and terminal heating units are placed on the standby power circuits. Control panels as well as the circuits for the distributed control unit transformers also need to be located on the standby power system. Note that redundant/nonconcurent power loads, such as primary/backup pumps, only need to be counted once in the total power calculation. Fuel tanks and service supply to fill tanks should also be taken into consideration when design emergency backup power. Location of fuel tanks are also critical, as underground tanks are problematic during a flood—water may float the tanks, breaking piping.

ARCHITECTURAL SYSTEMS

Though these systems are not normally within the scope of work of the mechanical designer, the designer needs to be aware of the issues and guide the design team toward appropriate solutions.

Building Envelope

Refer to Chapter 8 for a thorough explanation of building envelope design considerations for cold climates.

Roof Design

Where cold roof designs are used, be mindful to keep all heat and moisture sources out of the unconditioned cold space. Warm, moist air from sources such as a clothes dryer, cooking device, bathroom, or even just air from the conditioned space can create roof ice damming and condensation problems within the unconditioned space.

Exhaust and relief air terminations on the outside of the building also need to be coordinated with the cold roof intake openings. If eave vents are used along the edge of a roof, it is recommended to cover the roof vents on both sides of the exhaust penetration to minimize the intake of warm, moist air.

Exhaust and relief air ducts are not to be discharged into the cold space. Even combustion air openings can transfer heat by convection from the mechanical space into the cold space. Any mechanical penetrations going through the cold space, such as plumbing vents, boiler flues, and exhaust ducts, should be sealed tightly to the vapor barrier and be fully insulated through the cold space.

Damage from Snow and Ice

Snow and ice can accumulate on a roof under two conditions: natural gravity and wind-blown snow. When snow accumulates, it can cover up plumbing vents and get into ventilation openings, but the bigger concern is when snow and/or accumulated ice on a sloped roof breaks loose and slides off. Being mindful of how and where the roof pitches is an important part of cold-climate design. Pitched roofs, especially metal roofs, will slide snow and ice with great force onto whatever lies below. In addition to property damage, this can also lead to significant personal injury risk. The design team should endeavor to avoid locating the following items at the outfall of sloped roofs:

- **Building Entry and Exit Doors.** This includes exterior maintenance doors into mechanical rooms. The snow will block door swings and access.
- **Vehicle Parking.** Vehicles can easily be damaged by falling snow and ice.
- **Public Sidewalks.** Pedestrians can be injured by falling snow.
- **HVAC Hoods, Sidewall Exhaust Fans, and Equipment Sidewall Venting.** The snow can damage these items and even rip them off the side of the building.
- **Condensing Units.** Often located near a building, falling snow and ice can cause damage to a unit including the loss of the refrigerant charge.

There are methodologies for protecting openings at the bottom of roof slopes, such as doors. Use the help of an architect experienced in cold-climate design with any of the following options:

- Locate canopies above entries and walkways. Be sure that the structural design for the canopies takes into account the extra weight and impact of the snow from the roof above.

- Use shingled roofs or other roofing systems that are not as slippery as metal roofing systems. This will not eliminate the danger of snow falling from the roof, but should minimize how much releases at once.
- Add snow guards or rails on sloped metal roofs to improve snow retention and minimize the amount of ice and snow that is released at once. This solution will not eliminate the possibility of damage, but will help to minimize the chances.

HVAC and other equipment mounted on sloped roofs is also susceptible to damage from sliding snow and ice. If not properly designed, a plumbing vent or other ducted penetration can be greatly damaged or even shear off of the roof. Design considerations to help minimize this include the following:

- Locate roof penetrations as high up on the roof slope as possible.
- Use *crickets*, wedge-shaped snow deflectors, upstream of the penetration to deflect sliding snow and ice around the protrusion.

On flat roofs, the accumulation of snow on a roof can also create problems. Openings into the building, including louvers and hoods, should be elevated above the anticipated drifting snow elevation on the vertical surface. The accumulation of snow should also be considered when defining the height of equipment curbs. Ideally, the top of the curb should be higher than the anticipated snow height to minimize the potential of melting snow getting inside the building through the seams along the top of the curb/equipment interface. Traditionally, this has been 18 in. (0.5 m) above the roof surface.

Piping routed exposed across roofs and on grade should be elevated so that the piping and its accessories are kept clear of compacted snow and ice. A minimum 5 in. (0.13 m) is recommended. This will reduce corrosion as well as physical damage from moving ice or being tripped over. In-line gas regulators should be elevated so that they are above the anticipated snow level so that their vents are not plugged.

Figure 9.29 A parking lot sign warns drivers of the potential of falling snow and ice from the roof above.

Figure 9.30 Roof cricket in Valdez, AK, protecting a roof-mounted exhaust fan.

Though light-colored, high-albedo roofs noticeably reduce cooling loads in the summer, they also create icing conditions in the winter and shoulder seasons. Some extreme cold areas prefer using dark-colored roofing systems to facilitate the melting and evaporation of ice and provide a safe walking area for their maintenance personnel. One compromise is to use light, high-albedo roof coverings with black walking mats from the roof access to and around equipment.

HOW COLD IS IT?

It is so cold ... the 3D printer keeps spitting out igloos!

10 Controls

Controls and control systems are an essential part of a building's HVAC system. Control systems ensure that the HVAC systems operate in a safe and efficient manner while maintaining the required building environmental conditions such as temperature, humidity, pressure, and ventilation. The intent of this chapter is to share best practices and lessons learned as they pertain to the implementation of HVAC control systems in cold climates.

The complexity of building control systems range from the very simple (e.g., a thermostat controlling baseboard radiation) to the very complex (e.g., a direct digital control (DDC) system controlling air-handling units (AHUs), individual room terminal units, and the heating plant and cooling plant, all monitored by a graphic front end) and every type of control system in between. There is no standard control system just as there is no standard HVAC system. The building control system is custom designed for each building to meet the unique requirements of that building's HVAC system. It is this unique custom-design requirement that can make HVAC control systems appear complex, and is perhaps the reason why control systems are so poorly understood in the industry.

All control systems provide common functionality—they turn HVAC equipment on and off and regulate the outputs of the HVAC equipment to maintain the proper building environment as required. Understanding control systems starts with understanding the fundamental building block of HVAC control systems—the control loop. All control loops consist of three basic elements:

- A *sensor* measures the value of the controlled variable such as temperature, humidity, pressure, etc.
- The *controller* receives the information from the sensor, compares the actual value to the required set point value, and calculates an output signal to send to the controlled device.
- The *controlled device* (heating valve or damper) is the component used to directly modify the controlled variable, such as temperature, humidity, or duct pressure.

A control system can be as small as one control loop, or it can consist of many, many control loops. For example, an AHU can have three or four control loops to regulate the outputs of the unit.

Control systems are classified by the type of power they use and by the type of input and output signals they use (analog or digital). Most control systems are a combination of the five basic control system types:

- Self-powered (thermostatic)
- Electric
- Electronic
- Pneumatic
- Microprocessor-based or DDC

Before understanding the implementation of HVAC controls in cold climates, it is important to understand the basics of HVAC control systems and the application of controls to particular HVAC systems. However, it is beyond the scope of this book to cover these control system fundamentals and system applications, especially when these topics have been addressed so well in other ASHRAE documents and independent publications. The following are excellent resources and can provide a good general understanding of HVAC control systems.

HVAC CONTROL SYSTEM RESOURCES

ASHRAE Handbook—Fundamentals, Chapter 7—Fundamentals of Controls

Chapter 7 of *ASHRAE Handbook—Fundamentals* (2021a) discusses control fundamentals and terminology; the physical components such as sensors, control devices, controllers, and auxiliary control devices; communication networks; and commissioning, among other topics.

ASHRAE Handbook—HVAC Applications, Chapter 47—Design and Application of Controls

Chapter 47 of *ASHRAE Handbook—HVAC Applications* (2019e) discusses the design and implementation of control systems for specific HVAC applications such as hot-water systems, chilled-water systems, air systems, and terminal units. It provides a schematic of the HVAC system indicating placement of sensors and controlled devices with a brief description of the system control strategy.

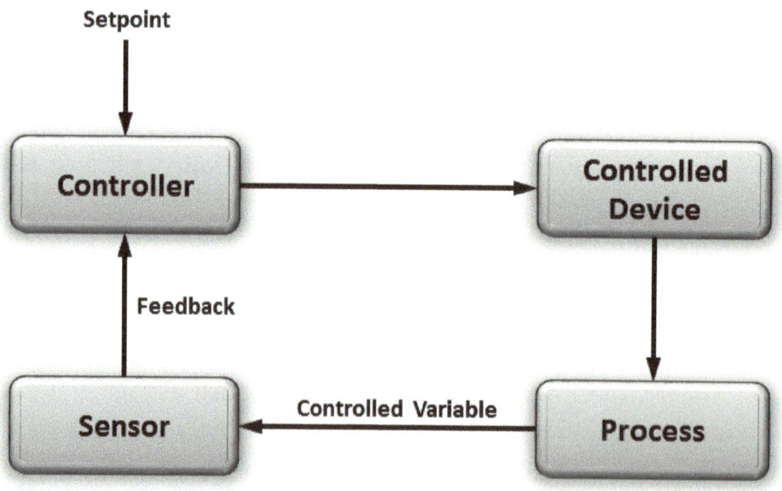

Figure 10.1 Control system.

Courtesy DLB Associates

CIBSE Knowledge Series—*Understanding Controls*

Published by The Chartered Institution of Building Services Engineers (CIBSE), London, UK, *Understanding Controls* (2005) provides an excellent overview of HVAC controls. It also provides a list of questions to consider when implementing or retrofitting a control system.

Manufacturers

HVAC controls manufacturers and vendors are also a very good source of information. Many vendors have design guides and engineering manuals covering all aspects of HVAC control systems.

HVAC CONTROL SYSTEM STRATEGIES

Climate-developed strategies that are routinely used in every type of cold-climate control system from the very simple range to the very complex. These strategies are generally applicable to all control systems types, but some strategies listed here can only be implemented with DDC systems. This is by no means a complete list of strategies to use in a cold climate. Widely varying ambient conditions present different problems, and different solutions are frequently developed.

Here, in no particular order, are some lessons learned from control contractors and building owners.

Do Not Let Things Freeze

The underlying concern when designing control systems for cold climates is to consider the effects of cold on the building, the building occupants, and the HVAC equipment and systems. In simple terms—do not let things freeze. The designer must always be aware of the outdoor air temperature and what effect it will have on the HVAC system during both system start-up conditions and system steady-state operating conditions.

Valve Fail-Safe Action

Space zone control actuators on valves and dampers should fail to the full-open position in the event of a system failure, to ensure that the building is heated. A system failure could be a controller failure or a power failure. Floating control valves are driven both open and closed, and they fail in the last-commanded position. The use of floating control zone valves for a particulate application should be evaluated to determine if their use would represent a risk. What is the likelihood that all the valves are commanded to a closed position when the control system failed? If a majority of valves were closed, would this be a problem? Would heat be available if the valves were open?

AHU heating coil control valves should be commanded to the open position when the unit is off and should fail to the open position. This ensures that there is flow through the coil and that the plenum is heated.

However, one side effect of commanding the heating valve to full-flow position when the AHU is off is that the AHU plenum can heat up significantly, making it difficult to achieve stable control of the unit on start-up. One approach that has been used to prevent overheating the plenum when the unit is off is to control the heating valve from the mixed-air temperature sensor (using a proportional-only control routine) to maintain, for example, a mixed-air temperature of 59°F (15°C). This opens the heating valve to inject heat when the plenum is getting cold but closes the valve to prevent overheating as the temperature rises in the plenum. This makes the AHU much easier to control on start-up because there is less waste heat to dissipate before the heating coil control routine reaches steady-state control.

The steam control valves on a steam/glycol heat exchanger should fail closed. The excess heat will cook the glycol/water mixture, turning it into a thick sludge.

Damper Fail-Safe Action

The AHU mixed-air damper actuators should be spring-return with a minimum of one actuator per damper. The exact number of actuators required per damper is determined by the torque requirements of the damper section.

The actuators should be installed such that, with no power to the actuators, the outdoor air intake and exhaust air discharge dampers are closed and the return air damper is open—a fail-safe position that preserves heat in the unit. The mixed-air dampers should return to a fail-safe position or their normal position when the AHU is off or when the AHU has failed to operate. Dampers (and valves) are described as being normally open or normally closed. The normal position is the position the actuator assumes with no energy source (electric or pneumatic) applied to the device. Outdoor and exhaust air dampers are typically normally closed and the return air dampers are normally open.

It is best to operate the outdoor, exhaust, and return air dampers in unison from the same control signal to ensure that when the fan is running there is always an airflow path through the unit. If the outdoor, exhaust, and return air dampers are controlled by separate control signals, there is a possibility that all dampers could be driven closed starving the fan for air. Note that, in some cases, the exhaust air damper is operated from a separate control signal as part of exhaust air volume tracking and/or exhaust plenum pressurization routines.

Cold Air Inrush

Start-up of AHUs with mixed-air dampers can be very difficult when the outdoor air temperature is below freezing. In these cases, the controller should slowly ramp open the mixed-air dampers to their control point to prevent excessive overshoot of the control point. If a ramp function is not used on start-up, the control routine could drive the dampers to 100% outdoor air to dissipate the heat in the unit. When the outdoor air temperature is below freezing, the temperature in the unit will drop rapidly and the controller will drive the dampers to the fully closed position. The temperature will rise again and then the dampers will open—this is known as *hunting* or *overshoot*. This hunting action can be difficult to stabilize; it can cause the freezestat to trip, therefore shutting down the unit, and also places equipment downstream of the unit at risk due to a burst of freezing air being injected into the AHU. A ramp function slowly opens the dampers and eliminates or minimizes the wild swings in the damper position on unit start-up.

In some cases, the ramp function is not sufficient to get the unit started and other strategies are required on start-up. Consider using proportional-only control on start-up of the AHU in addition to the ramp function, and enable the integral functions of the proportional-integral-derivative loop after a period of time. This is applicable only to DDCs. Note that use of the derivative function was not mentioned, because it can be extremely difficult to tune a loop with the derivative function enabled and the end result may not be that much better than proportional and integral control.

Another option if the ramp function is not successful is to start the AHU with the dampers in recirculation mode and the heat on full. Operate the AHU this way for several minutes in order to bring the heating element (electric or hot water) up to temperature and preheat the unit. After several minutes of preheating the AHU (the time period is adjustable and unique to each device) the dampers are very slowly ramped open to their control point while the heating is very slowly ramped down until its control point is reached.

Freezestat

A low-temperature thermostat, also known as a *freezestat*, *low-limit cutout switch*, or *safety low limit* is a very commonly installed device used to protect hot-water coils from freezing. These devices are commonly referred to as *freezestats;* this name implies that they can prevent freezing when, in fact, they cannot. No matter what they are called, they are an important part of the AHU control system when installed properly.

The freezestat consists of the device head installed outside the unit near the top of the coil and a 20 ft (6 m) long sensing element, a vapor-filled capillary installed below the level of the device head across the downstream side of the coil. The capillary must be installed in a horizontal serpentine pattern across the face of the coil, not vertically, with no sharp bends or kinks in the capillary. If the temperature of any 1 ft (0.3 m) section of the capillary tube falls, the vapor condenses, and the pressure in the capillary tube drops, tripping the switch. A 1 ft (0.3 m) of the capillary covers 1 ft^2 (0.92 m^2) of coil, so a 20 ft (6 m) sensing element will cover 20 ft^2 (1.8 m^2). If the coil is bigger than 20 ft^2 (1.8 m^2), multiple freezestats should be installed. It is not uncommon to have three or four freezestats installed on a heating coil.

The freezestat can be manual- or auto-resetting type and, when tripped, the line voltage contact is hardwired to shut down the AHU. Some models have a second set of low-voltage contacts to provide a signal to the DDC system. When multiple freezestats are installed, the normally closed line voltage contacts of all the freezestats wired in series to shut down the AHU, and the normally open low-voltage alarm contacts (if available) are wired in parallel to the direct digital controller or alarm circuit. It is important to ensure coverage across the bottom of the coil, because this is the area most vulnerable to freezing. It is important to follow the installation directions provided by the manufacturer.

A freezestat can also be a very troublesome device, causing "nuisance" shutdowns and problems dropping out units on start-up. Often, when a coil has burst and the cause is investigated, it is discovered that a freezestat was installed but was deactivated because of nuisance trips. Before resorting to deactivating the freezestat, try to correct the issue causing the freezestat to trip. This can be difficult, because the No. 1 priority is to restore heat, not determine why the freezestat tripped. It can be difficult to simulate the problem later when things are less hectic, because the outdoor air conditions are not as extreme.

Proper installation of the device can significantly reduce the occurrence of nuisance trips. There are several other options available to achieve the same functionality the freezestat provides but without the problem of nuisance trips. A time delay function can be wired into the freezestat circuitry to override trips on system start-up; however, care must be taken when setting the time delay so that a true freeze condition is not overridden.

When DDCs are used, a software low-limit control routine can be used to monitor the temperature on the discharge of the coil and shut down the AHU when required, taking into account start-up conditions. These delays are typically only used when there are other strategies protecting the heating coil, such as glycol in the heating system or a coil pump to maintain minimum fluid velocity through the heating coil.

Steam Coil Control

Steam heating coils require a freezestat installed on the downstream face of the coil to protect against freezing. When the outdoor air (or mixed air, depending on the unit configuration) entering the coil is below freezing, the steam coil valve should be fully open. To control the temperature downstream, face-and-bypass dampers are modulated to maintain the discharge temperature at set point. The face-and-bypass dampers work in unison but in opposite directions—when the face damper is open, the bypass is closed, and as the face damper drives closed, the bypass damper drives open. The face-and-bypass dampers mix two airstreams—the air (very hot) flowing through steam coil and the air (below freezing) bypassing the steam coil. Stratification of the air downstream from the coil can be an issue. The location of the discharge air temperature sensor should be selected to provide the most uniform temperature. Additional measures, such as mixing baffles, may be required to ensure proper mixing downstream of the face-and-bypass dampers. When the outdoor air temperature is above freezing, the face-and-bypass dampers are positioned to provide full airflow through the steam coil, and the steam valve is modulated to maintain the discharge air temperature at set point.

Temperature Sensor Alarm Limits

When using DDCs, it is always a good practice to program low- and high-temperature alarm limits on the AHU discharge temperature sensor to act as a backup for the freezestat (in case the freezestat device has failed or has been disabled).

Low- and high-temperature alarm limits should also be programmed on all the space temperature sensors to provide a warning of potentially dangerous low-temperature conditions. Alarm limits on each sensor can identify localized potential freeze conditions that could be caused by inadvertent sources of cold air, such as doors that have been left open or windows cracked open and left overnight or for a weekend.

Humidity Control

Thermal comfort in a space is affected by temperature and humidity. In cold climates, it is often necessary to add humidity to the space during the winter months. The desired humidity level is always a balance; it is a trade-off between achieving a humidity level sufficiently high for occupant comfort and yet low enough to prevent condensation on the building envelope.

Humidification can cause problems, such as condensation on windows and flooding of supply ducts, if the proper control devices and strategies are not implemented. The control elements for a duct-mounted humidifier on an AHU include, at a minimum, a discharge humidity sensor mounted downstream from the humidifier, a space-mounted humidity sensor located in an area representative of the building conditions, an outdoor air temperature sensor, and an interlock with the AHU. The humidifier is controlled from the space humidity sensor to maintain the humidity at set point. The discharge air humidity sensor acts as a high limit to keep the supply air below the moisture condensing point.

The outdoor air temperature sensor is used for compensation or reset control of the space humidity set point. As the outdoor air temperature falls, the space humidity set point is decreased to prevent, or at least reduce, condensation on the windows. A typical reset schedule is shown in Table 10.1.

The actual numbers used for the reset schedule are unique for each building and may be very different from those shown in this example. The values selected are impacted by the building envelope, the insulation value of the windows and the unique operating parameters of the building. The high- and low-humidity set point limits may have to be

Table 10.1 Outdoor Reset of Space Humidity Set Point

Outdoor Air Temperature, °F (°C)	Space Humidity Set Point, % rh
50 (10)	40
−22 (−30)	25

adjusted through trial and error from the first few seasons of operation to find the best balance between maximizing humidity and minimizing condensation.

Device Temperature Rating

Select field devices rated for use at the winter design temperature. A device rated for use to –22°F (–30°C) (e.g., a cat whisker end switch) can freeze and fail to work properly at –40°F (–40°C). If a device cannot be found rated for use at that design temperature, there are a couple of installation options. Consider installing the device inside the heated building whenever possible. This simplifies the access to and maintenance of this device. Look to the manufacturer for accessories devices that could extend the ambient operating range of the device. For example, a direct-coupled damper actuator has an operating temperature range of –22°F to 122°F (–30°C to 50°C). To use this device at a location with a winter design temperature of –40°F (–40°C), some accessories are required. These accessories include a heater with integral thermostat that is attached to the actuator or a weatherproof enclosure, or a combination of the two, which can extend the device's ambient operating range. In extreme situations, even these accessories may not be sufficient, and creative solutions must be found. A custom enclosure complete with heat-trace tape can be fabricated and used to protect the device.

Direct Digital Controller Location

DDCs should be located in a heated indoor space with easy access to the enclosure. Installing the DDCs outside in a heated enclosure is not recommended. Although the heated enclosure may be able to maintain the enclosure temperature within the operating range of the direct digital controllers (typically 32°F to 122°F [0°C to 50°C]) under normal conditions, as soon as any maintenance or troubleshooting is required and the enclosure is opened, the temperature will quickly equalize with the ambient temperature. At – 22°F (–30°C), the direct digital controller will not function for very long.

Night Cycle and Night Setback

One important function of control systems is to turn the HVAC equipment off as an energy-saving strategy when the building is not in use. However, precautions must be taken in cold climates when implementing occupancy schedules, to protect the building and the equipment in it from freezing.

The building space temperature is monitored using one or more temperature sensors. When any one of the temperature sensors falls below the unoccupied set point(e.g., 59°F [15°C]), the HVAC equipment associated with that sensor is cycled on and off to maintain the space temperature at the unoccupied set point.

A dead band of 4°F (2°C) is typically used to prevent excessive cycling of the unit; that is, the unit will turn on when the space temperature falls below 15°C (59°F) and turn off when the space temperature rises above 17°C (63°F). This is referred to as *night-cycle* or *fabric* protection.

In lieu of turning systems off per an occupancy schedule, the system can be operated continuously but at a reduced set point during the unoccupied period. This is referred to as *night setback*. This ensures that the building space temperature does not fall too low during the unoccupied period and reduces the time it takes to bring the system back up to required occupied set point. A word of caution: using night setback in spaces maintained at higher humidities increases the risk of condensation on any surface reaching its dew point as a result of lower space temperature.

In some buildings, a combination of the two unoccupied strategies can be used. For example, a building with perimeter radiation and AHUs would, during the unoccupied period, turn off the AHUs and go to an unoccupied set point (for example, 59°F [15°C]) to control the perimeter radiation. If the space temperature could not be maintained by the perimeter radiation, the AHU would be started and controlled to maintain the space temperature at the night-cycle set point (e.g., 54°F [12°C]). Thus, during the unoccupied period, the AHUs act as a heating backup for the perimeter radiation.

During a long stretch of winter design-day conditions, turning the HVAC equipment off during unoccupied periods or operating at night setback conditions may not be feasible, because the building may never be able to recover to the required occupied temperature set point. During a long cold snap, HVAC equipment may have to operate at the occupied set point 24/7. Provisions should be made in the control system to easily allow this operational change.

Outdoor Temperature Reset Control

There are a number of control loops that can benefit from adjusting the control set point based on the outdoor air temperature. This is called *reset control* or *compensation control.*

Perimeter Hydronic System. The hot-water supply temperatures for perimeter radiation and space reheats loops can be reset to provide the maximum supply temperature (e.g., 194°F [90°C]) on the design days and a lower supply temperature on the more moderate days. The hot-water distribution system must be designed to provide different loops for various functions and provide a method to control the supply temperature in these loops.

Hot Deck Discharge Air Temperature. The supply air temperature on an AHU hot deck can be adjusted higher when the outdoor air temperature is below freezing and adjusted lower when the outdoor air temperature is above freezing.

Hot-Water Boilers. The boiler discharge temperature can be reset based on the outdoor air temperature to provide a higher supply water temperature when the outdoor air temperature is below freezing (i.e., when the heating load is high). Before implementing a reset schedule for a boiler hot-water discharge control, consult the manufacturer for the recommended minimum boiler inlet temperature. Some boilers are at risk of shocking at boiler inlet temperatures below 140°F (60°C).

Space Humidity. The space humidity set point is adjusted based on the outdoor air temperature to prevent condensation on the windows. As the outdoor air temperature falls below freezing, the space humidity set point is decreased. Refer to the Humidity Control section for further details.

Valve Sizing

The sizing of the heating coil and the heating coil control valve is critical to a responsive operation of the heating control loop. Oversized coils and valves are extremely difficult to control to a stable operating state, especially at low-load conditions.

There are many factors to consider when selecting and then sizing a control valve. These variables include valve characteristics, fluid characteristics, fluid flow rates, system pressure, valve close-off requirements, and pipe size and configuration.

Globe valves, ball valves, and butterfly valves are commonly used in HVAC applications, with globe valves being the most popular for modulating applications.Valves are sized according to the flow coefficient C_v (I-P units) and K_v (SI units). The flow coefficient C_v is defined as the amount of flow in gallons per minute of 60°F water that will flow through a fully open valve with a 1 psi pressure drop across the valve. The flow coefficient K_v is defined as the amount of flow in cubic meters per hour of water (5°C to

30°C) that will flow through a fully open valve with a pressure drop of 1 bar. To calculate the required C_v or K_v for a particular flow rate, use the following formula:

$$C_v = Q\frac{\sqrt{SG}}{\sqrt{\Delta P}}$$

where
Q = flow, gpm
SG = specific gravity of the fluid (water = 1)
ΔP = pressure drop across the valve, psi

$$K_v = Q\frac{\sqrt{SG}}{\sqrt{\Delta P}}$$

where
Q = flow, m^3/h
SG = specific gravity of the fluid (water = 1)
ΔP = pressure drop across the valve, bar (100 kPa = 1 bar)

$$K_v = 0.865\ C_v$$

The required pressure drop across the valve can be difficult to determine. In theory, it should be equal to or greater than the pressure drop across the coil. However, in practice, this information may not be available when sizing the valve. A good rule of thumb is to use a pressure drop across the valve of 3 to 5 psi (21 to 35 kPa).

Select the valve size that most closely matches the calculated C_v. The calculated C_v should not exceed the actual C_v by more than 10%.

For HVAC applications, the valve size is selected to provide the maximum design flow at 100% of the valve stroke range (i.e., when the valve is fully open). This is in contrast to industrial or process control applications where the recommendation is to select a valve that will provide the design flow in the range of 60 to 80% of the valve stroke range. Using this selection guideline will result in an oversized valve for HVAC applications; HVAC heating coils rarely operate at the max load conditions. The required system flow rate can vary from 0 to 100% of the maximum flow rate. The valve rangeability is defined as the ratio of the maximum to minimum controllable flow. A smaller valve will have a smaller minimum controllable flow, making the system easier to control at the low-flow rates.

At the maximum flow rate (winter design days), the goal is not controlling to a precise flow rate but rather just achieving full flow through the valve. Therefore, the valve that will provide the lowest controllable flow while still allowing the maximum flow when fully open is the preferred valve.

Industrial applications tend to control to a specific design flow rate with a very narrow variation in the flow. These applications require a control valve that will provide that narrow range of flow rate over the linear operating range of the valve. Therefore, the valve that will provide the design flow rate over the linear range of the valve (60 to 80% stroke) is the preferred valve.

A rule of thumb used to validate the valve selection is that typically a globe valve for a coil should be one size smaller than the pipe size. For example, a heating coil with a 2 in. (5 mm) pipe typically would require a 1.5 in. (40 mm) valve. This rule of thumb however is not a substitute for calculating the required C_v or K_v and selecting the proper size valve.

Readers are advised to study the principles of valve authority and contents of the *ASHRAE Handbook—HVAC Systems and Equipment*, Chapter 47 (ASHRAE 2020a).

Building Management System Alarm Notification and Remote Access

Consider implementing a building management system in the building to take advantage of its alarm notification and remote access features. A building management system integrates the monitoring, operation, and management of a building's microprocessor/direct-digital-control-based HVAC control systems to a computer "front end" and presents real-time information to the user typically in graphical format. The building management system can broadcast alerts and alarms to other devices such as cell phones and computers. The building management system can be accessed from alternate locations within the building and from locations remote to the building via Internet connectivity.

This allows the building operator to be alerted immediately to a problem situation and to access and manage the HVAC control system remotely to take corrective action. Remote access is a time and money saver, not only in remote locations where the costs to fly a technician in to troubleshoot and correct an issue can be extremely high, but also in urban centers where a quick response by a service technician is required and expected.

On a very simple electric control system where the advanced alarm notifications features of a building management system are not available, consider providing an alarm light mounted on the outside of the building. This can be very effective in remote locations and small communities.

Dirty Power

Dirty power conditions such as brownouts, spikes, and other power abnormalities can cause big problems for DDC systems in urban, rural, and remote locations alike. Dirty power conditions have been known to lockup or reset the direct digital controllers and, in extreme cases, corrupt the program in the controller. If dirty power conditions are common, a little planning can help alleviate the impact of these conditions.

A small uninterruptible power supply (UPS) unit with surge protection installed in the direct digital controller enclosure, or a larger central UPS unit providing the power to all direct digital controllers will keep the direct digital controller operating smoothly during power loss, brownouts, or power spikes that may be inherent to the local utility electrical supply or caused by transitions to generators.

Simple Control Loops

Keep the control loops simple and quick responding. For example, control mixed-air dampers from the mixed-air sensor and the heating valve from the discharge air temperature sensor. An overly complicated sequence of operation or control strategy can be very difficult to understand and troubleshoot. When there are problems—real or perceived—with a complicated system, people tend to override the controls in attempt to solve the problem, because they do not understand how the controls are supposed to work. Overriding the controls can be dangerous and can lead to serious equipment damage. The principle of simplicity is always important to keep in mind.

Sensor Location

Locate temperature sensors on interior walls well away from drafts or open doors that could adversely impact the temperature sensor reading. An often overlooked source of temperature sensor error in cold climates is cold air migrating through conduit and or wiring chases to the space sensor junction box. Without proper sealing at the junction box, the space sensors will show errors, even on interior walls.

Stratification

When there is a large temperature difference between two airstreams, stratification is likely to occur with the cold air on the bottom and the warmer air on the top of the duct. This can make placement of temperature sensors difficult and the value measured by the sensor inaccurate. There are several ways to deal with stratification. An averaging temperature sensor should be used in the mixed-air section to attempt to compensate for the stratification. AHUs may need special mixing sections to better mix the airstreams and reduce the stratification. Moving a temperature sensor further downstream from the point of mixing may allow the airstreams to mix more completely, providing a more uniform measurement. Stratification can also occur in water systems and duct-mounted humidification systems.

Air-to-Air Heat Recovery Systems Frost Control

Regardless of type of heat recovery system used, condensation occurs when the warm airstream is cooled below its dew point. Frost or ice may form if this airstream is also cooled below freezing. For most air-to-air heat exchanger types (heat pipes, glycol run-around loops, plate heat exchangers, heat wheels, and so on) frost buildup on heat exchanger surfaces can impede airflow and heat transfer. The threshold temperature at which frost or ice buildup in a heat exchanger becomes problematic is a function of the warm airstream temperature and humidity, the cold airstream temperature, the type of heat recovery equipment used and its heat recovery effectiveness. To address frost and ice problems, designers may select or design systems that prevent frost formation or apply strategies that periodically remove frost from the heat exchanger.

Frost prevention strategies prevent the warm airstream from being cooled below its dew point temperature during cold weather by preheating the cold airstream (or warm airstream) before it enters the heat exchanger or by reducing heat transfer between airstreams by, for example, bypassing a portion of the cold airstream around the heat exchanger (face and bypass dampers), tilting heat pipes, or slowing heat wheels. Recommended temperatures at which to activate frost prevention strategies and methods of control may be provided in equipment manufacturers' literature.

Frost prevention strategies are simple in concept and application, but impose an energy penalty in cold climates because the maximum amount of energy that can be recovered from the warm airstream in cold weather is limited to the difference in enthalpy of the entering warm airstream and that air near its dew-point temperature. In moderate climates (e.g., less than 4000 DDC/7200 DDF), the lost heat recovery potential tends to be modest. However, in colder climates, frost prevention strategies significantly reduce the potential amount of energy that can be recovered. Depending on how well the frost control strategy is optimized, a quarter to half the potentially recoverable energy is unavailable for recovery in 5000 DDC/9000 DDF climates, and 40% to 80% is unavailable in 10,000 DDC/18,000 DDF climates (Phillips et al. 1989; Phillips et al. 1992). Be aware that implementing electric preheat for frost control may use more source energy than ventilating without heat recovery.

Most manufactures of packaged heat recovery ventilators (HRVs) and energy recovery ventilators (ERVs) employ defrost strategies that, during cold weather, periodically warm the heat exchanger core to melt ice and frost accumulations. Typically, the cold airstream is interrupted while warm airflow continues through the heat exchanger. This may be done on a time schedule which varies the frequency and/or duration of defrost with outdoor air temperature, or may be controlled by a pressure differential switch that senses pressure drop across the warm side of the heat exchanger. More detailed descriptions of defrost strategies employed in HRVs and ERVs can be found in the literature (Phillips et al. 1989; Phillips et al. 1992; ASHRAE 2020a). At any rate, the designer may have to adjust ventilation rates during cold weather to account for reductions in ventilation related to frosting and operation of the defrost mechanism.

If air-to-air energy recovery equipment selected for a project does not include controls to address frosting and freezing in the heat exchanger core, consult the heat recovery equipment manufacturer for recommendations for frost prevention and defrost strategies specific to their equipment.

HOW COLD IS IT?

It is so cold ... Sunday communion was served with iced wine and toasted biscuits!

11 Residential

THE RESIDENTIAL CLIENT

Unlike for other buildings, residential clients have different objectives for their homes and private spaces. Dr. Nancy Wells, an environmental psychologist in the Department of Design and Environmental Analysis at Cornell University has demonstrated, "how natural and built environments affect people's mental and physical well-being throughout their lives" (Cornell University 2010). From the proceedings of an international workshop published by the World Health Organization (WHO 2010), "Housing and built environments have a profound impact on human health … exposures and health risks in this private setting are of crucial relevance." Residential dwellings are occupied for more hours than other archetypes. As stated in *Climate Change, the Indoor Environment, and Health* (NAS 2011), according to research by Berstein et al. (2008), of the 92% to 94% time spent indoors, the general population spends approximately 69% of its time at home; for infants the time at home varies by age, ranging between 66% and 76% (EPA 2009), and for the elderly the time at home is between 82% and 95% (Berry 1991; EPA 1996). It is not clear from research how these numbers relate to climate zones, but one could speculate that colder climates would see higher values. The nature of the occupant/home relationship, being far more intimate than that of a place of work or learning, means designers should thoroughly explore potential client needs and wants by asking such questions as:

1. Is the house a starter home where application of minimum codes is acceptable even if they result in relatively higher heating energy use and greater levels of winter discomfort? If so, the consequences of code-built homes should be explained to the client (Bean 2012).
2. Are the homes, apartments, or condos investment properties where hazards caused by snow, ice, and frost could become an accessibility issue for tenants? If so, solutions such as snow/ice melting or frost prevention strategies need to be explained to the client (Figure 11.17).
3. Is the home a retirement asset with greater emphasis on reduced maintenance and better indoor environmental quality and reduced operating costs through energy-efficiency measures? If yes, the designer needs to take the client through a comprehensive personal, environmental, energy architectural, mechanical, and electrical needs and wants assessment.
4. Will the house eventually become (as many do) a satellite extension of the institutional healthcare system and need special considerations for the aged,

Figure 11.1 Residential homes in cold climates have to deal with heat, sleet, rain, winds, snow, frost, and ice. They are occupied more than any other type of building and have an impact on human health and global energy use.

Robert Bean

injured, and/or infirm choosing to remain at home during palliative care? If yes, the principles of "lifetime" housing need to be evaluated with the client, including the ergonomics of the indoor environment for the occupants and health care equipment (Bean 2015).

Answers to these and other questions of a more technical and financial nature will provide early guidance to the designer in deciding strategies and tactics for meeting the residential clients' needs.

INDOOR ENVIRONMENTAL QUALITY

Residential versus Nonresidential

Where occupants in commercial, institutional, and industrial buildings can express indoor environmental quality (IEQ) dissatisfaction to a third party (management, lessors, building owners), the homeowner, for the most part, bears the brunt of dealing with their own criticisms of the property (or those of their tenants). For example, consider a stale air complaint and occupant needs for outdoor air at –40°F (–40°C). It is not a simple matter of running the energy recovery ventilator, as it likely is in defrost mode and not actually ventilating the home. Nor could one open a frozen window to obtain outdoor air. Such an exercise during a winter storm can be difficult and increases the risk of introducing vehicle emissions and traffic noise, affecting air and sound quality. It also introduces drafts of cold air, quickly affecting thermal comfort. Windows left open also destroy energy efficiency and frequently lead to frozen pipes and water-related damages. Consider also the common complaint of cold floors, for those with disabilities such as arthritis, diabetes, multiple sclerosis, or Parkinson's disease, it may not be a simple adaptive task of putting on more socks. Nor is it easy for vulnerable populations to adapt to the discomforts or being homebound due to extreme cold temperatures and/or snow/ice accumulation on walkways, driveways, and streets. For these and many other reasons, designers are advised to work with clients and CIBSE/REHVA/ASHRAE standards, guidelines and position documents as tools in designing above code cold-climate residences.

IEQ AND THE HOMEBOUND

Standard protocol for residential designers is to ignore the subjectivity and needs of the occupants, focusing strictly on the objectivity of code compliance. However, being indoors for long periods by necessity should require designers to pay stricter attention to the sensory physiological processes of the IEQ of sound, thermal conditions, air, lighting, odors, and vibrations. Likewise, it is important to recognize the psychological process of perceiving the environment and resulting wintertime stresses related to indoor environmental ergonomics. Often, occupants find wintertime spaces cool/cold, dim/dark, stuffy/stale, and dry/staticky. Together the sensation and perception of the IEQ will ultimately determine the energy used, as occupants use architectural, electrical, and mechanical systems to compensate for discomfort. At the same time, architects, engineers, and interior designers need to respect the positive tensions existing between the human factor principles of homeostasis and alliesthesia (Rodolfo 2000, Cabanac 1988) and ought not to design residential indoor climates so bland that the sensory system remains unstimulated for extended periods during winter months.

Adaptive Comfort: Not Always an Option

Though there is a global shortage of research looking at the adaptive habits of cold-climate dwellers, it is reasonable to say that outdoor conditions common in warmer climates (moderate outdoor temperatures, breezes, and abundance of natural lighting) are not always accessible to those in cold areas. As such, thermal adaptive measures are by default restricted to choices in clothing. On the surface, clothing adaptability seems like a simple solution, but such adaptations are not easily adopted by certain inhabitants; for example, infants and the elderly. Adaptability is also difficult for those with physiological and/or physical challenges, and those with mental health and/or cognitive disabilities (WHO 2017). For this reason, residential buildings should be designed using the analytical compliance method for comfort as prescribed by such standards as ASHRAE Standard 55 and/or ISO 7730 (ASHRAE 2020b, ISO 2005).

ARCHITECTURAL AND HVAC OBJECTIVES

Prime Objective: Control the Radiant Exchange

Air temperature is used in homes as a proxy for comfort conditions. In part, this is because of building code minimum requirements that specify dry-bulb control to approximately 72°F (22°C). Unfortunately, what typically happens in housing is minimum code practice in code becomes maximum practice in the field. Hence, 72°F (22°C) dry bulb has become the definition for comfort. Additionally, it is standard strategy in non-European countries such as Canada and the United States to use a single air-based thermostat located centrally in a home to serve as the occupant's "ambassador" to the HVAC systems. These tactics serve the need for code compliance but frequently prove to be poor and inadequate strategies for serving the needs of the occupants as demonstrated by surveys and current research (Decision Analyst 2016; Kapedan et al. 2017).

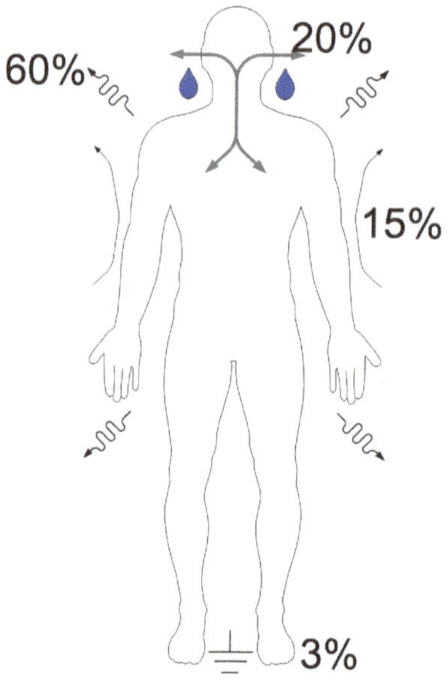

Figure 11.2 Approximately 60% of sensible body heat is lost via radiation at normal activities, clothing, and low air velocities. Respiration/evaporation accounts for 20%, convection for 15%, and conduction for 3%.

According to medical literature and ASHRAE Handbooks, approximately 60% of sensible body heat transferred to the environment is accomplished via radiation not convection[1] (Betts et al. 2013; ASHRAE 2021a). That is to say, the challenge to overcome in homes is more of a radiant exchange problem than an air temperature problem. The industry-wide disconnect between how body heat is primarily regulated and how houses are controlled is a root cause for thermal discomfort in homes. For this reason alone, cold-climate architecture should first enable the retention of body heat through the reduction of energy lost via radiant transfer. Reducing radiant transfer is accomplished with high-performance enclosure design and radiant-based HVAC heating systems (ASHRAE 2017; NRCAN 2010).[2]

Solving the Radiant Exchange Problem

Promoting warm indoor surfaces includes conservative use of window surface areas and liberal use of higher-performance glass/frames, doors, ceilings, walls, and floors. Achieving higher-performance residential enclosures also includes the use of simpler envelope geometries and exterior insulation to reduce thermal bridging and infiltration occurring from complicated framing systems more characteristic of homes designed for more moderate climates (Figure 11.3).

Fenestration

Of particular challenge for cold-climate architecture is residential fenestration systems, specifically finding a balance between window-to-wall ratios (WWR), solar heat gain coefficients, visible light transmittance, and U-factors. Ideally, sufficient amounts of

1. When wearing clothing typical of occupants in residential environments at normal activities and air velocities.
2. Radiant-based HVAC systems first consider the enclosure as a means of regulating body heat supported by the use of mechanical or electrical radiant heating systems.

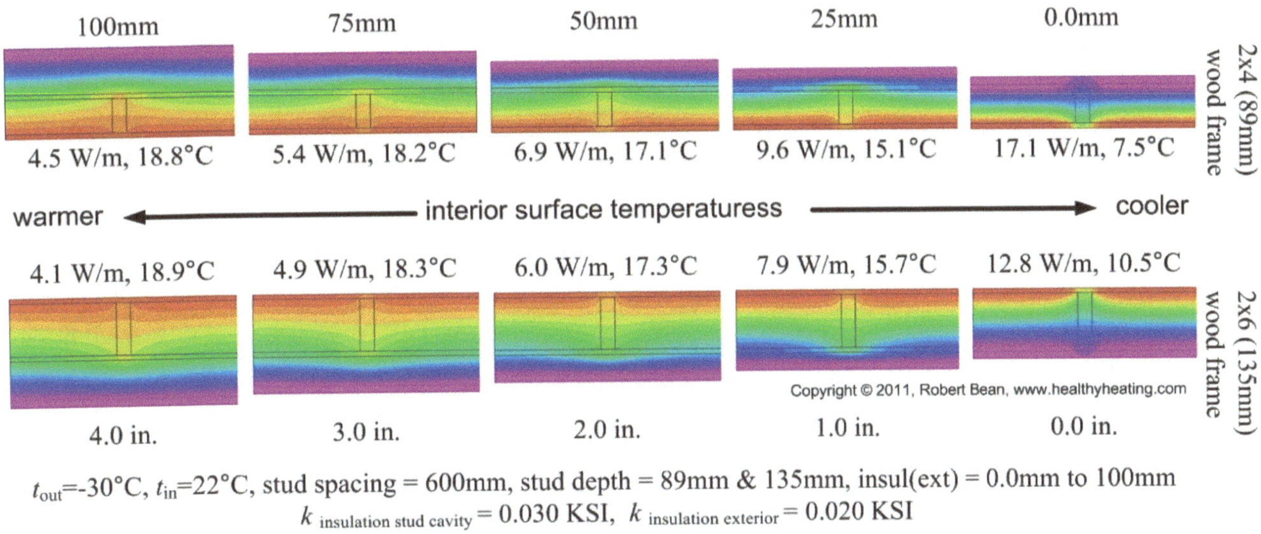

Figure 11.3 Designs must promote warm indoor surfaces through enclosure engineering, interior design choices, and selection of mechanical system types. Compared above are interior surfaces temperatures resulting from the addition of exterior insulation on wood-frame wall construction. Observe how the thicker exterior insulation keeps the colder temperatures toward the outdoors.
Robert Bean

shortwave (solar) energy should be permitted to enter a home during winter months to offset loads and utility heating cost. However, shortwave energy should also be reduced during the summer with careful selection of window films and use of exterior shading to prevent overheating and unnecessary cooling cost. Gases and U-factors should be selected to keep indoor glass surfaces warmer in winter and cooler in summer. Visible transmittances should be selected to optimize natural daylight during darker winter months.

Refer to Table 11.1. Even high performance glass at 0°F (–18°C) has an inside surface temperature of 63°F (17°C), almost 25°F (14°C) lower than mean skin temperature. At –40°F (–40°C) outdoor temperatures, a window with a U-factor of 0.25 has an approximate indoor surface temperature of 56°F (13°C). The designer should appreciate that the radiant transfer will be highly motivated with a 35°F to 45°F (19°C to 25°C) occupant-to-window differential. These large surface temperature differentials are why people feel cold when next to cold surfaces even when the thermostats reads 72°F (22°C) (ASHRAE 2017).

Fenestration and Humidity

Tied directly to window performance is humidity control. It should be managed within a range beneficial to the occupant and their moisture-sensitive assets (heirlooms, furniture, paintings, musical instruments, and so on) without promoting moisture condensation on cold glass (LBNL 2006). The ideal range is 35% to 55% rh (±5%), but condensation concerns at cold temperatures prevent the higher and more preferable ranges, to the detriment of IEQ and, by connection, detriment to the occupants. Figure 11.4 illustrates condensation potential on various window assemblies as a function of relative humidity and outdoor temperature. It should be noted, condensation is more typically an edge of glass problem, unless warm edge spacers and insulated/low conductivity frames are used. Air at windows is cooler than at the center of a room, so relative humidity near windows is higher than at the center of a room, especially when curtains or blinds are closed.

Table 11.1 Indoor Surface Heat Transfer Coefficient h_i in Btu/h·ft²·°F (W/[m²·K]), Vertical Orientation (Still Air Conditions) (ASHRAE 2017)

Glazing ID[a]	Glazing Type	Glazing Height, ft (m)	Winter Conditions[b]			Summer Conditions[c]		
			Glazing Temp., °F (°C)	Temp. Diff., °F (°C)	h_i, Btu/h·ft²·°F (W/[m²·K])	Glazing Temp., °F (°C)	Temp. Diff., °F (°C)	h_i, Btu/h·ft²·°F (W/[m²·K])
1	Single glazing	2 (0.6)	17 (–9)	53 (30)	1.41 (8.04)	89 (33)	14 (9)	1.41 (8.10)
		4 (1.2)	17 (–9)	53 (30)	1.31 (7.42)	89 (33)	14 (9)	1.33 (7.64)
		6 (1.8)	17 (–9)	53 (30)	1.25 (7.10)	89 (33)	14 (9)	1.29 (7.41)
5	Double glazing with 1/2 in. (12.7 mm) air space	2 (0.6)	45(7)	25 (14)	1.36 (7.72)	89 (35)	14 (11)	1.41 (8.30)
		4 (1.2)	45(7)	25 (14)	1.27 (7.21)	89 (35)	14 (11)	1.33 (7.82)
		6 (1.8)	45(7)	25 (14)	1.22 (6.95)	89 (35)	14 (11)	1.29 (7.58)
23	Double glazing with e = 0.1 on surface 2 and 1/2 in. (12.7 mm) argon space	2 (0.6)	56 (13)	14 (8)	1.31 (7.44)	87 (34)	12 (10)	1.38 (8.20)
		4 (1.2)	56 (13)	14 (8)	1.23 (7.00)	87 (34)	12 (10)	1.31 (7.73)
		6 (1.8)	56 (13)	14 (8)	1.19 (6.77)	87 (34)	12 (10)	1.27 (7.50)
43	Triple glazing with e = 0.1 on surfaces 2 and 5 and 1/2 in. (12.7 mm) argon spaces	2 (0.6)	63 (17)	7 (4)	1.25 (7.09)	93 (40)	18 (16)	1.45 (8.73)
		4 (1.2)	63 (17)	7 (4)	1.18 (6.72)	93 (40)	18 (16)	1.36 (8.20)
		6 (1.8)	63 (17)	7 (4)	1.15 (6.53)	93 (40)	18 (16)	1.32 (7.94)

a. Glazing ID refers to fenestration assemblies in Table 4.
b. Winter conditions: room air temperature t_i = 69.8°F (21°C), outdoor air temperature t_o = 0.4°F (–18°C), no solar radiation
c. Summer conditions: room air temperature t_i = 75.2°F (24°C), outdoor air temperature t_o = 89.6°F (32°C), direct solar irradiance E_D = 248.2 Btu/h·ft² (748 W/m²), $h_i = h_{ic} + h_{iR}$ for I-P units $= 0.3(\Delta T/L)^{0.25} + \varepsilon\sigma(T_i^4 - T_g^4)/\Delta T$ [for SI units $= 1.46(\Delta T/L)^{0.25} + \varepsilon\sigma(T_i^4 - T_g^4)/\Delta T$], where $\Delta T = T_i - T_g$, °R (K); L = glazing height, ft (m); T_g = glazing temperature, °R (K), σ = Stefan-Boltzmann constant, and ε = surface emissivity.

Radiant-Based HVAC Systems

The tendency for heating-dominated homes to have cool to cold interior surfaces suggest architects, builders and mechanical designers consider when necessary the benefits of low-temperature, radiant-based HVAC systems to solve mean radiant temperature (MRT) challenges and related discomfort metrics (Bean 2017). An example of a home that has a MRT challenge is shown in Figure 11.5. This project and many like it in cold climates have MRT challenges due to large WWRs even when using high-performance materials and assemblies. Code-built enclosures can expect a higher probability of standards noncompliance with moderate WWR. The results from an ASHRAE Standard 55 thermal comfort analysis from a residential home in ASHRAE Climate Zone 7 are shown in Figure 11.6.

The results between using forced air heating and radiant floor heating are compared, where the former resulted in a MRT of 67.3°F (19.6°C) and noncompliance with ASHRAE Standard 55 at 24% predicted percentage dissatisfied (PPD); and the latter with a MRT of 74.4°F (23.6°C) and compliance at 9% PPD (see MRT solution in Figure 11.7). Progressively increased PPDs and PMVs (more discomfort) will result when buildings like this are constructed in progressively colder climates.

Thus, mechanical designers should work with architects and interior designers for integrated solutions such as reducing heating fluxes; using conductive, low-volatile organic compound (low-VOC) finishes; and where possible, increasing heat exchanger surface areas and fin efficiencies.

INTEGRATED SOLUTIONS

Reduce Heating Fluxes

Heating flux is simply the heating load in a zone divided by its floor area in units of Btu/h/ft² (W/m²). For relative comparison using Canada's voluntary R2000 housing

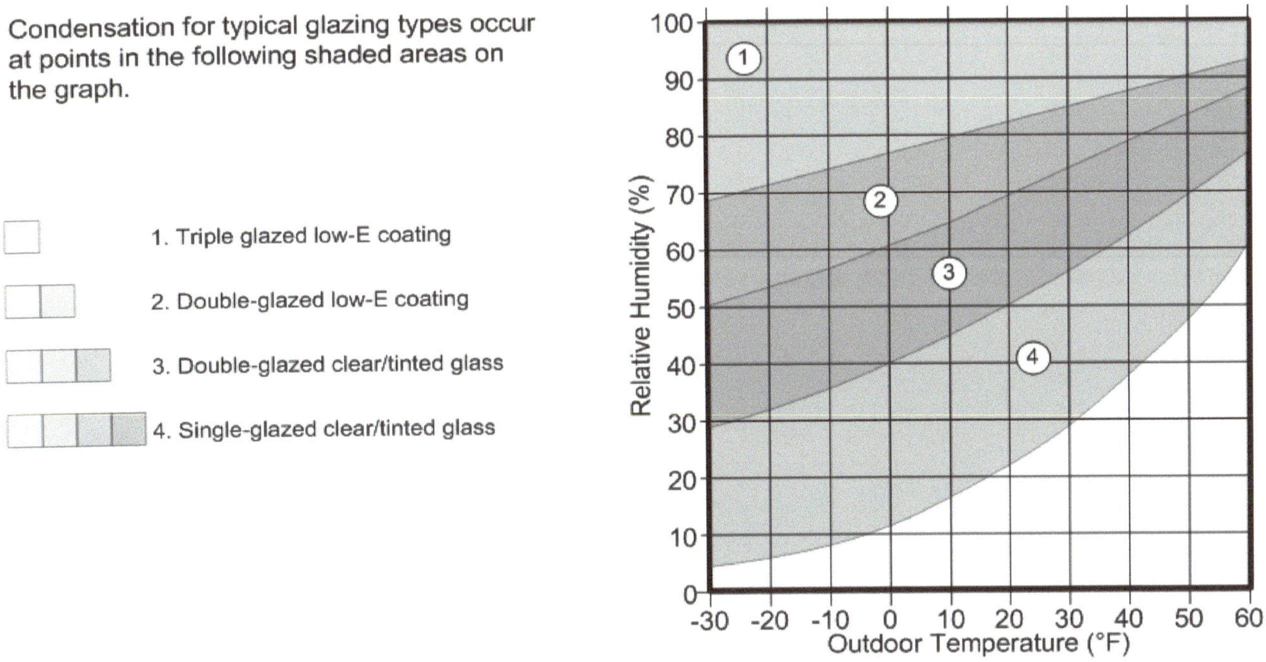

Figure 11.4 Condensation potential on glazing (center of glazing) at various outdoor temperatures and indoor relative humidity conditions (LBNL 2006).

Image credit/adapted from NFRC/LBNL Open Source

Figure 11.5 Even when an enclosure is designed and fabricated with high-performance materials and systems, a large WWR will result in low mean radiant temperatures, high radiant temperature asymmetry, and aggressive downdrafts resulting in noncompliance with thermal comfort standards.

Robert Bean

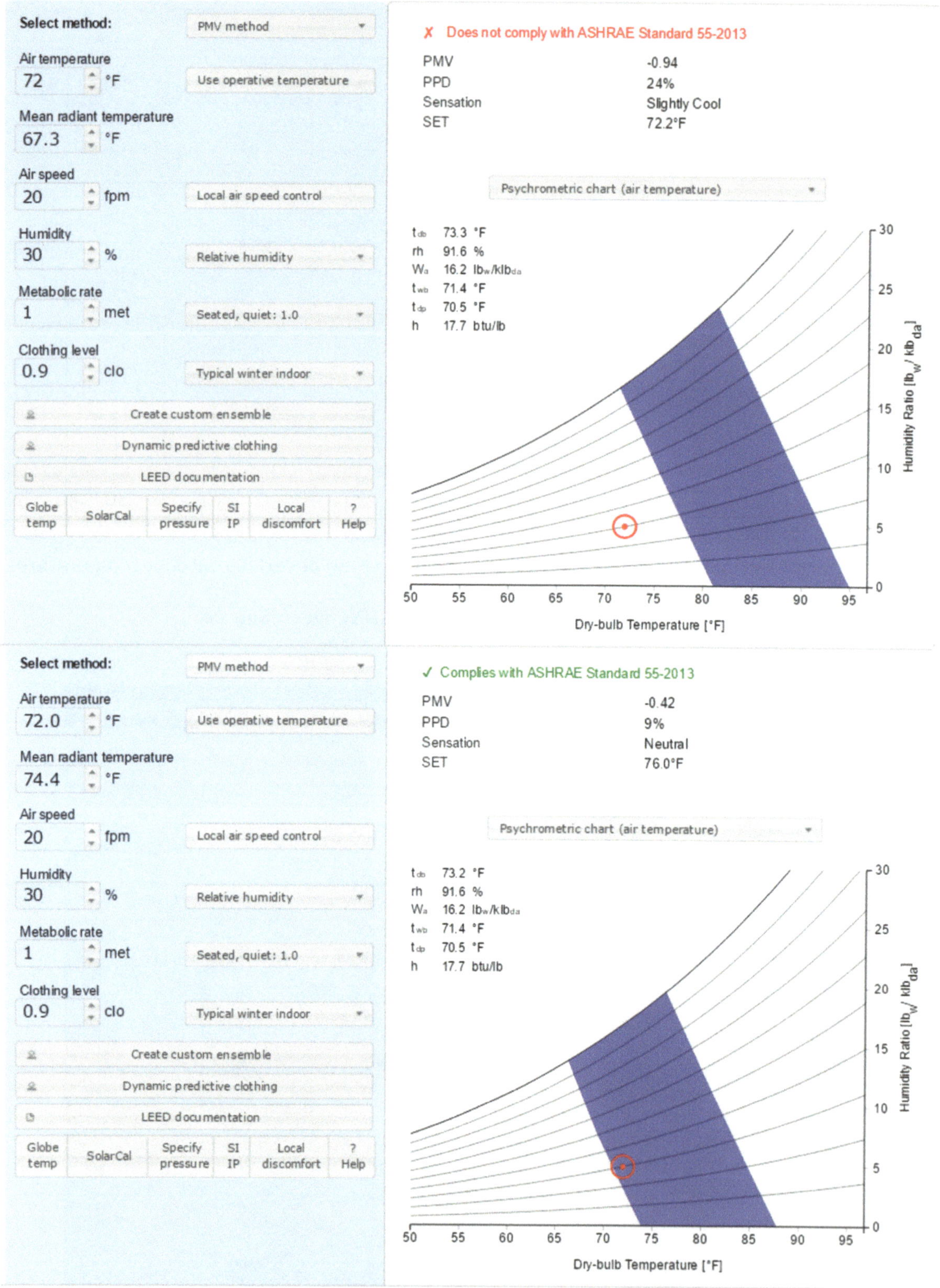

Figure 11.6 Air-based noncompliance system (top) versus radiant floor heat compliant system (bottom) based on influencing the MRT. Circle and shaded area on psychrometric chart represents occupant relative to the comfort zone.

LBNL (2020)

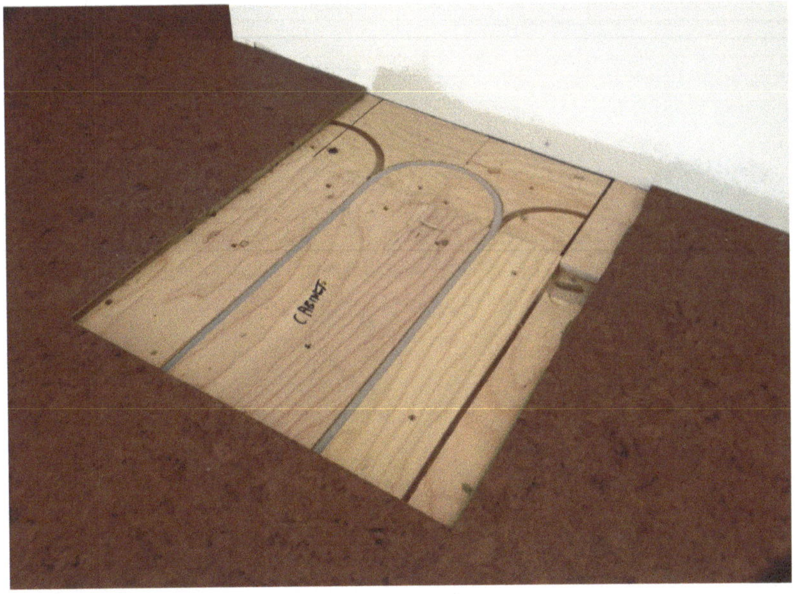

Figure 11.7 Radiant heating (shown) was the compliant MRT solution for the space shown in Figure 11.5 and as modeled in Figure 11.6. These low-temperature solutions are quieter relative to air-based systems, have lower vibration issues, enable the use of low-VOC finishes and dedicated ventilation systems while promoting energy efficiency in boilers, heat pumps, geoexchange, geothermal and solar systems.

Robert Bean

program in comparison to the Canadian National Building Code, high-performance residential enclosures in cold climates will have fluxes typically less than 10 Btu/h/ft^2 (32 W/m^2), whereas poor-performance enclosures have fluxes exceeding 30 Btu/h/ft^2 (95 W/m^2). Reducing heating fluxes in part enables the use of lower fluid temperatures in the heating systems, promoting IAQ through reduced outgassing and improved efficiency in the heating appliance. This is important as it is common for high-efficiency boilers and furnaces to be purchased according to their efficiency rating but not operated in a system designed to achieve the published value. Reducing heating fluxes through enclosure design is therefore one of several key metrics within the integrated design process to achieve both acceptable IEQ and energy efficiency (see Figure 11.3).

Use Conductive, Low-VOC Finishes

Given the extended exposure to interior finishes in cold climates, often without the benefits of natural ventilation, it makes sense that interior designers and HVAC engineers should evaluate the outgassing from interior finishes such as floor coverings, paint, and cabinetry. Consider that conductive. low-VOC finishes tend to have characteristics of natural materials. For example tile, slate, stone, terrazzo, and stained concrete fall into the category of masonry-like materials. These are known for their low- to no-outgassing issues and their stability. Additionally, these finishes do not have the same composition as synthetic materials that may contain such compounds as polybrominated diphenyl ethers (a common fire retardant), phthalates (plasticizer used in such things as flooring and shower curtains), and formaldehyde (used in glues and resins). All of the latter are known health hazards and known for their contribution to poor IAQ from outgassing in the presence of increasing temperatures and humidity. Unlike many synthetics, masonry-like

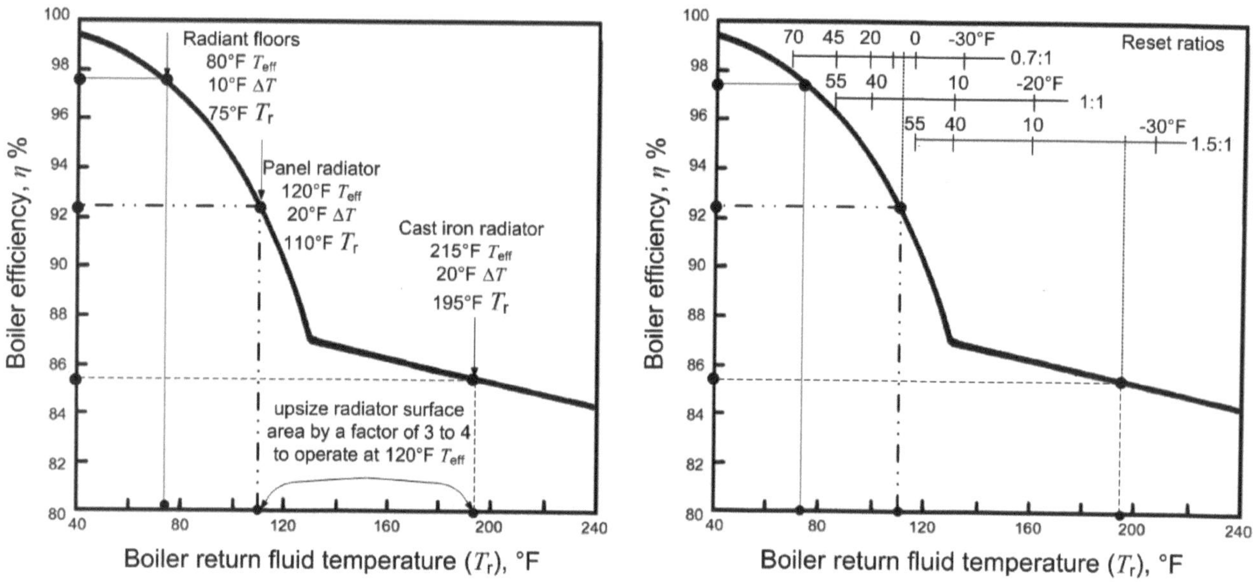

Figure 11.8 Integrated design delivers energy efficiency and IEQ. Shown are improvements to boiler efficiency based on lower return temperatures from different types of heat terminal units.
Robert Bean

materials also tend to be more conductive. This means when they are used with, for example, low-temperature floor heating systems, they enable improved IAQ, thermal comfort, and efficiency gains through the use of lower return temperatures in the heating appliance (see Figure 11.8). Residential ventilation standards, including CSA F326 and ASHRAE Standard 62.2 (ANSI 2019, ASHRAE 2019a), have provisions for source control. Designers are encouraged to adopt the implied principles around the choice of building and finishing materials.

Increase Heat Exchanger Surface Areas and Improve Fin Efficiency

Heat exchanger design is a key component in energy efficiency and thermal comfort. Regrettably, designers in the housing sector do not typically apply this design principle. For example, selecting default forced-air furnaces and boilers means the designer cannot manipulate the appliances exchanger design. However, designers using hydronic systems (including heat pumps, solar, geoexchange, and geothermal) can manipulate the secondary exchangers (radiators, convectors, fan/coils, floor heating, and so on) employed in conditioning the occupants and space (CSA Group 2021, ISO 2019). Additionally, using a separate comfort system enables the use of a dedicated outdoor air system (DOAS) such as a heat recovery ventilator. This feature and benefit should not be underestimated in extracting the maximum energy and exergy efficiency from renewable energy sources (Zmeureanu and Wu 2006).

Consider typical processes for residential designers in selecting panel radiators, fan-coils, or convectors. The default strategy is to select combinations of "off the shelf" manufactured equipment and ignore the resulting deficiencies. However, skilled designers have learned to work with boilers, heat pumps, earth loops, and solar systems' ideal return temperatures to maximize efficiency (see Figure 11.8). From this approach the engineer will discover that, for example, the exchanger coils require a larger surface area and an increase in the tube/fin density (see Figure 11.9, Coil B) (Bean 2016b). This extra

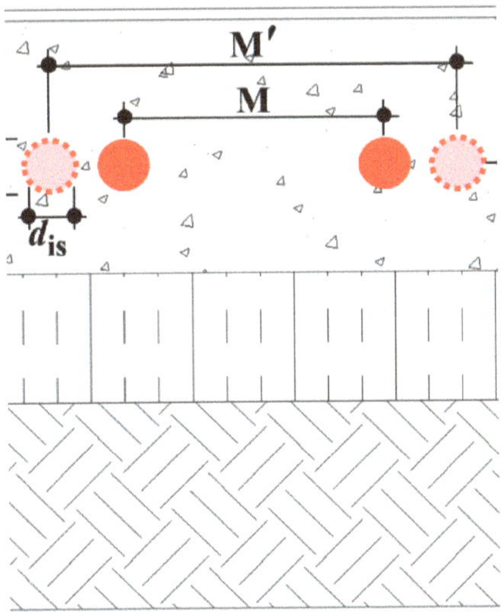

Figure 11.9 Radiant floor heating. By reducing the spacing of the tubes (M' to M) and increasing the pipe diameter (d_{is}), the pipe density and internal heat exchanger surface area is increased. Increased surface areas enables the system to run at lower temperatures. Low temperatures promote safety, improved air and thermal comfort quality, and efficiency in the heating system (Bean 2017).
Robert Bean

capital cost is a one-time expense with a lifetime of efficiency benefits. This principle also applies to convectors, radiators, and radiant panels where their geometries can be selected for large surface areas and their effectiveness improved by increasing conductivity, fin density, and tube spacing (fin efficiency) (see Figures 11.10 and 11.8).

These integrated strategies are safer and promote better IAQ, thermal comfort, and low return temperatures, enabling residential heating equipment and systems to achieve their rated efficiencies (CCHRC n.d., Bean 2013). Additionally, gaining control of the enclosure resulting in lower fluxes enables greater control over drafts, flows of humidity, and temperatures across the occupant. Such regulation serves to reduce discomfort due to evaporative and convective losses from the body (representing approximately 35% of total body heat; see Figure 11.2) (OpenStax 2013).

ENCLOSURE DESIGN

Differentials in moisture, temperatures, and pressures cause mass and energy to flow back and forth across an enclosure (ASHRAE 2011). The enclosure by default then becomes a filter, sponge, and capacitor (see Figure 11.12). The flow becomes the transport mechanism for destructive elements that affect building materials and systems and, by association, could then affect the occupants. Though similar materials and assembly process are used to mitigate effects in other building types, the design, assembly, and inspection process at the residential level often does not happen with the same understanding or rigor as other building types. For this reason housing authorities, designers, manufacturers, trades and inspectors should pay strict attention to the materials and methods of construction used in cold climates. Specific attention should be paid to 1) moisture control, 2) heat/temperature/ultraviolet light, and 3) air leakage/ventilation.

Moisture

Uncontrolled moisture from precipitation, ground water, and water vapor released by occupants and occupant activity leads to decay of wood, paper-faced finishes, fabrics, and

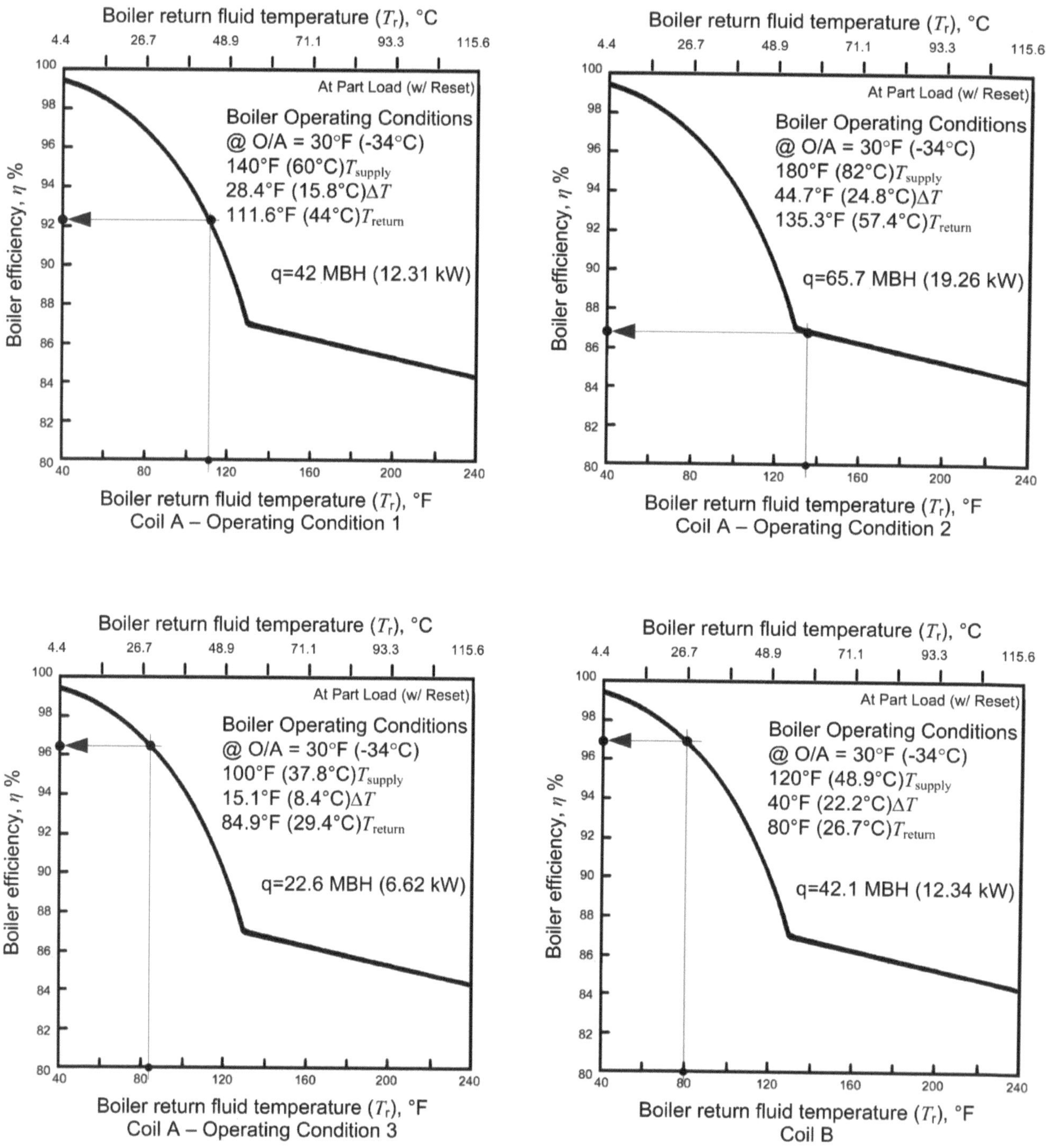

Figure 11.10 The effects of heating coil selection on boiler efficiency. Coil A is shown under three operating conditions—note the wide range of performances. Note the benefit of coil B on efficiency and performance. Coil B has a slightly larger face area (2.92 versus 2.25 ft^2 [0.27 versus 0.21 m^2]), two more rows (4 versus 2), a corrugated fin pattern versus sine wave fin pattern, and is built with smaller diameter tubes (1/2 versus 5/8 in. [12.7 versus 16 mm]). This extra capital cost is a one-time expense with a lifetime of efficiency benefits.

Robert Bean

Figure 11.11 Low-temperature, radiant-based, high-efficient hybrid heating and heat recovery ventilation system. Includes kitchen exhaust/makeup air and domestic water-heating systems for large residential home. Note the preheat coil on the DOASs and associated glycol makeup tank (bottom left). Coil surface area and radiant floor based on a low-temperature log mean temperature difference.

Robert Bean

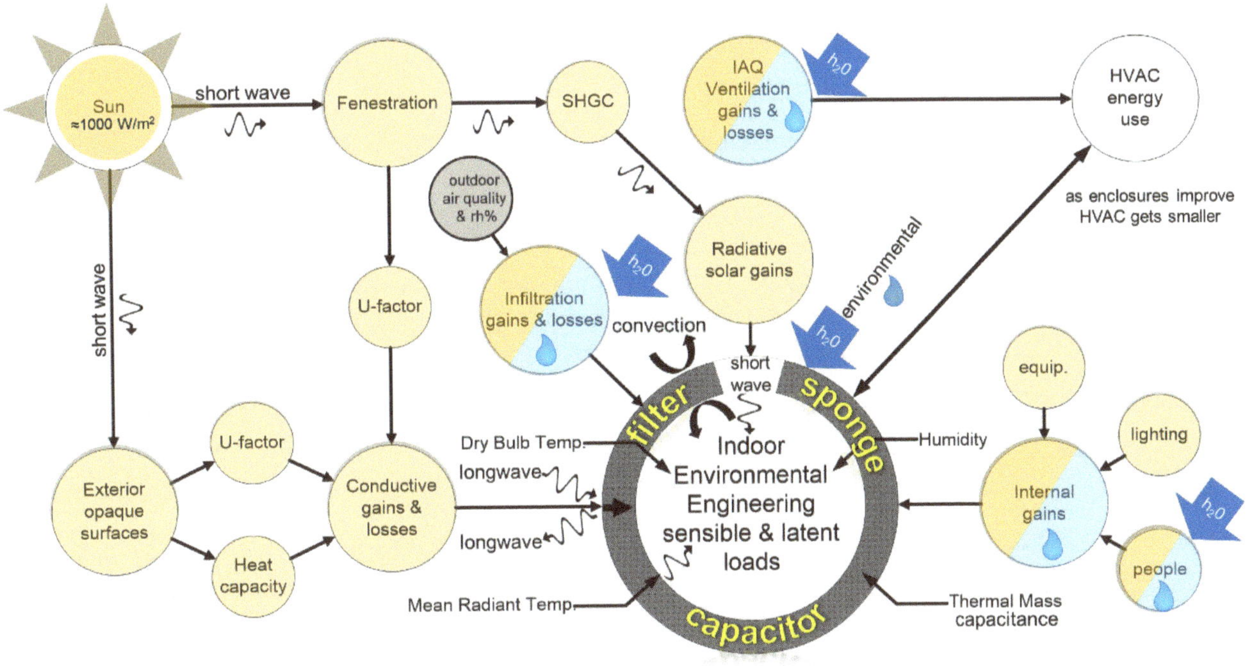

Figure 11.12 Differentials in moisture, temperatures, and pressures cause mass and energy to flow back and forth across the enclosure. The flow becomes the transport mechanism for destructive elements (ASHRAE 2011).

Adapted from Figure 3-2, ASHRAE (2011)

similar materials. It enables growth of mildew and molds (and thus odors) on host surfaces, and enables corrosion of fasteners and components found in framing and mechanical and electrical systems. Particular to cold climates are moisture risks leading to ground heaving, condensation, frosts, and freeze-thaw cycles on construction materials and assemblies. The following are key reasons why moisture ought to be controlled in residential buildings (Bean 2016c):

- To control hydrolysis (VOC emissions)
- To control microbes (virus, bacteria, molds, mites, some insects)
- To support hydration (affects cognition and wound healing)
- To enable physical comfort in mucous membranes (respiratory, eyes, skin)
- To enable positive perceptions of thermal comfort
- To enable positive perceptions of IAQ
- To enable positive perceptions of indoor odor quality
- To maintain dimensional stability in hygroscopic materials (woods)
- To prevent condensation on hydrophobic materials (glass)
- To prevent condensation in hydrophilic materials (drywall)
- To preserve moisture-sensitive artifacts/photographs/collectibles/musical instruments

Heat and Temperature

All the concerns of excessive temperatures found in hot and mixed climates frequently apply to homes in cold climates. For example Canada, considered one of the coldest countries on the planet, has recorded highs of 108 °F (42.2°C) in Manitoba, similar to extremes found in Florida (109°F [43°C]). But unlike Florida, it has also recorded low temperatures of –54.0°F (–47.8°C). At both ends of the temperature spectrum there is increasing degradation potential from ultraviolet (solar) radiation, heat, and frost and freeze-thaw cycles.

Infiltration/Exfiltration/Ventilation

Airflow across the enclosure and through a home is a function of leakage paths subject to wind pressures, stack effects, and effects from mechanical systems. When controlled, airflow can be treated to the benefit of the occupant. Uncontrolled inbound airflow introduces particulate matter, moisture, odors, sounds, and cold air. Uncontrolled outbound air takes with it energy and moisture accounted for in utility bills. For this reason, homes should target tightness levels less than 1.5 ACH50 and use balanced ventilation systems with filtration and heat recovery (and moisture recovery where appropriate).

Kitchen Range Hoods

Of special concern in cold-climate homes is the use of high-capacity kitchen range hoods that have the ability to place the home into a deep negative pressure. Such events exacerbate the introduction of cold, untreated air through enclosure flaws and penetrations used for venting, exhaust, and makeup air for HVAC appliances, central vacuums, bath exhausts, and so on. Additionally and of greater importance is the potential for carbon monoxide poisoning from natural draft appliances back-drafting due to the negative pressures created by these devices. Wood stoves, fireplaces, attached garages and any other source of combustion-based fumes are high-risk offenders. It should also be noted that some power-vented, sealed combustion appliances are at risk under deep depressurization and designers are advised to consult with the manufacturers for their threshold values. Regardless of absent or inadequate code provisions, large-capacity exhaust hoods

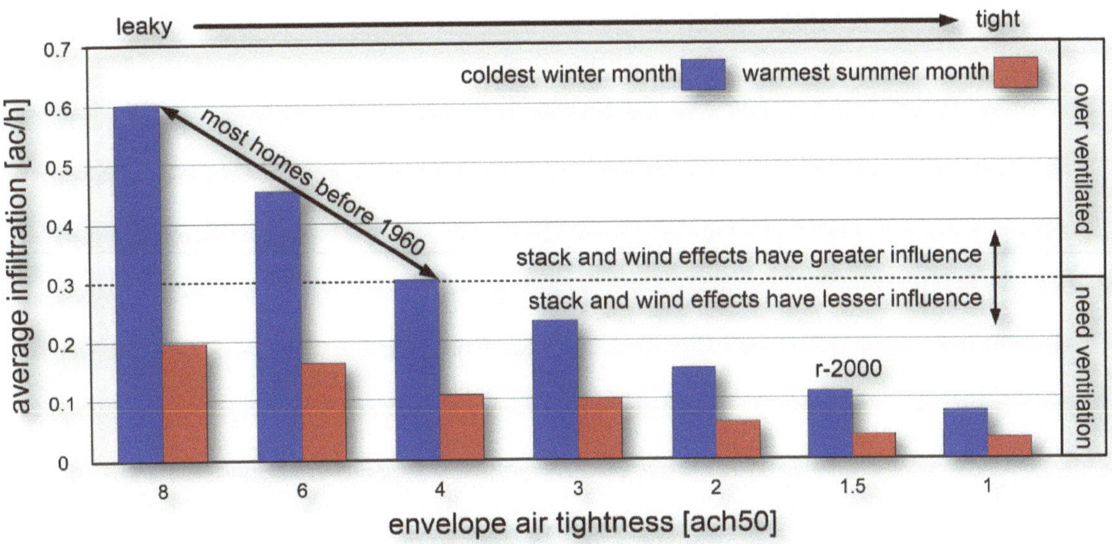

Figure 11.13 Homes should target tightness levels of 1.5 ACH50 or lower, and use balanced ventilation systems with recovery of heat (and where appropriate moisture recovery).

Bean (2017)

should be matched with proper makeup air systems that treat entering cold outdoor air to acceptable conditions. Additionally, carbon monoxide sensors must be installed wherever exhaust appliances share a house with combustion appliances and/or attached garages.

In general, designers for cold-climate homes are advised to seek out construction details that exceed minimum code requirements and use materials and methods proven to withstand the extremes of moisture, heat, and air pressure differentials.

Due to the uncertainty of climate change consequences, engineers and architects are also cautioned to avoid design obsolescence strategies when selecting materials and methods of construction. Rather, one should consider the design choices made in the present and their implications of changing environmental conditions in the future (CIBSE 2005). These considerations should also include passive cooling and heating solutions (Bean 2016a).

CONSTRUCTION CONSIDERATIONS

The intent of this section is to highlight a few unique considerations encountered during construction of cold-climate homes.

Land and Lot Development

When laying out lots and placing buildings on those lots, especially consider the following:

- Consider prevailing winds, snow drifting, and melt/runoff/freeze conditions when locating building entries.
- Remember that as one moves away from the Equator, sun angles are lower and shadows from adjacent buildings, vegetation, and land forms are longer. This can affect the effectiveness of passive solar heating strategies and incident solar on photovoltaic panels.
- In areas with intermittent permafrost, permafrost can be very localized. There can be spots with and without permafrost under the footprint of one house.
- Site drainage systems need to be functional during freezing conditions.

Figure 11.14 A large range hood exhaust in cold climates needs treated makeup air to prevent depressurization of the home, leading to poor IAQ risks, and to prevent thermal discomfort from untreated cold air being introduced to the space through registers and enclosure flaws.
Robert Bean

Foundations

Climate and soil conditions drive foundation choice. Special care must be taken when designing foundations for buildings located on permafrost and intermittent permafrost. Significant movement and differential settling can occur in foundations resting on permafrost when permafrost thaws, especially if thawing is uneven. Similarly, frost heaving and frost jacking can occur when expansive clays and silts under and against foundations and piles go through freeze/thaw cycles. In these cases, proper site drainage is an important preventative element. Designing foundations for buildings in locations where frost penetrates the soil by more than 2 ft (0.6 m) requires knowledge of the soil conditions of the building site and of appropriate foundation designs for the proposed building and those soil conditions.

Heat loss through ground-connected building elements is a concern in cold climates, particularly if these elements include heating ducts, pipes, or cables. Local energy-efficiency agencies and heating equipment suppliers can provide guidance on appropriate insulation strategies for below-grade building elements.

Items to consider for below-grade and on-grade slabs include the following:

- Soil characteristics including soil type, moisture content, and density below and adjacent to slabs, moisture storage function, porosity, conductivity and specific heat capacity.
- Water table characteristics, including historical data (i.e., risk assessment for change in the water table).

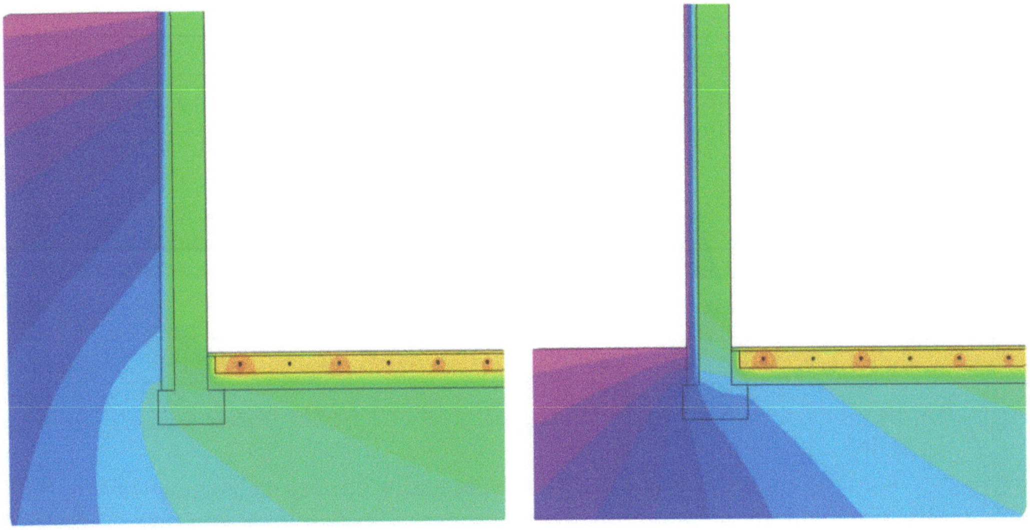

Figure 11.15 FEA modeling heat flow through ground for a basement (left) and slab-on-grade (right). Insulation values should consider the items from 2a-2h.
Robert Bean, produced with FlexPDE (PDE 2021)

- Temperatures above and below grade.
- Temperature desired for space conditions.
- Temperatures in embedded heating pipes and in ground heating ducts.
- Duct, tube, or cable spacing and slab conductivities (upward and downward)
- Tube (and cable) patterns (i.e., warmest fluid versus coolest fluid in relation to edge).
- Slab characteristics including thickness, conductivity, and elevation in relation to grade.
- Enclosure performance above and adjacent to the heated slab, including effects from solar gains. The greater the performance above the slab, the more significant the slab heat loss.
- Aspect ratio of building, either the width-to-length ratio or area to perimeter ratio (i.e., rectangular buildings versus square buildings).

Moisture entry through foundations is a greater concern in cold climates than in warmer ones because of the potential for moisture evaporating from below-grade foundations condensing in or on cold above-grade building elements. Good foundation designs prevent moisture in the ground from migrating to the interior of buildings.

Walls and Roofs

The requirements for exterior walls and roofs to control heat, moisture, and air flows are covered in Chapter 8, as are other cold-climate building envelope issues such as ice dams. Exterior walls and roofs in cold climates can see huge seasonal temperature variations and are subject to numerous freeze-thaw cycles. Design the exterior envelope to shed water and choose materials that dry quickly, that can survive thermal variations, and can accommodate freeze-thaw cycles.

Snow can accumulate to great depths on roofs, particularly at roof elevation changes and around rooftop equipment. Potential snow loads need to be calculated and considered in the structural design of roof trusses and supporting walls.

Windows

Designers must consider whole-window performance (frame and glass) when evaluating loads and comfort conditions, and the poorest-performing element of the window (often the edge of the glass) when assessing condensation resistance. When closed, curtains and blinds can result in the local environment around the window being cooler than the ambient conditions in the room, increasing susceptibility to condensation. In cold weather, indoor humidity needs to be controlled so that the dew point of indoor air is below the temperature of the coolest elements of the windows.

Mechanical Systems

Heating, cooling, humidity control and ventilation systems appropriate for residential applications are discussed in Chapter 1 of *ASHRAE Handbook—HVAC Applications* (ASHRAE 2019e).

Humidity control in cold weather generally implies dehumidification, which is usually achieved by ventilation. Refrigerant-based mechanical dehumidification used to dehumidify spaces in warm weather is generally not suitable in cold weather, because dehumidification coils typically ice up when relative humidity levels are pulled down about 50% at normal indoor comfort conditions. Refrigerant-based mechanical dehumidification with heat recovery may be appropriate for spaces with swimming pools, but not for general living spaces.

Where spaces are humidified for comfort or other reasons in cold weather, keep space air dew point temperatures below the temperature of the windows and other cool surfaces to prevent condensation. This has been noted as a concern when using setback thermostats in humidified spaces in cold climates. Also, where humidification is installed, exhaust systems should not be independently operated to control humidity in the space, or else humidification and dehumidification systems will compete, wasting energy.

Of particular concern is the affect of exhaust devices on combustion appliance vent operation. In energy-efficient, airtight houses, modest exhaust airflows can create negative indoor to outdoor pressure differentials that may cause some combustion appliances to backdraft. Devices of concern are not just exhaust fans, but also clothes driers and central vacuum systems. This life safety concern can be mitigated by the following:

- Selecting sealed or power vented combustion appliances
- Locating spillage-susceptible appliances in spaces not subject to depressurization
- Installing makeup air systems to offset exhaust airflows of concern
- Installing carbon monoxide sensors in spaces with spillage-susceptible combustion appliances

Heat recovery ventilators are commonly used in cold climates to lessen the energy requirements of conditioning ventilation air (see Figure 11.16). It is common for moisture in the exhaust air stream to condense in the heat exchanger core as the exhaust air is cooled and, if the air is sufficiently cooled, for frost and ice to form. Frost and ice formation adversely affects exhaust airflows and heat recovery. A variety of frost-control strategies are available for both packaged and built-up heat recovery ventilation systems. Not all are appropriate for all climates or applications. For example, electric preheat frost prevention systems may be appropriate for very mild winter climates, but, in a cold winter climate, may be both more expensive and more energy consumptive than an exhaust-only system (i.e., one that pulls makeup air through the building envelope), if source energy is considered.

Snow and ice melting (Figure 11.17) are a specialty systems that serve cold-climate residential projects to ensure safe movement of pedestrians and vehicles. Their use in private homes, apartments, and condominium complexes replaces the use of corrosive salts, chemicals, and noisy combustion-based snow removal equipment. They should be considered where hazards pose a life and safety risk, such as emergency egress for inhabitants, where vehicles could slide over ice into pedestrians, or where access to services is necessary.

This chapter provides general discussion on residential buildings in cold climates. For further details readers are advised to search out guidance published by local authorities and use the list of resources below:

- ANSI/ASHRAE Standard 55, *Thermal Environmental Conditions for Human Occupancy* (ASHRAE 2020b)
- ANSI/ASHRAE Standard 62.2, *Ventilation and Acceptable Indoor Air Quality in Residential Buildings* (ASHRAE 2019a)
- ANSI/ASHRAE/IES Standard 90.2, *Energy-Efficient Design of Low-Rise Residential Buildings* (ASHRAE 2018a)
- ASHRAE Guideline 10, *Interactions Affecting the Achievement of Acceptable Indoor Environments* (ASHRAE 2020a)
- ASHRAE Guideline 24, *Ventilation and Indoor Air Quality in Low-Rise Residential Buildings* (ASHRAE 2015a)
- ASHRAE Position Documents: www.ashrae.org/about-ashrae/position-documents
- Cold Climate Housing Research Center, Alaska: www.cchrc.org
- Natural Resources Canada, Energy Efficiency for Homes: www.nrcan.gc.ca/energy-efficiency/homes/20546
- *Keeping The Heat In*, Chapter 6: Basement Insulation (NRC 2021)
- Building America Solution Center: https://basc.pnnl.gov/

Figure 11.16 Dedicated heat recovery ventilators extract heated stale air from the home and use the outbound energy to preheat incoming outdoor air. Note condensate drain line.
Robert Bean

Figure 11.17 Installation of a snow melting system in a sidewalk.
Courtesy Uponor

HOW COLD IS IT?

It is so cold ... I can't even get into a heated argument with my spouse!

12 Seasonal and Off-Grid Applications

INTRODUCTION

Managing seasonal and off-grid buildings in a cold environment is exceptionally challenging. These facilities lack standard municipal infrastructure including road access, municipal sewer systems, electricity from the grid, and a reliable treated potable water supply. Solutions for energy, potable water, and wastewater are challenged by local environmental conditions, access to the site, sustainability issues, rising costs, and resource availability.

CONSTRUCTION

Reduce the Need for Heat Energy

Reducing the need for heat energy is a good policy for any building in a cold climate, but of particular importance to off-grid buildings because they are often located in remote areas and incur a transportation cost for fuel. If a new facility is planned, the use of architectural technologies that reduce the need for heat energy is one of the most important decisions to make. A tightly constructed building that is well insulated and has high-performance windows will require less energy to heat, resulting in significant cost savings from energy source transportation and use over the lifespan of the building. Building techniques and features that reduce the need for heat energy include site orientation and window placement, tight building envelope, superinsulated building envelope, high-performance windows, passive solar heating, heat recovery ventilation, solar thermal air collectors, and airlock vestibule entries.

Controlling Moisture

As with all buildings, moisture management involves controlling moisture entry, accumulation, and removal. Limiting moisture entry involves controlling bulk moisture movement (by controlling building envelope openings and managing the driving forces [gravity, momentum, capillary suction, surface tension, air pressure]); capillary suction; airborne moisture; and vapor diffusion. Controlling moisture accumulation involves limiting condensation and providing a means for the building materials to dry. For more information, refer to Chapter 8.

Condensation is of particular concern with seasonal and off-grid cold-climate design. These buildings may be used intermittently and may experience lower indoor temperatures when they are unoccupied; the level of heating varies and steady-state thermal con-

Figure 12.1 Alpine hut under construction.
Courtesy Alpine Club of Canada

ditions do not exist. In addition, ventilation may be significantly reduced during unoccupied times. Condensation will occur when moisture held in warmer air and materials meets a colder surface. Strategies to control condensation include

- increasing the condensing surface temperature by introducing a heat source to vulnerable areas (windows),
- adding more insulation to make the interior surface of exterior walls warmer,
- reducing or eliminating thermal bridging,
- keeping furnishings away from exterior walls so that warm air can reach the interior surface of the exterior wall,
- incorporating an airlock vestibule entry,
- using airtight construction methods,
- installing a vapor barrier on the warm side of the insulation to keep moisture generated indoors out of the building envelope,
- using ventilation to remove moisture generated indoors,[3] and
- limiting the amount of moisture generated indoors by controlling the source.[4]

3. As with all well-insulated, airtight buildings, ventilation is essential to prevent the formation of condensation on the interior surface of the exterior walls (NRC 2006).
4. Boiling water as a potable treatment method and drying wet ski gear indoors are two activities that introduce a significant amount of moisture not normally encountered in the built environment.

Indoor Air Quality

Indoor air quality (IAQ) is the condition of the air inside buildings and is evaluated by its effects on human health and well-being. Comfort parameters (temperature, relative humidity, airflow) and the presence and concentration of contaminants affect IAQ. IAQ is described in more detail in Chapter 3. Issues particular to off-grid cold-climate buildings include:

- A higher risk of condensation brings a higher risk of fungi (mold growth) on building materials.
- Respirable particulates may be released by wood-burning stoves.

ENERGY

Energy is needed at off-grid locations for heating, cooking, lighting, waste management, ventilation, and communications. Typical sources of energy include solar, micro-hydro, wind, propane, diesel and biodiesel, and wood-burning (biomass). Helicopters and other forms of motorized transportation used to access the site also represent significant sources of energy use. Energy in general is discussed in Chapter 4. Some issues and considerations particular to off-grid buildings are as follows:

- There may be no meteorological data for a particular remote site and, since the selection of a wind turbine is based on wind characteristics, meteorological data should be monitored for a year prior to installation of a wind turbine.
- Remote sites are difficult to access, resulting in issues with installing and maintaining wind turbines. Consider mounting wind turbines on a tower that can be lowered for maintenance so that helicopter servicing is not required.
- The primary use for propane in an off-grid facility is as a fuel for cooking and lighting. Propane may also be used to provide heat for the building.
- Wood burning as a source of heat at an off-grid facility is reliable, culturally acceptable, and easy for users to operate. In some locations, the fuel is renewable and readily available. However, if the cost of transportation and the inefficiency of burning wood are considered, the cost is relatively high. Wood-burning is also a source of airborne particulates.

POTABLE WATER TREATMENT METHODS

Drinking water at off-grid facilities comes principally from meteoric precipitation (surface water from snow, snowmelt, and rain) that collects in lakes or flows in ephemeral streams. Some facilities use groundwater from springs or subgrade water from aquifers. Wells can be drilled into water-bearing rock and water can be pumped out.

In the winter, water is often obtained directly by melting snow. Larger facilities with a greater demand for water will take lake water from a nearby source during the summer, treat it, and store the water in large heated tanks for use in the winter.

Potability refers to whether water is safe to drink. Water is an ideal breeding ground for a variety of organisms. Many are harmless, others have a mild nuisance value, and still others are a source of disease. There are four types of organisms with disease-causing potential: bacteria, protozoa, worms, and viruses. The presence of certain varieties of these organisms can cause infectious diseases such as typhoid fever, dysentery, cholera, jaundice, hepatitis, undulant fever, and tularemia.

High-altitude water sources contain a higher concentration of fine suspended solids from glacial meltwater (turbidity). Fine sediments can protect microorganisms from the effects of disinfection and stimulate the growth of bacteria (WHO 1997).

Figure 12.2 Wall-mounted solar panel and wind turbine.
Courtesy Alpine Club of Canada

Heavy metal concentrations have been found in snow, lakes, and streams, as well as subgrade aquifers. It is not uncommon for subgrade aquifers to have high iron levels or even arsenic. All water sources should be routinely tested to ensure the proper treatment process is provided.

The following sections are overviews of the different types of sanitation methods. This is not an all-inclusive guide on water treatment and the designer should conduct further research on methods and practices, including consultation with engineers who specialize in the treatment of raw water systems.

Boiling

Boiling kills all enteropathogenic germs except Hepatitis A virus. Hepatitis A virus is very rare at high altitude.

Procedure: At sea level, water should boil with bubbles in it (a rolling boil) for at least one minute to kill harmful microorganisms. Because water boils at a lower temperature at elevation, water should be boiled for an additional minute for every 3281 ft (1000 m) of elevation (WHO 1997). For example, at 9840 ft (3000 m) elevation, water should be boiled for four minutes.

Advantages: Simple method, nearly no failure.

Disadvantages: Takes time and consumes energy. This procedure also does not remove heavy metals or minerals such as arsenic.

Table 12.1 Bleach Drops Required to Treat Potable Water

% Strength Chlorine Solution	1%	4% to 6%	7% to 10%	unknown
	Number of drops* of bleach per L of water			
Clear Water	10	2	1	10
Turbid Water	20	4	2	20

** Drops should be the size of those that come from a typical eye dropper.*

Chemical Treatment

Chemicals such as sodium hypochlorite (chlorine) or calcium hypochlorite kill microorganisms that contaminate drinking water. Iodine should not be used because it causes side effects (Küpper et. al 2012). At concentrations used in drinking water, chlorine and its by-products are considered to have low toxicity to humans and its use as a disinfectant is very effective (Health Canada 2011).

Procedure: A sufficient amount of disinfectant must be added to the water. Shake well. Wait the appropriate disinfectant time, usually 30 minutes. Disinfection time increases with turbidity.

Advantages: Can be used anywhere. Does not require energy.

Disadvantages: Takes time and taste can be impaired. Failures are possible:

- Cold water takes longer to disinfect
- Neutralizing of cysts and eggs require higher concentration of disinfectant
- Water with higher concentration of organic material requires more disinfectant

Procedure:

1. Prefilter cloudy or turbid water through clean cloths or coffee filters.
2. Read the label on the bottle of unscented chlorine bleach to determine its concentration (most bleach sold for household use is 5% solution of sodium hypochlorite). Use a dropper to the bleach to the water. The amount of bleach is doubled for turbid water (see Table 12.1).
3. Stir the water and let it stand for 30 minutes. For water that has too strong of a chlorine taste, let it stand open for a few hours without a lid to allow the chlorine to off-gas.

Filtration

Pathogens, turbidity and chemicals can be removed from the water by several physical characteristics such as their size relative to the filter's pores or hydrophobic or electrostatic interaction between the impurity and the filter material.

Filtration includes several levels of technology including ultrafiltration, microfiltration, and nanofiltration. Reverse osmosis uses applied pressure to overcome osmotic pressure and a semipermeable membrane to remove water impurities.

Filtering Particulates

Particle filtration is a process that removes suspended particles from water, ranging in size from sand to clay. Glacial meltwater should be filtered to remove fine, suspended particles before additional treatment.

Filters function in two general modes:

1. Surface or screen filters remove the particles at or very near the filter surface. Ceramic filters are porous ceramic cylinders that filter at the surface. They are expensive but long-lasting, may be cleaned, and provide precise filtration.
2. Depth filters have a thick filter medium. Particles are retained throughout the thick filter mat and these filters may be used for a wide range of particle sizes. There are several types of depth filters including string-wound, spun-fiber, and pleated-fiber.

   Sand filtration systems are a type of depth filter that provides relatively safe, clean, and clear drinking water. These systems filter out 97% of pathogens and bacteria. This does not meet Canada's standard of 100% (Health Canada 2017) and so the water will require further treatment such as boiling or the addition of chlorine. Sand filters would be a good method for removing fine particulates from glacial meltwater.

Procedure: Water will pass through a pore size of 0.2 μm or larger.

Advantages: Simple procedure.

Disadvantages: Equipment must be handled with care (ceramic material may break). Ceramic filters are high-tech products and manufacturers vary. Clogging is a frequent problem and filters must be cleaned or regularly backwashed through controls, which typically requires electricity.

Ultraviolet (UV) Treatment

UV treatment is the disinfection process of passing water by a special light source. This method of treatment is growing in popularity because it does not require the addition of chemicals or the amount of energy needed to boil water. UV systems alone are neither intended to treat water that is visually contaminated nor intended to convert wastewater to safe, microbiologically potable water. A point-of-use system is a small, portable device that attaches to a faucet and rests on or under the counter. Handheld devices that can be placed in a glass of water are also available.

Procedure: Immersed in the water in a protective transparent sleeve, the special light source emits UV waves that inactivate harmful microorganisms including bacteria, viruses, Cryptosporidium oocysts, and Giardia cysts. The UV rays, similar to the sun's but stronger, alter the DNA of microorganisms so that they cannot reproduce.

Advantages: Does not require as much energy as boiling. Does not alter the water chemically.

Disadvantages: An energy source (electricity or battery) is required to power the light. Inactivated microorganisms are not removed from the water. UV treatment does not remove dirt and particles. The technology is intended to be used in visually clear water, not colored, cloudy, or turbid water. UV does not work well in turbid water. The bulb gradually loses its disinfecting capabilities over time.

GRAY-WATER TREATMENT METHODS

Gray water is used kitchen wastewater and wash water. It contains soap, food particles, oil, and grease. It can also contain nutrients, bacteria, viruses, and a variety of chemicals including endocrine disruptors[5] associated with detergents and personal care

5. An endocrine disruptor is a synthetic chemical that, when absorbed into the body, either mimics or blocks hormones and disrupts the body's normal functions. This disruption can happen through the alteration of normal hormone levels, halting or stimulating the production of hormones, or changing the way hormones travel through the body, thus affecting the functions that these hormones control.

Table 12.2 Comparison of Water Treatment Methods for Removing Pathogens *(Adapted from Küpper et al. 2012)*

Procedure	Safe For				Remarks
	Viruses	Bacteria	Cysts (Giardia, Amoebic) and Helminth Eggs	Cryptosporidium	
Boiling	+[a]	+	+	+	Consumes fuel
Chemical disinfection[b]	+	+	(+)	+[c]	Does not work well in very cold water[d]
Ceramic filter	(+)[e]	+	+	+[f]	Type specific failures / limitations
Chemical disinfection and ceramic filter	+	+	+	+c,f	The only absolute safe procedure at high altitude
UV	+	+	+	+	Does not work well in turbid water
Sand filter	—	(+)e	(+)[g]	n.d.	Fine sand and low flow necessary
Charcoal filter	—	(+)e	(+)g	n.d.	Low flow necessary
Sand and charcoal filter	—	(+)e	(+)g	n.d.	Fine sand and low flow necessary
Textile filter	—	(+)e	(+)g	n.d.	The tighter the textiles, the better the filter effect

+ = *safe*
(+) = *safe with some limitations*
— = *not safe*
n.d. = *no data available*

a. *Hepatitis not completely inactivated*
b. *Using chlorine*
c. *High concentration of chemical necessary*
d. *Requires higher concentration of disinfectant or longer disinfection time*
e. *Not safe, but reduces concentration of germs*
f. *Pore size <1 μm necessary*
g. *Nearly safe*

products (such as sunscreen). If dumped outdoors, it may attract wildlife and pollute the local environment. Food particles, oil, and grease in various stages of decomposition may also present a health hazard.

Objectives of gray-water management are to

- avoid the creation of bad odors where untreated gray water may be deposited,
- prevent contamination of groundwater and surface water,
- prevent eutrophication of surface waters through the introduction of nitrates and phosphates, and
- disperse gray water on site without attracting wildlife.

Grease Interceptor

A grease interceptor is a method of removing food particles, oil and grease from kitchen wastewater. Units can be purchased and placed within the plumbing system under the sink. Food particles will sink to the bottom and oil and grease will rise to the top. Water is plumbed away from the middle. The grease interceptor must be regularly cleaned and should be installed so it is easily accessible. An under-sink grease interceptor will not be suitable if the indoor temperature drops below freezing when the facility is unoccupied.

If the facility freezes during unoccupied times, a low-cost, low-tech solution for removing food particles, grease, and oil from gray water is to plumb the gray water to a grease interceptor located in a barrel buried outdoors. Gray water from the kitchen is passed through a preliminary sieve under the sink to remove larger food particles. The sieve is cleaned daily. The gray water is plumbed outside through a buried and insulated line. The water flows into a barrel where the food particles sink to the bottom and the oil and grease float to the top. Water is taken out of the barrel from the middle and is passed through a secondary sand filter before it is allowed to disperse on-site. The grease interceptor must be insulated so that is does not freeze in winter. Maintenance involves periodically removing the barrel for cleaning. Grease and food particles must be handled as hazardous waste and taken to an appropriate waste treatment facility or landfill.

Dispersion Field

Another low-tech method of gray-water treatment is the dispersion field. Kitchen wastewater is passed through a sieve to remove large food particles and is collected under the sink in a bucket. The bucket is carried out by hand and the contents are poured into one of a number of PVC pipes that are placed vertically in the ground and located over a constructed bed of sand and gravel. Half of the pipe is buried to prevent wastewater from being dumped onto the surface. The sand and gravel filter out the grease, oil, and remaining food particles; the resultant effluent either sinks or emerges to the surface. Bacteria on the sand and gravel decompose the organic material. The system is shallow enough that the decomposition process remains aerobic.

Container Treatment Systems

Higher-technology off-the-shelf container and treatment systems are available to handle larger quantities of gray water. These multistage containers combine filtration and microbial digestion to clean the gray water sufficiently so that the effluent can be safely dispersed on site. Key treatment processes may include grease removal, solids capture, dissolved solids removal, nutrient removal (denitrification, dephosphatation), and dechlorination.

Soaps and detergents found in gray water can be biodegraded in a gray-water container and treatment system. Soaps and detergents are organic macromolecules composed of carbon, oxygen, and hydrogen. If exposed to certain microorganisms, they will biodegrade into water and carbon dioxide. Phosphates and phosphate-containing detergents will precipitate out.

The low-tech gray-water treatment methods described in the previous subsections are currently accepted practices for removing food particles, oil, and grease; however, they do not remove soap, detergents, or disinfectants such as bleach. Only a gray-water multistage containment system is capable of treating soaps and detergents.

BLACK-WATER TREATMENT METHODS

Black water (human sewage) consists of fecal waste, urine and toilet paper. The primary objectives of a black water waste management system are to control pollution onsite and prevent disease through human contact.

What most operators of off-grid facilities are looking for is a waste management strategy that

- is functional (it works) and reliable,
- uses the least amount of energy,
- requires the minimum amount of attention,
- has minimal safety risk for users and maintenance staff during servicing,
- is robust enough to withstand periods of high use and long periods of vacancy,

- has a low environmental impact,
- is user friendly, and
- is affordable.

Options for managing black water at remote sites include pit toilets, barrel fly-outs, incinerating, carry-out, vermicomposting, urine diversion, and flush to tertiary waste water treatment.

There are advantages and disadvantages to each system and the application of each technology should be assessed according to the following:

- Access
- Environmental conditions (soils, temperature, proximity to surface and ground water)
- Amount of use (seasonal, number of visitors)
- Energy requirements
- Budget
- Availability of resources (custodian, access to transportation, level of skill required to operate the technology)
- Health and safety codes
- Land manager requirements

Pit Toilets

A pit toilet consists of a hole in the ground that is deep enough to hold a significant amount of waste but not deep enough to penetrate the water table. The pit has an outhouse structure over top containing the toilet seat to provide privacy. When the pit is full, it is covered with soil and the outhouse is moved or rebuilt over a new pit.

Pit toilets should not be used in areas without a significant layer of biologically active soil, because it is the microorganisms in the soil that eventually break down the waste. Without the microorganisms, the waste will not decompose. Alpine areas generally have too little topsoil and are too cold, which causes slow decomposition. In addition, the rocky substrate makes it difficult, if not impossible, to dig a hole.

Pit toilets are very reliable, easy to maintain, and result in low risk to users and operators.

Consider using a ventilation fan powered by a solar panel to control odors.

Barrel Fly-Out

The barrel fly-out system involves collecting the waste in barrels. When one barrel is full, it is moved off to the side and sealed and a new barrel is placed under the toilet chute. The full barrels are removed by helicopter and taken to a staging area with road access where a pumper truck extracts the waste and transports it to a proper waste processing facility.

There is a certain amount of technology involved in the design of an outhouse structure that is built to accommodate the barrel fly-out system. The toilet seat is positioned above the barrel. The barrel storage area must have easy access to the flight deck where barrels are loaded into a long line from a helicopter. Barrels must have tight-fitting lids. Plastic material such as arena board is placed on the floor of the barrel storage room to make the barrels slide easier during replacement and servicing.

The advantage of this system is that it removes all waste from the site, resulting in absolutely no adverse affects on sensitive local ecosystems. There is low risk to users except when changing the barrels, at which time they should wear protective gear (gloves, clothing). Barrel fly-out is easy to maintain.

Disadvantages are that helicopters are expensive, noisy, use fossil fuels, and disturb wildlife and people. In addition, there is a health risk for maintenance personnel handling raw sewage and a small risk potential for spills during transport.

Incinerating

Incinerating toilets collect waste in a chamber below the toilet seat. During their burn cycle the liquids are evaporated through boiling and the solids are converted to a sterile ash. There are both electric and gas (propane or natural gas) burning models available. Each unit requires a flue that exhausts to the exterior. They can be used indoors and outdoors, even under freezing conditions.

Some models are designed to burn after each use; some hold 40 to 60 uses before a burn cycle is initiated. They are manually operated. The electric models have a push-button start. The gas-fired models start similar to lighting a barbecue grill. Depending on the model and volume of waste that is being burned, burn cycles typically last between 1 to 4.5 hours. At elevations greater than 4000 ft (1219 m) it is recommended to run the burn cycle for 6 hours (EPA 1999). A 100 lb (45 kg) propane cylinder will burn 16 capacity loads (approximately 960 uses).

Incineration methods have two clear advantages when compared with other methods. First, there is no waste to remove from the site when the process is complete, except a few cups of ash. Second, pathogen destruction is almost certain (some models perform better than others). There is low risk to maintenance operators.

The main disadvantage is that incineration uses fossil fuels. Odors are also cited as a problem, however air pollutants and odors can be minimized by filtering the smoke within the incineration chamber through an odor control catalyst (much like the one found on an automobile exhaust system) before it goes out the exhaust vent. Odors are present between burns, much like an unvented pit toilet.

Carry-Out

The carry-out method is suitable for individual use away from a roofed accommodation, such as single- or multiday mountaineering trips. This method is also suitable for low-use shelters that receive less than 100 overnights per year. More use than that and the barrel fly-out method may become more economical, although this depends on helicopter flight distance.

User carry-out involves the users to carry out their own waste in puncture-proof, sealable bags. There are many manufacturers of these commercial products. Some bags come with a chemical powder that gels urine to prevent leakage. Some come with enzyme powders to help biodegrade and stabilize solids so the bags can be disposed in regular trash intended for a landfill, much like baby diapers. Some bags are biodegradable. Some systems use double bags to protect from leakage. Some systems come with hard containers such as tubes or small barrels.

Carry-out bags made from biodegradable material can be deposited in a pit toilet or composting toilet. Flush toilets will be damaged if plastic or biodegradable bags are flushed. One could progressively take one's waste contents out of its bag and flush it in a domestic toilet in small quantities; however, this option is not appealing to most people and the waste bag still requires proper hygienic disposal. It is not hygienic to bury human waste in a garden or yard. Unless it is the type of bag that contains a stabilizing enzyme, the bag should not go in the regular trash. Do not use RV dump stations or pump-out vault toilets as these have pumping systems that can be blocked by plastic and biodegradable bags. The bags have also been known to plug up septic systems.

User compliance is key for the carry-out method to succeed. Noncompliance may result in randomly placed waste, which can pose health risks. This problem is becoming more apparent along popular mountaineering routes. Education is required to encourage users to change old habits, and initially, policing will also be required. Hazards to users are high.

Vermicomposting

The idea of composting human waste seems like a sound environmental solution to providing remote area toilets. In theory, composting human waste is a feasible waste management option and has garnered strong support. However, composting has also fueled disillusionment and even inspired active opposition, making composting the most controversial human waste management option.

The only way to ensure that pathogens are destroyed by composting is to ensure temperatures reach 131°F (55°C) (British Columbia Regulation 2002). Composting toilets do not reach these temperatures and the body of literature indicates that they are unreliable in the production of compost suitable for discharge into the environment (Hill and Henry 2013).

Another method of composting involves urine diversion and the introduction of worms, bacteria and invertebrates: vermicomposting. The worms facilitate all essential actions in decomposition necessary to render human waste safe, stable, and mature. These actions include: aeration (anaerobic fermentation is smelly, but aerobic fermentation has little odor), mixing, pathogen destruction, biochemical nutrient oxidation to plant fertilizer forms, and metal extraction through bio-accumulation. Naturally occurring bacteria in the worm's digestive system feed on the pathogenic human germs in the feces resulting in pathogen destruction. The naturally occurring bacteria also reorganize the elements contained in the feces by breaking molecules and building new ones. The result is a stable material (humus or compost), which no longer ferments.

Urine Diversion

Urine diversion can be used in conjunction with other waste management technologies.

Urinating is the body's primary method of excreting toxins, chemicals, and drugs. In addition to nitrogen and ammonia, urine also contains phosphorous, inorganic salts, amino acids, proteins, enzymes, carbohydrates, vitamins, hormones, and pharmaceuticals such as antibiotics. The nitrogen-phosphorous-potassium ratio is similar to commercial fertilizer; humans have used urine as fertilizer for many generations. The pH of urine is neutral (pH = 7).

If deposited on site, the nutrients in urine may change the quantity and quality of the natural flora. In addition, urine can support the growth of microorganisms in the soil by providing a source of moisture and nutrients. If soils are inadequate to take up the urine, odors associated with ammonia may develop. Environment Canada has guidelines for on-site urine disposal. The practice is acceptable provided it can be done in a manner that ensures the protection of public health, does not harm the local environment, and does not negatively impact aesthetics (Rollins and Young 1992).

Requirements for urine diversion include the following:

- The waste liquid must be free of fecal coliform bacteria.
- The soil mantle conditions (including slope, depth, and carrying capacity of the regolith) are such that neither surface water nor groundwater quality are adversely affected.

- The soil absorption is adequate to prevent odor and other nuisance conditions from developing.
- Frequency of freezing conditions and seasonality of use can affect urine gathering.

Urine diversion is currently used at many mountain facilities. Urine diversion combined with vermicomposting is a viable waste management alternative. (For more information see www.toilettech.com.)

The potential to treat urine on-site could lead to enormous savings in transportation and reduced risk in subsurface disposal. The most promising technique involves the evaporation of all the water in urine (about 95% of the total volume), thus creating a stable, low odor, safe to handle solid material (Bethune 2015).

Flush to Tertiary Wastewater Treatment System

Tertiary wastewater treatment systems are small-scale municipal sewage treatment systems that treat gray water along with black water. They involve multistage treatment

Figure 12.3 Ice-cold water.

David Lima

processes that treat the waste biologically, chemically, and physically. Tertiary wastewater treatment produces a high standard of effluent before it is discharged into the environment.

The first stage of treatment involves removal of particulate matter in settling tanks. Oils and grease from gray water floats to the top and solids from gray water and black water sink to the bottom. The solids are strained off and the sludge is treated in an anaerobic sludge-digesting tank. Anaerobic digestion involves microorganisms that break down the organic matter (proteins, lipids, and cellulose) into smaller substances that can be metabolized by other organisms. Any remaining sludge is incinerated or buried in a landfill.

During the second stage of treatment, microbial populations in the liquid waste are reduced by aerobic digestion. The fluid waste is aerated and passed through a filter where colonies of bacteria, fungi and protozoa remove organic materials. The microorganisms metabolize organic compounds and convert them to carbon dioxide, sulfate, phosphates, nitrates, and other ions. This removes 99% of microorganisms. Liquid waste can also be treated in an activated digester after it has been aerated. Several hours of treatment will significantly reduce the microbial population.

In the third stage of treatment, phosphates and nitrates are precipitated out by combining them with calcium or iron. Ammonia is released by oxidizing it to nitrate in the nitrification process. Adsorption to activated charcoal removes many organic compounds such as polychlorinated biphenyls (PCBs) and odor producing gases. Final disinfection treatment can be achieved with chlorine, ultraviolet radiation, or ozone.

There are several manufacturers of packaged waste treatment systems. They tend to be expensive, have complex systems with many components, and require electricity to operate. They require professional help in design, installation, and careful daily monitoring. Problems have arisen with turbid effluent for which ultraviolet radiation is not effective. They are most suitable for larger operations. Cost varies with number and type of components.

HOW COLD IS IT?

It is so cold ... we parked the dog sled on the tread mill to keep the dogs running!

13 Oil and Gas Applications

This chapter focuses on the unique challenges associated with HVAC systems in oil and gas facilities located in cold climates. Cold-climate considerations play a major role in the design of oil and gas HVAC systems along with the selection, operation, and maintenance of HVAC equipment. Additionally, the unique nature of requirements and design practices specific to oil and gas facilities also influence HVAC systems and equipment selections.

CODES, DESIGN, AND SAFETY CONSIDERATIONS

HVAC systems are an integral part of an oil and gas facility's hazard reduction and life safety design. Protection of occupants and assets while ensuring the continuity of operations are the main objectives in designing HVAC systems for oil and gas facilities. The design of these systems needs to account for present hazards and their potential impact on occupants and assets.

Codes and Standards

HVAC systems used in cold-climate oil and gas applications follow normal industry codes and standards, such as those from NFPA, SMACNA, ASHRAE, local building codes, and so on. There are additional codes and standards that are unique to or have greater applicability to oil and gas HVAC system designs. Typical codes, standards, and recommended practices for determining ventilation requirements for hazard-reduction applications in oil and gas hazardous (classified) locations are NFPA 70, NFPA 497, API RP 500, and API RP 505 (NFPA 2020, NFPA 2021b, API 2012, API 2018). NFPA 496 is used to identify pressurization requirements with respect to hazardous environments. Additionally, local code requirements may address specific ventilation requirements for oil and gas facilities. An example of this in the State of Alaska, which follows the ICC series, and has created an additional Appendix LL to the IBC which addresses specific oil and gas criteria under the Alaska Administrative Code (AAC 2017).

Design Considerations

Enclosed oil and gas processing and drilling facilities located in cold climates place particular focus on indoor ventilation rates and pressurization for both fugitive emissions[6]

6. Fugitive emissions are leaks and other irregular releases of gases or vapors from pressurized containment.

and hydrocarbon containment. The design of HVAC systems and equipment should be simple, reliable, and easy to operate and maintain.

Some hazards associated with oil and gas operations are fugitive emissions, overpressure (blast) events, fire, flammable and toxic gas release, and equipment failures associated with ambient conditions. Occupants are faced with hazardous materials in the workplace during normal operations, impact of hydrocarbon release in abnormal or upset operating conditions, and IAQ and temperatures outside of normal ranges.

Oil and gas applications can also include asphyxiants, such as those associated with nitrogen generators and distribution pipework (nitrogen is used to purge pipelines and analyzers to maintain a nonflammable environment). Asphyxiants are of particular concern in enclosed facilities associated with cold climate design.

Hazardous materials present in daily operations need to be accounted for. Both acute and long-term exposure limits with respect to federal occupational health guidelines must be evaluated. Point-of-source capture hoods, fume hoods, or local exhaust air systems can be used to mitigate local hazards. Proper ventilation rates and outdoor air dilution along with adequate air distribution to eliminate dead-air spaces and mitigate stratification need to be evaluated. Local building codes, API RP 500, and API RP 505 provide guidance with respect to minimum ventilation rates and adequate ventilation (API 2012, 2018).

Both heavier-than-air and lighter-than-air fumes and vapors need to be evaluated and accounted for. Using exhaust or relief assemblies mounted both near the floor and the highest point of the facility are common solutions in oil and gas facilities. Insulated cold-air traps should be used at both locations. Care must be taken to avoid cold, dense air from leaking into the building at lower intakes, which can disrupt intended design airflow rates and distribution. Backdraft dampers have been used in the past, but are currently discouraged as the backdraft damper is not a positive isolation damper under actuator control. Inspection of existing backdraft dampers have shown that the linkages are prone to failure and sticking in a fixed position, thus rendering it ineffective for its original design purpose.

Redundant or backup HVAC systems are typically provided in safety- or operation-critical oil and gas facilities. Redundant applications are considered $N + 1$ systems that use the same type of HVAC equipment to ensure continuous operation. Backup applications are considered systems that use different HVAC equipment: for example, a duty AHU with direct-expansion (DX) coil for ventilation and cooling and a backup split system DX unit for space cooling only. Backup equipment ensures space design criteria are met for each operational mode. Some typical operating modes can be normal, emergency, uninterruptible power supply (UPS), shutdown, or high rate ventilation. Controls need to be provided to automatically energize redundant or backup equipment and isolate duty equipment upon failure.

When designing oil and gas HVAC systems, complexity is not desirable because of the potential impact on safety. Experience has shown that complex systems have a high failure rate because of the harsh operating environment. Simplified equipment and control sequences, along with inherently safe design, result in more reliable HVAC systems. Proven technology, as opposed to cutting-edge equipment relying on a high level of controllability or calibration, is more successful in cold-climate oil and gas applications. Sufficient maintenance space needs to be provided for all equipment along with access to HVAC equipment. While maintenance space may be provided, access to the space is often overlooked due to the size of module and the congested nature of processing equipment.

Design analysis and calculations must account for high heat rejection equipment associated with oil and gas applications in enclosed facilities. Electrical, telecom, instrument, and control rooms are examples of spaces with high heat rejection. A sufficient safety factor should be used during design to account for additional heat gain associated with future electrical equipment and field modifications of process control equipment.

Calculations should also take into account the construction quality associated with oil and gas facilities in cold climates. While buildings with higher U-factors are desirable, it is common to encounter non-insulated buildings along with significant air gaps and infiltration associated with remote, irregularly occupied buildings and equipment enclosures. A sufficient safety factor for heating systems should be used.

Oil and gas processing applications often include loud, noise-producing equipment (such as compressors, generators, turbines, and large-diameter pipework) that can have a potential impact on HVAC systems for occupied facilities. Sound attenuation on HVAC systems needs to be thoroughly evaluated through acoustical analysis. Verification that noise exposure and sound pressure level requirements for individual spaces are met should be provided.

Hazards and Electrical Area Classification

The processing of hydrocarbons associated with oil and gas facilities is an inherently hazardous operation; hydrocarbons by nature can be both flammable and explosive. NFPA 70, NFPA 497, API RP 500 are API RP 505 are examples of oil and gas industry documents that provide guidelines for establishing hazardous area classifications and boundaries (NFPA 2020, 2021; API 2012, 2018). These documents contain normative references, detailed information, and common applications establishing electrical classification. The following definitions are per API RP 500:

- Unclassified locations (nonhazardous location) are locations determined to be neither Class I, Division 1 nor Class I, Division 2; or any combination thereof.
- Class I location are locations in which flammable gases or vapors are, or may be, present in the air in quantities sufficient to produce explosive or ignitable mixtures.
- Class I, Division 1 locations are locations in which ignitable concentrations of flammable gases or vapors are expected to exist under normal operating conditions or in which faulty operation of equipment or processes might simultaneously release flammable gases or vapors and also cause failure of electrical equipment.
- Class I, Division 2 locations are locations in which flammable gases or vapors may be present, but normally are confined within closed systems; are prevented from accumulating by adequate ventilation; or the location is adjacent to a Division 1 location from which ignitable concentrations might occasionally be communicated.

API RP 505 outlines similar definitions for electrical area classification, but uses a Class and Zone (0, 1, or 2) definition, as opposed to the Class and Division (1 or 2) found in API RP 500. API RP 500 is aligned with NFPA 70 Article 500, while API RP 505 is aligned with NFPA 70 Article 505.

Hazardous gas-air mixtures, termed *lower flammable limit* (LFL) or *lower explosive limit* (LEL) for hydrocarbon gases, have been established and can be found in oil and gas industry documents. The upper bound of hazardous gas-air mixtures are termed *upper flammable limit* (UFL) or *upper explosive limit* (UEL). Gas-air mixtures below LFL are too lean to support flammable conditions while gas-air mixtures above UFL concentration are too rich to support flammable conditions.

Maximum equipment temperature ratings (T rating) for components that produce heat (i.e., fan motors, heating elements, and so o) and a material group (Group A, B, C, or D), which identifies the specific hazardous material, can also be found in NFPA 70. An example of an electrical classification would be Class I, Division 2, Group D, T-3.

Minimum ventilation rates for hazardous areas are established in local and national industry documents. Ventilation rates such as 1 cfm/ft^2, 6 air changes per hour, or adequate ventilation sufficient to keep gas-air concentrations below 25% LFL are all code, standard, and recommended-practice phrases used to describe ventilation requirements in hazardous areas. Specific ventilation rates are dependent upon each unique location, adopted codes, and application.

Ventilation of hazardous areas can either be by natural or mechanical ventilation. Natural ventilation is considered an inherently safer option, as there is no reliance on electrical power. However, natural ventilation is not always achievable because of environmental conditions, space congestion, or hazardous areas located within buildings, partial enclosures, or behind wind walls. Mechanical supply ventilation providing dilution and exhaust ventilation in enclosed process facilities or congested process areas are used to mitigate buildup of hazardous vapor-air mixtures associated with fugitive emissions.

Recirculation of air is often used in a mixed air configuration to reduce outdoor air heating loads in cold-climate oil and gas design. API RP 500 and API RP 505 detail when and how this can be implemented (API 2012, 2018). In summary, recirculation can be used if continuous gas monitoring is provided incorporating a fail-safe system designed to automatically sound an alarm, stop recirculation, and provide full exhaust to the outdoors if gas-air concentrations over 25% LFL are detected.

Ventilation systems can be used as justification to reclassify an electrical area classification, provided the ventilation system is in accordance with API RP 500 or API RP 505. This allows a Class I, Division 1 area to be reclassified as Class 1, Division 2, which has less stringent construction criteria for electrical equipment. Class I, Division 2 spaces cannot be reclassified to unclassified spaces.

Vaportight barriers, as illustrated in Figure 13.1, can be used to create a separation between unclassified areas and adjacent hazardous classified areas. API RP 500 defines a vaportight barrier as "a wall or other obstruction that will limit the passage of gas or vapor at atmospheric pressure, thus preventing the accumulation of vapor-air or gas-air mixtures in concentrations above 25% of their lower flammable (explosive) limit" (API 2012). Degradation of vaportight barriers due to environmental conditions or future penetrations from conduit or piping are common, making the integrity of vaportight barriers difficult to maintain. Additionally, the presence of a door between hazardous and unclassified areas will negate the benefits of a vaportight barrier. Providing positive pressurization per NFPA 496 in unclassified spaces adjacent to hazardous classified areas, regardless of the presence of vaportight barriers, is considered best practice within the oil and gas industry. Typically, an unclassified area such as an electrical or instrumentation room that is adjacent to a process area will require a supply fan for cooling, so providing positive pressurization is relatively easy and cost-effective. Positive pressurization provides a greater level of assurance that no hazardous gas-air mixtures will infiltrate the unclassified area and contact a potential ignition sources. Other methods for maintaining separation between hazardous and nonhazardous areas may be possible, but these will be unique fit for purpose solutions based on site conditions and operator requirements along with code and standard interpretation requiring multiple engineering discipline interfaces.

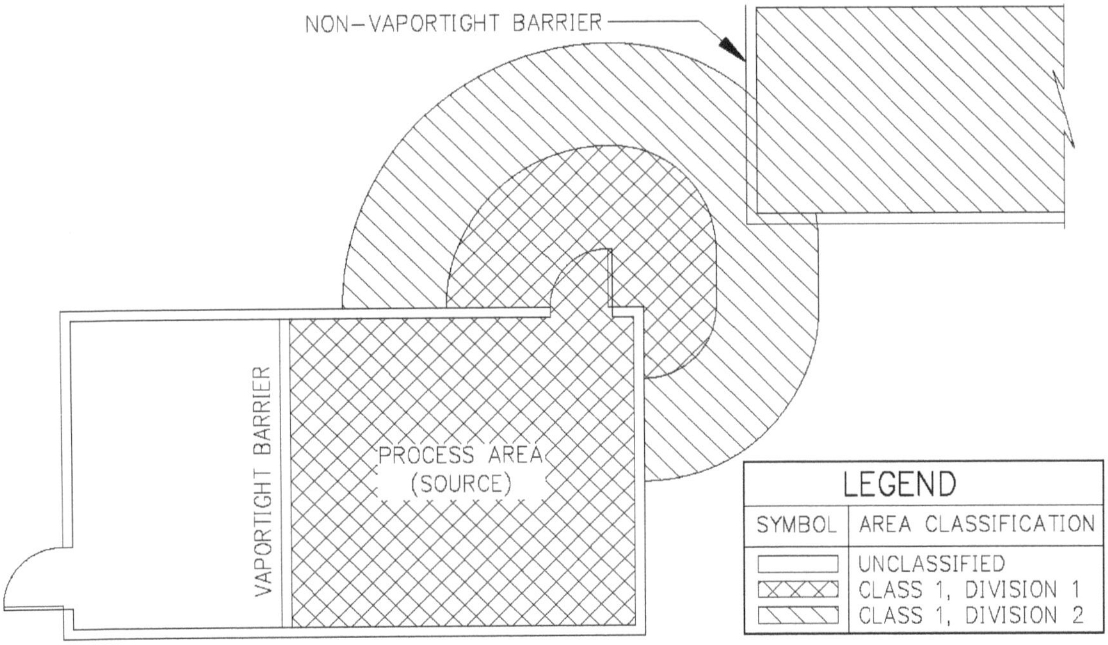

Figure 13.1 Electrical area classification designation example from API RP 500. The hazard source can be either from inside of an enclosed space, or outside.

Pressurization

Pressurization of oil and gas facilities using mechanical ventilation reduces the risk of gas and smoke ingress along with infiltration of the external environment and consequent heat gain or loss to the space. Supply, exhaust, and relief systems need to be balanced with respect to each other to ensure sufficient flow rate and to ensure that positive space pressurizations are maintained under normal wind conditions. There are some specific instances of facilities being operated at negative pressure, such as battery rooms and hazardous areas, but the general cold-climate concept is to maintain positive pressurization with respect to ambient conditions but maintain these spaces at a lower pressure than adjacent nonhazardous or unclassified spaces.

Passive or active pressurization control can be used to control facility pressure. Passive pressurization control is generally the preferred approach, but complex facilities in hazardous areas may require active pressurization control.

Passive pressurization is achieved by designing supply and exhaust air flows to be in excess of one another (volumetric offset) to obtain a positive or negative pressure differential, respectively. A pressure-relief damper combined with exfiltration and exhaust sources are typically used to maintain positive-pressure spaces. Exhaust fans are typically used to maintain negative-pressure spaces.

Active pressurization is achieved by means of automatic pressure control dampers, fan variable-speed drives, or fan inlet radial vane dampers to modulate airflows in response to monitored space differential pressure.

Purging and pressurization to protect unclassified spaces in hazardous oil and gas locations is governed by NFPA 496. The following highlight the main requirements found in NFPA 496 for pressurizing unclassified spaces (NFPA 2021a):

- Minimum positive pressure shall be 25 Pa (0.1 in. wg) of water with all doors closed.

Table 13.1 Building Pressurization Criteria

Location Area (Hazardous or Nonhazardous)	Facility Classification	Pressure, in. w.g. (Pa)	Envelope Integrity (Leakage Performance Criteria)[a]
Hazardous	Unclassified	+ 0.1 (25) with respect to adjacent hazardous area	0.25 ach @ 0.2 in. w.g. (50 Pa)
Hazardous	Classified	+ 0.05 (15) with respect to atmosphere	N/A
Hazardous	Division/zone boundaries (i.e., Division 1 to Division 2)	+ 0.05 (15) with respect to higher hazard[b]	N/A
Nonhazardous, but within potential gas cloud dispersion	Unclassified	+ 0.1 (25) with respect to atmosphere	0.25ACHR @ 0.2 in. w.g. (50 Pa)
Nonhazardous	Unclassified	+ 0.05 (15) with respect to atmosphere	N/A

a. *Verification of performance standard for envelope integrity is normally demonstrated through positive and negative blower door testing.*
b. *Pressurization between Division 1 and 2 and Zone 0, 1, and 2 spaces is used to mitigate air transfer from higher- to lower-hazard spaces.*

- Minimum outward velocity shall be 0.3 m/sec (60 ft/min) through all openings, measured when all doors are opened. A temporary drop in pressure (i.e., 30 seconds) below 25 Pa (0.1 in. wg) is permitted in this condition.
- Failure of pressurization system shall activate alarm at constantly monitored location.

Pressurized spaces must have building envelope integrity to mitigate infiltration and ingress. Building pressurization is closely linked to envelope integrity. Evaluation of space envelope integrity can include HVAC intakes, outlets, and dampers, as well as wall construction, door and window seals, cable entries, lighting, plumbing vents, sanitary drains, piping, and similar space penetrations.

The use of vestibules and airlocks to maintain building pressurization to mitigate infiltration of hazardous gas is considered best practice. Differential pressure gages should be installed at each normally used doorway to indicate space pressure for local monitoring.

Typical facility pressurization in oil and gas applications is given in Table 13.1.

Intake and Isolation

Air intakes for ventilating hazardous and nonhazardous areas need to be taken from nonhazardous locations where toxic gas, noxious contaminants, or objectionable odors will not be introduced into the ventilation system. Air outlets from hazardous areas need to be located where they will not create additional hazards by inadvertently extending the hazardous area.

The following should be accounted for when determining the elevation and location of air intakes and outlets:

- Prevailing wind direction and environmental conditions.
- Electrical area classification boundaries.
- Output from gas dispersion studies. Facilities with lighter-than-air gas hazards need to evaluate air intakes at low elevation, i.e., underneath facilities or platforms. Conversely, facilities with heavier-than-air gas hazards need to evaluate air intakes at high elevation.
- Process vents, chimneys and flues; power generation exhaust, ventilation exhaust; plumbing and drainage vents; other known hazards and contaminants.

Air intakes are typically located a minimum of 10 ft (3 m) from a hazardous area boundary in oil and gas applications.

Isolation dampers need to be provided at penetrations of buildings or spaces that have gaseous fire suppression systems installed regardless of the building classification or location in hazardous or nonhazardous areas. The isolation dampers are used to contain the gaseous fire suppression agent within the space. Gaseous fire suppression systems typically are only used in process, control, electrical, telecom, or instrumentation spaces.

Fire and Gas Dampers

Fire dampers are provided at HVAC penetrations of fire-rated barriers to maintain the fire barrier integrity. In oil and gas applications, fire-rated barriers can be both internal or external fire-rated walls (i.e., the building envelope). It should be noted that external fire-rated walls are not common and are typically not provided for onshore oil and gas facilities. External fire-rated walls are typically only provided when a hazard analysis indicates that a fire scenario is the governing hazard. Hazard analysis and similar evaluations are typically conducted by a process safety engineer or specialist and are outside the normal realm of an HVAC engineer's expertise.

Fire dampers can either be actuated through the facility safety control systems or be provided with a fusible link, and in many cases are provided with both. Specific closure requirements will be governed by local building codes. Fire damper integrity levels are based on the barrier construction in which they will be located (i.e. a 1.5 hour fire damper in a less than 3 hour resistance rated construction, or an A60 fire damper in an A60-rated construction). Fire dampers for oil and gas facilities can require approvals from different certifying authorities based on the applicable code or standard, which will be based on the local building codes and damper type (i.e., 1.5 hour damper or A60 damper). Dampers with 1.5 or 3 hour terminology are typically based on ratings from UL 555 (UL 2006a), while dampers with A, H, or J ratings are typically based on ratings from an approved certifying authority (e.g., Lloyd's Register, Det Norske Veritas, American Bureau of Shipping, Bureau Veritas) as designated by local regulations.

Internal smoke dampers can be required in buildings with internal smoke barriers and partitions. Specific application should be coordinated and verified with local building code requirements and building operational modes. Oil and gas buildings with critical application equipment and instrumentation for asset protection may utilize smoke dampers to isolate an area of incident. Control and electrical rooms can fall into this category, based on site-specific oil and gas operational philosophies.

Gas dampers are provided at ventilation intakes and outlets to protect against ingress of toxic or flammable gas. Unclassified buildings located in hazardous areas or in a gas dispersion cloud from a credible release source need to use gas dampers. Buildings can be in a nonhazardous area and be unclassified but still be within a gas dispersion cloud. Buildings in hazardous areas also need to be NFPA 496-complaint for gas ingress protection.

Gas dampers are typically rated to UL 555S, *Standard for Safety—Smoke Dampers* (2014), using Class 1 for leakage. UL 555S is used as there currently is no corresponding rating specifically for gas leakage.

Gas dampers are typically actuated through the facility safety control systems, with input based on gas detectors. Gas detection philosophies specific to the facility dictate gas detectors locations (e.g., area detection, inlet ductwork detection, detection inside of the facility). A typical gas damper closure concept is to provide sufficient distance between inlet ductwork gas detectors and a downstream gas damper to allow closure of the gas damper before gas enters the facility. Cold-climate issues with gas detection on inlet ductwork include frosting and icing of the detector, so care needs to be taken on determining the location of the detector. Based on site and environmental conditions, a gas damper may be located upstream of gas detectors. In these applications, evaluations need to be conducted to investigate concentration and quantity of gas entering the facility. Determining gas detector locations, detector spacing evaluations and required gas damper closure time are analysis typically conducted by a fire and gas engineer, process safety engineer or specialist and are outside the normal realm of an HVAC engineer's expertise.

Gas dampers can use either pneumatic or electric actuators, but in either case dampers are typically required to close within a few seconds of actuation. Actuators are required to be rated for the electrical area classification in which they are installed. Pneumatic actuators offer an advantage over electric actuators due to no electrical components and the requirement for a hazardous area rated enclosure over the electric actuator. However, maintenance of pneumatic actuators has become increasingly difficult due to lack of maintenance knowledge and advances in electric actuator enclosures have led to electric actuators starting to become the oil and gas industry norm.

Blast Protection

HVAC inlets and outlets on blast-protected structures in oil and gas applications are typically fit with blast-protection devices, such as blast dampers or blast attenuators. Blast dampers use damper blades held open by a spring that automatically close at a set high pressure, established by an overpressure analysis. Blast attenuators (also called *static blast dampers*) use multiple changes in direction to reduce the overpressure. Blast attenuators are physically longer in length than blast dampers, with the size generally based upon the blast attenuator pressure reduction per unit length and the over pressure analysis. HVAC intake and outlet penetrations on blast-protected structures should be located where the overpressure load is minimized.

Blast protection can be omitted at HVAC intakes and outlets if an overpressure analysis concludes the overpressure load will not cause adverse effects to occupants or equipment. An overpressure analysis is not an HVAC analysis; this analysis is typically conducted by a process safety engineering or specialist.

Where possible, HVAC equipment should be contained within the blast-protected structure. An exception to this is equipment that requires air cooling, such as air-cooled condensing units, or where space prohibits interior location of HVAC equipment. Externally located equipment needs to be anchored and secured to prevent significant displacement during an overpressure event.

Hazardous Area Equipment

In oil and gas applications, equipment upstream of a gas damper is typically considered as being in a hazardous area, requiring hazardous area protection and equipment approvals and listings. Equipment downstream of gas dampers is typically considered unclassified, as the gas damper is part of the protective barrier.

When selecting HVAC equipment, ensure that it is rated for operation in the electrical area classification in which the equipment will be located. Class I, Division 1; Class 1, Division 2; and unclassified ratings all have differing construction and approval criteria. Additionally, local and regional criteria may have requirements for testing and approval of electrical components. An example is criteria outlined in the Code of Federal Regulations (CFR) for Nationally Recognized Testing Laboratory (NRTL) approval; a list of current NRTLs is available on OSHA's website.

Fans located in hazardous areas or drawing air from hazardous areas typically are AMCA 99 Type A or B spark resistant construction (AMCA 2016). Type A construction requires all nonferrous components, while Type B construction only requires nonferrous impeller and ring about the opening through which the shaft passes. The choice of Type A or B construction typically is dictated by end user requirements. The impeller, bearings, and shaft are all required to be adequately restrained to prevent rubbing of components due to a lateral or axial shift in this component. Brass rubbing plates are commonly provided around shafts, on the casing, or on belt guards.

Explosion-proof or intrinsically safe requirements may be required by the end user. These requirements are electrical in nature, while the AMCA 99 spark-resistant construction requirements are mechanical in nature.

Explosion-proof design typically includes an enclosure capable of containing an explosion originating within its housing and prevent sparks from within its housing from igniting vapors, gases, dust, or fibers in the air surrounding it (APA 2012). Therefore, *explosion-proof*, when referring to electrical enclosures, does not mean that it is able to withstand an exterior explosion. Instead, it is the enclosure's ability to prevent an internal spark or explosion from causing a much larger blast.

Intrinsically safe equipment is equipment that should not be capable of releasing sufficient electrical or thermal energy, under normal or abnormal conditions, to cause ignition of a flammable or combustible atmospheric mixture in its most easily ignitable concentration.

It is also important to ensure that all packaged equipment, components, ductwork and devices are electrically grounded to eliminate potential ignition source from static electricity discharge.

Even though commonly used in cold-climate HVAC design, air-to-air heat recovery is not suggested for use in hazardous areas or used with hazardous exhaust due to the potential for cross-contamination with the hazard.

SITE-SPECIFIC CONSIDERATIONS

Cold-climate oil and gas facility projects have many geographical, seasonal, and environmental limitations that require additional consideration depending on their specific design application. The following are a few potential limitations to consider:

Gas Dispersion Effects

The processing of hydrocarbons associated with oil and gas facilities is an inherently hazardous operation; hydrocarbons, by nature, can be both flammable and explosive. This makes the location of potential ignition sources vital to facility location, layout, and design. An ignition source is defined as a process or event, such as arcing-sparking equipment, open flames, or hot surfaces, that can ignite a gas-air mixture and cause a fire or explosion.

When facilities contain hazardous processes with the potential for indoor gas releases, it is imperative that the proper electrical area classification for the space be identified and that all equipment installed within the space meets the minimum electrical area classification requirements to avoid creating an opportunity for ignition of a gas-air mixture. As mentioned previously in this chapter, ventilation and proper air distribution are key components when determining the electrical area classification of a facility. They are also key components when considering gas dispersion both inside and outside of a facility. Providing proper air distribution inside a facility will keep pockets of high gas concentrations from forming in dead-air spaces, which reduces the likelihood of an unnecessary and potentially undetectable safety hazard from being created. Exterior exhaust and relief air hood termination locations need to account for potential nearby ignition sources such as exterior lighting not rated for hazardous locations or nearby unclassified facilities located downstream of the prevailing wind. These exhaust/relief air hood termination locations will also have a hazardous electrical area classification boundary around their termination that needs to be coordinated with the electrical engineer.

Many facilities will be located near potential hazards, but will not contain hazardous processes and will otherwise be considered electrically unclassified. It is the location of these facilities relative to potential hazards that is of importance. In particular, outdoor air intake locations need specific consideration relative to potential gas release locations. A dispersion model of what a potential release would encompass may be desired. The predominant wind direction can play a large role in this dispersion modeling; see Chapter 5 for additional detail regarding predominant wind direction. Outdoor air intake openings should be oriented to avoid being within any hazardous electrical area classification boundary and should be located outside of any modeled gas release clouds.

Seasonal Construction Windows

Many cold-climate oil and gas facility projects will have seasonal construction windows that require careful consideration when determining the engineering execution strategy. An example of this would be a modular facility that is going to be elevated on vertical support members (VSMs, or *piles*), is being transported by ice road, and will rely on a campus-wide heating glycol system for building heat. A few of the seasonal considerations would be:

- **Facility Elevated on VSMs.** It may be advantageous to have these VSMs driven into the permafrost layer during the winter months if the facility location is on wet or exceptionally soft ground. This may be true of other buried utilities, such as electrical, as well. Completing this design effort ahead of facility transport may be beneficial.
- **Modular Facility Transport by Ice Road.** This transport method has a clear seasonal window. In many cases, these ice roads are only available for two to three months of the year. Depending on project schedule, it may make sense to have a separate modular facility design package that is managed to meet this seasonal window constraint to avoid missing the ice road and delaying the project a year. The project may also find that the modular facility will have to sit outdoors and cold soak without any heat for a prolonged period of time before being transported and installed on-site. The design will need to account for this and ensure that building materials and components are suitable for this application. An example of this would be the use of arctic-rated wiring in the facility and the facility components, such as an air-handling unit, to avoid costly rework in the field if the wiring becomes too brittle and fails.
- **The Heating Glycol Tie-In Scope.** It would clearly not be ideal to bring down the campus-wide heating system during the winter months, which limits the seasonal construction window to the short summer months. The project may consider having a separate tie-in package completed for summer install ahead of the facility shipment for final tie-in on-site upon facility arrival.

Emissions

It is important to define site-specific emission regulations and any client-specific emission requirements early in the project development to avoid improperly selecting fuel-burning equipment. One example of this could be a small, diesel-powered generation project design requested by a client. Some applications may require you to reduce emissions by providing a Tier 4 Final generator set while others may not. Tier 4 Final (4F) regulations were enacted by the United States Environmental Protection Agency (EPA) to reduce particulate matter and nitrogen oxide emissions from specific diesel-powered engines. These types of requirements will depend on location, air permit allowances, other adopted regulations in the area, and anticipated generator usage.

Wildlife Considerations

Many cold-climate oil and gas applications are in very remote locations and may be in environmentally sensitive areas. It is important to understand site-specific sensitivities for project scheduling, design considerations, and when visiting the areas. An example of this would be a project that is being designed in an area with the potential for high polar bear traffic. The designer will want to consider enclosing exits with cages that will allow occupants to safely scan the area prior to exiting the facility premises. This can affect facility design and layout depending on available space to install the facility.

Muster/Temporary Refuge/Shelter-in-Place

In cold-climate onshore oil and gas facilities there are seasonal windows where people may not be able to completely exit the accommodations during process safety or fire events. Muster points for temporary refuge (or shelter-in-place locations) are identified and the HVAC system designs for these areas need to consider the potential occupant density during one of the events to ensure sufficient IAQ to support the occupants. If the muster point is located within a potential blast or jet fire radius then additional details associated with any HVAC or utility system penetrations need to be considered, such as blast dampers or hydrocarbon rated (H-rated) fire dampers. When the muster point is considered a shelter-in-place location and there is a potential for the normal outdoor air source to be compromised during a process safety event, the facility may need to consider adding air scrubbers or oxygen for occupants to survive the anticipated event duration.

FACILITY CONSTRUCTION AND SIZE CONSTRAINTS

Many cold-climate oil and gas facility projects are installed in remote areas with limited available footprint. The facility footprint may be restricted by many factors such as available gravel pad space, facility siting, and available methods of transport. With remoteness comes the lack of readily available construction resources such as materials, labor, and nearby infrastructure to support the effort. On-site (or *stick-built*) construction becomes cost-prohibitive in many of these applications and transportable (or *modular*) facilities become necessary.

Modular facility size constraints are generally determined by the required mode of transportation. Modular facilities are typically hauled to the project site by truck or barged to the project as a sealift module. Trucked modules are limited in size by the road systems that they must travel. For instance, there may be overpasses, tunnels, or narrow bridges that can limit the height, width, and weight of these facilities.

It is important to understand facility size restrictions at the beginning of the project to ensure that enough space is allotted for each discipline. In some cases, multiple modules may be required to achieve the project-specific goal. See Figure 13.4, which shows multiple trucked modules being stitched together in the field.

The mechanical ventilation system is frequently one of the largest systems in oil and gas facilities and, many times, one of the last considerations. It is important to allocate space for the mechanical systems early in the project to avoid costly redesign efforts. Similarly, the construction execution plan should consider the size of the mechanical systems and determine if any of them need to be brought inside the module footprint prior to completing the module framing effort. See Figure 13.5 for an example HVAC plan that helps illustrate how limited space can be in these facilities. Note that the provided plan has all process piping and electrical infrastructure hidden in the background.

Figure 13.2 Blast-resistant control room module trucked arrival to arctic site. Seallift modules are generally limited by barge capacity and can be much larger than trucked modules.

Figure 13.3 Sealift-style manifold module in the arctic.

Figure 13.4 Blast-resistant control room construction.

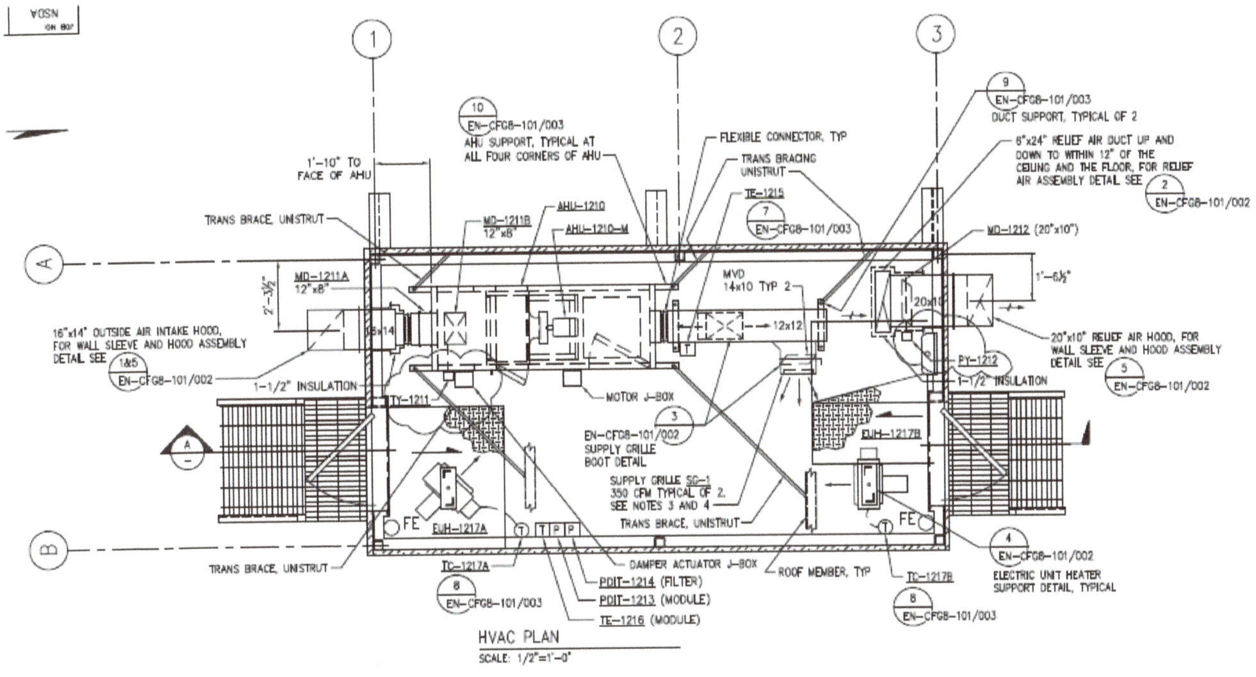

Figure 13.5 Example process module HVAC plan.

One additional design consideration that needs specific attention when considering modular construction is the transportation bracing. The forces associated with the mode of transportation (e.g., trucking, barging, or both) need to be identified and accounted for with proper bracing to ensure that the facility does not arrive on-site in pieces. In certain circumstances, it may make sense to reserve some installation tasks for on-site completion, such as acoustic ceiling grid installations.

This section covered oil-and-gas-specific strategic design considerations. See Chapter 5 for additional cold-climate strategic design considerations.

HEATING SYSTEM SELECTION

Proper heating system selection is critical to any cold-climate facility. While oil and gas installations may have similar utility cost concerns to any other industry, they may also have some advantages depending on the application. Understanding the available energy infrastructure is a key component of the heating system selection for oil and gas facilities. The following are heating system selection considerations for certain applications.

Central Facility Heating System Selection

For this discussion, the central facility is to be considered the primary oil and gas processing facility that collects and processes the pipeline fluid from producing wells.

Process Heating Glycol

Many cold-climate oil and gas facilities are remote and are required to install their own power generation infrastructure at their central facilities. In some cases, it may make sense to install heat recovery units on the generation equipment exhaust stacks to capture waste heat into a heating glycol distribution system to accommodate both oil and gas process heating demands and facility heating demands. When designing one of these systems, it is important to understand the required design and operating temperatures and pressures prior to selecting any piping materials and terminal heating equipment. It is likely that the oil and gas process heating demands will require these systems to operate at higher temperatures and pressures than more conventional hydronic heating systems. Process heating systems are typically designed in accordance with ASME B31.3 *Process Piping, ASME Code for Pressure Piping* requirements, thus these requirements may extend to the hydronic heating system (ASME 2021). Confirmation with the client on the required piping line classes and locations of line class breaks needs to be coordinated prior to starting detailed design. Additional scrutiny on the selection of hydronic heating equipment (e.g., unit heaters, heating coils, heat exchangers, and so on) should be provided to ensure that the selected equipment is fit for service and will not create a weak point in the system that could cause a process safety event. It is likely that standard hydronic heating equipment construction will not be suitable for direct connection to these systems and that special construction practices for the equipment will be required. An example of this is the selection of a glycol heating coil to be installed on a 60/40 triethylene glycol/water mixture system operating at 300°F (149°C) and 300 psig (2068 kPa). This heating coil may need to be fabricated with a Schedule 40 carbon steel header with Class 300 raised-face flange connections to meet the temperature and pressure needs of the system and the necessary hydrostatic testing requirements. Different heating equipment manufacturers may have different special construction options available to meet these demands.

Electric

Another typical heating medium selection for central facility design is to install electric heat and size the power generation system as needed to support this. If electric heating equipment is to be installed in a hazardous location, it is necessary to ensure that the selected equipment has been tested for use in that application and that it is listed and labeled for the service and location where it will be installed. In the United States, hazardous location electric heaters are tested by a federally approved Nationally Recognized Testing Laboratory (NRTL) such as UL, FM Approvals, or CSA Group to ensure that the equipment meets the minimum requirements outlined in UL 823, *Standard for Electric Heaters for Use in Hazardous (Classified) Locations* (UL 2006b). The equipment will then be listed by the NRTL and provided with a label outlining the applications it is suitable for. ATEX and the International Electrotechnical Commission (IEC) provide standards that are recognized on an international level for hazardous location electric equipment listings but may not be federally approved as an NRTL.

An example selection would be as follows: an electric unit heater is required to be installed in an oil and gas facility and will be located within a hazardous area that will see ignitable concentrations of hydrocarbon gas during normal operating conditions. Because the location will have flammable gases or vapors present in quantities sufficient to produce explosive or ignitable mixtures during normal operation, we can assume that the hazardous electrical area classification will be Class I, Division 1. Furthermore, we know that we have a flammable hydrocarbon gas that falls into a Group D category per NFPA 70. Finally, we need to identify the temperature code rating required for the hazardous gas that will keep the electric unit heater from reaching the auto-ignition temperature of a flammable gas-air mixture. For this example, we'll assume a T3 (200°C) temperature code will be sufficient. Thus, the electric unit heater selection will need to be Listed by a NRTL confirming compliance with UL 823 and labeled to meet Class I Division 1, Group D T3 requirements (UL 2006b).

Note: Some manufacturers build explosion-proof electric heaters built to withstand an explosion of a specified gas or vapor that may occur within the heater to prevent ignition of external gas or vapor by enclosing ignition sources and maintaining an external temperature below the auto-ignition temperature of the specified gas or vapor. These heaters are typically suitable for multiple hazardous location classes and groups.

Fuel Gas and Fuel Oil

If the central facility has available conditioned fuel gas, then gas-fired boilers and heaters may be considered for heating system selection like any other cold-climate facility design. The cost of fuel oil generally eliminates its use as a permanent heating source for central facilities that are planned to operate for many years. Keep in mind that fired equipment may not be installed within any hazardous electrical area classifications and may only be installed near these areas with proper engineering design consideration. The risk of having these potential ignition sources purposely located within the facility need to be considered during the final heating system selection.

Remote Drill Site Facility Heating Selection

In many cases, cold-climate oil and gas facilities will include remote drill sites that may be located miles away from the central facilities. The available infrastructure on the drill site will dictate the energy sources available for providing facility heating. The following are the most common energy sources available.

Electric

When sufficient electrical infrastructure is already in place or new power generation equipment is included in a project scope of work, using electric heating is typically the simplest and lowest upfront capital cost approach. The electric heating equipment selection needs to consider the hazards of the area served as discussed in the Central Facility Heating System Selection section earlier in this chapter.

Process Heating Glycol

The drill site may require a production heater, or line heater, to heat the well-stream of the pipeline fluid before transmitting down a pipeline. These line heaters are typically indirect-fired water or glycol bath heaters. With proper selection, the line heater glycol bath can be used to circulate heating glycol to the drill site facility heating equipment. Using this heating glycol energy source reduces the overall electrical load on the drill site and saves valuable electrical infrastructure for future production-related drilling activities.

Microturbine Heat Recovery

With the advancement of microturbine technology, some clients may choose to have their power generation or gas compression systems installed on-site. Many of these systems can be provided with a heat recovery system that can distribute heating glycol to the drill site facilities and significantly increase the energy efficiency of the overall system.

Fuel Gas

Fuel gas, primarily methane (natural gas), may be available on the drill site and may or may not be conditioned depending on its intended usage. If conditioned properly, the fuel gas can be used to supply gas-fired heaters or boilers like any other facility. In many of these remote locations, the fuel gas on the drill site is pulled directly from the ground, separated from the oil/water, and compressed to be reinjected as lift gas into the wellheads at high pressures. Before proceeding with one of these systems as a facility heating medium, there are quite a few challenges to overcome, starting with understanding the

Figure 13.6 Drill site production heater.

gas characteristics (heating value, specific gravity, temperature, impurities), potentially adding a gas letdown station, determining if freeze protection is necessary to deter hydrate formation problems, determining if gas odorization will need to be added to meet ASME B31.8 Gas Transmission and Distribution Piping Systems requirements, and placing any gas-fired equipment (i.e., open flames) in locations a safe distance away from any potential process gas release scenarios (ASME 2018).

Propane

At standard pressures and ambient temperatures below –44°F (42°C) propane will not vaporize into its gaseous state and will remain in liquid form. As a result, if propane is used in extreme cold climates, heat trace and/or heating pads may need to be used to allow for proper system function.

Fuel Oil

Fuel oil is a common heating medium in cold climates, but is generally not the most cost-effective approach when considering conventional remote drill sites that will be in operation for 20 or more years. If the drill site has a short lifespan and is accessible by road, then fuel oil as the heating medium may become more palatable. Fuel oil and diesel are extremely important in providing temporary heating for construction and for unplanned heating outages in extreme cold climates. Indirect-fired industrial grade heaters, both portable and skid-mounted, used for temporary heating are a common sight at remote cold-climate oil and gas facilities.

Photovoltaic and Wind Turbines

Photovoltaic (solar) and wind turbine power are not prominent power sources in the remote, extreme cold climate oil and gas facility designs—these areas lack sunlight during the heating season and the functionality of wind turbines is negatively affected by cold weather. Additional environmental concerns, such as migratory bird flight paths when potentially siting wind turbines, should be evaluated when considering these options.

Figure 13.7 Portable temporary heater.

Wellhead Shelter Heating Selection

Another unique aspect in cold-weather oil and gas design are shelters installed over wellheads and critical valves at drill sites. Wellheads are installed outdoors in most environments, but in this can be problematic in cold climates due to snow drifting, high winds, and extreme cold temperatures. These wellhead and valve shelters are typically heated enclosures that protect any instruments, valves, and fittings from the elements.

The shelters are typically provided with passive ventilation openings, one high and one low, that can be used during the warmer summer months for cooling the space. They are also typically heated to protect instruments and deter snow from building up in the enclosures. These shelters may be crane-lifted, removed, and set back in place during drilling and well workover activities such that their seams may not remain permanently sealed depending on the shelter construction. A higher-than-normal infiltration rate may need to be considered when sizing the heating system for these applications.

Two different heating approaches are generally considered for these shelters: electric unit heaters and wall-mount catalytic heaters when the drill site has lift or fuel gas available. Either approach requires special attention to the hazardous electrical area classification within the shelter. Typically, explosion-proof heating equipment is required.

The following are a few design details to consider if pursuing the use of catalytic heaters in these shelters:

- **Infrastructure.** Ensure that an appropriate gas source is available to supply the heater and that proper safety systems are in place in the fuel gas system. Keep in mind that the available gas source may be unconditioned and could contain

Figure 13.8 Wellhead shelter.

moisture, leading to freeze-up of small-diameter pipes if the system design does not take this into account. Coordinate any electrical connections with the electrical engineer.

- **Placement.** Catalytic heaters emit infrared energy and will not distribute heat throughout the space. The design needs to have the heaters pointed at the devices or locations that require heating. Proper spacing from piping and equipment must be accounted for because these heaters can emit significant heat with the potential to singe or ignite nearby combustible materials. Many of these heaters can be provided with an optional fan to help promote more uniform heat distribution throughout the space.
- **Venting.** The products of combustion from the catalytic process are water and carbon dioxide. Many times, these are vented directly into the space that they serve, trapping moisture within the space and potentially leading to early corrosion within the shelter. Vent hood assemblies are available that are sufficient for certain applications. Keep in mind that the typically provided vent hood snow cap may not be sufficient for extremely cold areas with high winds and that a simple downturn elbow may be more effective. See Figure 13.11 for an example of the damage that can occur with the manufacturer-provided snow cap. Consider snow drifting when placing the final vent termination.

VENTILATION SYSTEM SELECTION

The ventilation systems for cold-climate oil and gas facilities should be selected to be reliable, simple to operate, and easy to maintain. Most of these kinds of facilities are remote operations and may have minimal access to trained HVAC technicians and/or spare parts. Larger facility projects should consider standardization of equipment manu-

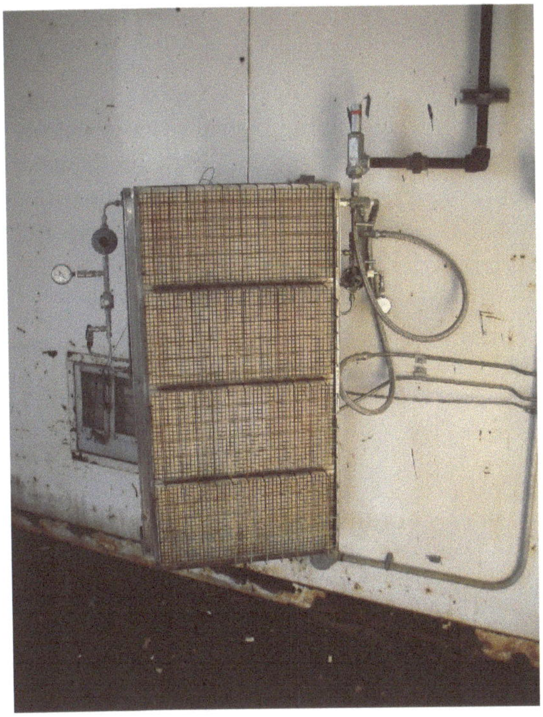

Figure 13.9 Catalytic heater (no vent hood).

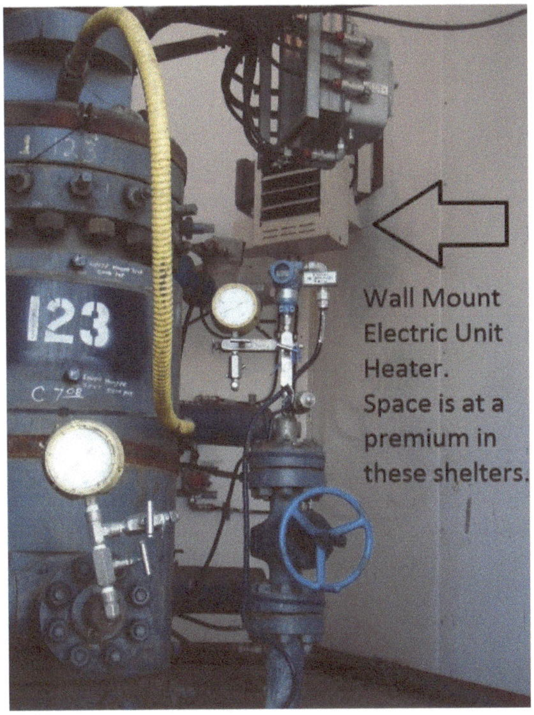

Figure 13.10 Explosion-proof electric unit heater.

Figure 13.11 Damaged catalytic heater vent hood.

facturers and types. It may be desirable to limit equipment and instrumentation manufacturers to a single type and potentially even a suitable model number to limit personnel training and spare part storage requirements. The ventilation system is an integral part of the process safety management of processing facilities and needs to be operable 24/7. Mission-critical facilities may require complete redundancy of the ventilation systems to ensure seamless operation.

The ventilation system selection for cold-climate oil and gas facilities depends on many factors. As discussed previously in this chapter, the ventilation system sizing basis needs to consider the following:

- Minimum code and standard requirements as adopted by the AHJ (similar to any other cold-climate facility)
- Hazards present inside the facility
- Hazards present adjacent to the facility
- Available facility footprint
- Available heating system infrastructure
- Internal heat generation within the space
- Fire and gas interactions (gas detection, fire suppression, gas inerting)
- Safety or operational criticality of the facility (redundancy requirements)

Hazardous Area Ventilation

Hazardous areas are locations where a flammable vapor, gas, fume, mist, or dust is present or may be present in concentrations exceeding 25% of the LFL of the substance at the expected room temperature. Many areas in cold-climate oil and gas hydrocarbon processing facilities inherently fall into this category. Similar to any other facility, these ventilation systems need to provide sufficient IAQ, cooling capacity, and heating distribution throughout the space while also taking into account process safety considerations, pressurization requirements, and potential electrical area classification reductions.

The minimum outdoor air ventilation rate for these hazardous areas is generally defined in codes and standards as either 1 cfm per ft^2 of floor area or 6 ach, whichever is greater. Codes and standards also discuss sizing the ventilation system such that flammable contaminants are diluted in uncontaminated air to maintain concentrations below 25%nt of the contaminants' LFL. This allows for a performance-based approach toward determining an appropriate minimum dilution outdoor air ventilation rate required to properly dilute potential contaminants to safe levels. Client specifications often outline predetermined minimum outdoor air ventilation rate requirements based on their typical process operating conditions and risk analysis. If no guidance is provided by the client, a fugitive emissions study should be completed by considering the specific leak rate of a contaminant during different common process release scenarios and determining the effectiveness of the ventilation system to dilute the atmosphere within the space to a safe level during these different process failure scenarios. In general, ventilation can be an effective safeguard against the smaller and more common gas release scenarios such as flange or valve gasket leaks. However, ventilation is not typically as effective at reducing a facility's risk profile for larger and less common gas release scenarios such as pipe failures or pump seal failures. Additional layers of safety beyond the ventilation system may need to be considered, such as a gas inerting system, depending on the client's asset risk analysis of the facility.

As discussed previously in the chapter, ventilation may also be used to assist in reducing the hazardous electrical area classification of the space from Class I, Division 1 to Class I, Division 2 when the space is considered adequately ventilated. Leveraging this allowance can reduce the cost associated with procuring and installing Class I, Division 1 rated gear and devices.

Constant-volume ventilation systems are typically used for oil and gas processing facilities because of their simplicity and reliability. Two methods are usually considered depending on whether or not a gas detection system is provided:

- When a gas detection system is not provided in the facility design, a ventilation system will need to be employed that provides the minimum code and standard ventilation rate requirements and considers the process safety dilution rate of ventilation along with any hazardous electrical area classification reductions. This approach will have one fixed outdoor air ventilation rate that it provides continuously. The system will be required to temper the complete outdoor air ventilation and is the least energy-efficient approach.
- When a gas detection system is provided in the facility design, a constant-supply air volume ventilation system with multiple outdoor air ventilation rates may be considered. As allowed by API RP 500 and API RP 505, recirculation of air is often used in a mixed-air configuration to reduce outdoor air heating loads in cold-climate oil and gas design (API 2012, 2018). This type of system consists of an air-handling unit with a mixing box with modulating outdoor and return air dampers to change the outdoor air ventilation rate for differing operational sce-

narios. This often leads to two outdoor air intake damper positions: a normal (minimum) ventilation rate position and an emergency (dilution or high-rate) ventilation rate position. The intent is that on positive gas detection the outdoor air damper will modulate open to the emergency ventilation rate position and dilute the hazardous atmosphere. The emergency ventilation rate is usually untempered in these scenarios and the heating system sizing only needs to accommodate the normal outdoor air ventilation rate position, making this approach more energy efficient. Keep in mind that if any cold-sensitive utility systems are employed in the facility, the emergency ventilation rate position may still be required to be tempered. An example of this would be a gaseous fire suppression system intended for both gas inerting and fire suppression that requires a minimum indoor temperature of 70°F (21°C) at discharge to operate effectively.

There are multiple system selection options available to meet applicable code requirements and design standards for hazardous area ventilation applications. The following are a few common approaches including their advantages, disadvantages, and insights to their application.

- Provide a constant-volume central air-handling unit sized for the emergency (dilution) ventilation rate, including a return air mixing box section, filter section, potentially a heating coil section, and a fan section. Provide either relief air dampers or exhaust fans.
  Advantages
  - Ability to provide higher outdoor air rates for both summer cooling and gas release scenarios while maintaining a constant fixed-air volume distributed throughout the space.
  - Allows a lower normal ventilation rate if gas detection is provided. Less facility heating required, assuming no utilities require tempered air during an emergency ventilation scenario.
  - Reduced load on a heating coil, if required, due to return air mixing. If the air-handling unit is sized appropriately to maintain a suitable supply air temperature on a winter design day, then a heating coil may not be required in the system and unit heaters in the space can pick up the heating load. This can be helpful in hazardous-area applications where an explosion-proof electric heating coil can significantly increase the size of the air-handling unit to a point that it may be difficult to fit into the available space without significant project effects.
  - Provides conditioned air to the space through a single filter section downstream of the mixing box.
  - Allows for suitable air distribution when distribution ductwork and diffusers are provided.
  - Positively pressurizes the space to reduce infiltration and ingress of snow, dirt, bugs, dust, and debris.

  Disadvantages
  - Redundancy, when required, will require a greater footprint than other options. Redundant fans in a single cabinet with a single coil and filter bank could be an option to reduce this disadvantage.
- Redundant 100% outdoor air makeup air units with relief air dampers or exhaust fans.
  Advantages
  - Redundancy with opportunity to increase ventilation by starting both units for both summer cooling and gas release scenarios.

    - Provides conditioned air to the space using summer and winter filter sections (one upstream of the heating coil in summer, one downstream of the heating coil in winter).
    - Allows for suitable air distribution when distribution ductwork and diffusers are provided.
    - Positively pressurizes the space to reduce infiltration and ingress of snow, dirt, bugs, dust, and debris.

    Disadvantages

    - Requires heating coils in the units without return air to help reduce the load on the coils. When using electric heat in extreme cold climates this can be problematic because it is difficult to find explosion-proof electric heating coils able to handle a 120°F (49°C) temperature rise.
    - Makeup air units are large because they require summer/winter filter sections and heating coils. Keep in mind that electric heating coils will have minimum upstream and downstream requirements that can significantly increase the size of the unit.
    - The increased cost and footprint of the system as compared to the other design approaches.

- Exhaust fans with tempered, unfiltered makeup air intake dampers.

    Advantages

    - Simple design with minimal footprint.
    - Easy to provide redundancy.
    - Lowest cost option.

    Disadvantages

    - Requires heating coils (glycol or electric) at the intake dampers to avoid freezing the facility. Without filtration, these coils tend to require a significant amount of maintenance due to plugging. If filtration is included upstream of the coils, it is important to remove the filters in the winter months when frost tends to plug them to avoid situations where the exhaust fans pull a significant negative pressure in the space, making it difficult to open/close doors.
    - Maintains a negative pressure in the space, pulling in dirt, dust, bugs, and debris while continually assisting cold-air infiltration into the facility. This can lead to cold spots within the facility that may include snow/ice build-up in areas with poor air distribution and a higher likelihood for pipe or equipment freezing.
    - Generally does not provide as effective air distribution to deter pockets of gas from accumulating during upset conditions.

The Constant-volume central air-handling units and the redundant 100% outdoor air makeup air units system selection approaches both use either relief air dampers or exhaust fans. In single-story facilities, the use of relief air dampers is generally acceptable. Many cold-climate oil and gas facilities have larger open areas that are two to three stories high and have the potential to contain both heavier- and lighter-than air-fumes or vapors. Thus, the exhaust or relief air must be pulled from both high and low points within the facility. In these instances, the use of exhaust fans is recommended because of the stack effect that occurs when the outdoor air becomes colder and denser. When exhaust fans are not used in these instances, the lower relief air damper openings tend to become intakes during the cold winter months and may not properly extract heavier than air fumes or vapors.

Whichever system is selected, it needs to be designed to provide proper air distribution within the facility to eliminate dead-air spaces, provide adequate mixing of module atmosphere, minimize stratification, and provide area temperature control. This includes the use of distribution ductwork with manual balancing dampers and diffusers and may include the use of ceiling destratification fans. If any ductwork is routed through a hazardous area to serve an adjacent nonhazardous or unclassified area, then special attention to duct sealing requirements is necessary; welded ductwork may be needed.

The ventilation systems should also consider potential asphyxiants. Nitrogen is a common agent at cold-climate oil and gas facilities and is usually generated on-site and distributed throughout the facility or facilities through a piping network. Depending on the quantities of an asphyxiant, and the overall room volume of the space that the nitrogen or other asphyxiant serves, O_2 sensors should be considered to alert individuals of a potentially oxygen-deprived enclosed space and to potentially increase the outdoor air ventilation rate as required to increase O_2 levels. the U.S. Occupational Safety and Health Administration (OSHA) sets limits on allowable oxygen levels of between 19.5% and 23.5% within the workplace (OSHA 2011). When completing the dilution calculations to determine anticipated oxygen levels, remember that air is roughly 78% nitrogen to begin with.

All equipment selected for use in hazardous areas will need to be listed and labeled for the appropriate service. For example, an air-handling unit may be required to be suitable for Class I, Division 2, Group D areas with a T3B temperature code. This unit will still be listed and labeled under UL 1995 (UL 2015), but will be required to be custom-fabricated to meet the hazardous electrical area classification requirements that it will be installed in. Confirm with the intended manufacturer that they are capable of this type of construction prior to finalizing the equipment manufacturer selection.

Facility Pressurization

Cold-climate oil and gas facilities tend to have a range of hazardous electrical area classifications housed within them. The application of facility pressurization to maintain higher-hazard areas at a negative pressure relative to adjacent lower-hazard areas is an important aspect of any ventilation system design. Facility purging and pressurization in conformance with NFPA 496 is required when less hazardous areas are adjacent to (i.e., shares at least one wall with) more hazardous areas. NFPA 496 provides the minimum requirements for properly purging and pressurizing less-hazardous areas relative to more-hazardous areas. It outlines three different pressurization approaches that are important to understand when selecting the ventilation systems (NFPA 2021a). The three approaches are summarized as follows:

1. Type X Pressurization: Enables the use of equipment suitable for unclassified locations within the protected enclosure that would otherwise be required to be suitable for Division 1 locations. (e.g., unclassified to Class I, Division 1)
2. Type Y Pressurization: Enables the use of equipment suitable for Division 2 locations within the protected enclosure that would otherwise be required to be suitable for Division 1 locations. (e.g., Class I, Division 2 to Class I, Division 1)
3. Type Z Pressurization: Enables the use of equipment suitable for unclassified locations within the protected enclosure that would otherwise be required to be suitable for Division 2 locations. (e.g., unclassified to Class I, Division 2)

Each of these approaches has specific requirements outlined in NFPA 496; for example, there is a requirement for Type X pressurization to de-energize all circuits and equipment within the protected enclosure that are not identified for Division 1 upon failure of the protected gas supply to maintain positive pressure. There is an exception that allows power to be continued for a short duration if immediate loss of power would result in a more hazardous condition and if both audible and visual alarms are provided at a constantly attended location. Relying on Type X pressurization between an unclassified space that communicates directly into a Class I, Division 1 area can be a big coordination problem; the process safety impacts of de-energizing all devices within the unclassified space that are not rated for a Class I, Division 1 environment may outweigh the benefits of the simplified pressurization system.

Another approach to consider for an unclassified space that communicates directly into a Class I, Division 1 area could be to use Type Y and Type Z pressurization. This approach would include a Division 2 airlock between the two areas that allows Type Y pressurization between the airlock and the Division 1 area, and Type Z pressurization between the control room and the airlock. This will allow the systems to be installed without the Type X pressurization requirement to de-energize all circuits within the unclassified space upon failure of the positive pressure air system.

The main requirements for purging and pressurizing to protect adjacent unclassified spaces are mentioned previously in this chapter. Two common methods for pressurization control are to use pressurization fans with variable-speed drives or to use outdoor/return air or relief air damper control. These methods provide active pressurization control that accounts for wind fluctuations (infiltration/exfiltration), particulate filter loading, and opening/closing of facility doors. Figure 13.12 depicts a few common approaches that can be considered when beginning to design one of these systems.

If the positive pressure ventilation system is required to be installed outside of the general-purpose area, ensure that all the equipment and devices selected are suitable for the area classification that they are going to be installed in. The incoming air streams of these systems are typically monitored by a gas detector that will shut the air system down upon positive gas detection. This detection could occur in the intake ductwork or as far down stream as the first supply air register depending on the client's gas detection philosophy.

Fire and Gas Interactions

The ventilation system actions must be coordinated with the project- or client-specific fire and gas philosophy. This philosophy should outline what actions are taken with respect to detection, alarming, and protection, and will usually be captured in a cause and effect matrix. An example of this would be a gas detection philosophy that increases the normal outdoor air ventilation rate to the emergency outdoor air ventilation rate on 10% LEL gas detection, but then shuts down the ventilation system on 50% LEL gas detection or on fire detection while a gaseous fire suppression system activates. The controls associated with these actions are all integral to the ventilation system selection.

COOLING SYSTEM SELECTION

Outdoor air economizer cooling is commonly used in both hazardous (classified) area and general-purpose (unclassified) area cold-climate oil and gas facilities. When an emergency (dilution) outdoor air ventilation rate is used for hazardous-area systems, it can also be used as a cooling ventilation rate during the warmer summer months.

In general-purpose areas with high heat rejection equipment, such as large variable-speed drives that support downhole electrical submersible pumps, it may be possible to cool the space to a practical indoor design temperature with outdoor air alone, but it could also require the use of a DX refrigeration system for summer month operation only to achieve the appropriate indoor design conditions. Using outdoor air for "free" cooling is typically preferred where practical. The use of DX systems is commonly seen in control rooms, server rooms, and computer rooms that require more precise temperature control. If a condensing unit is going to be installed outdoors, it should be provided with a low

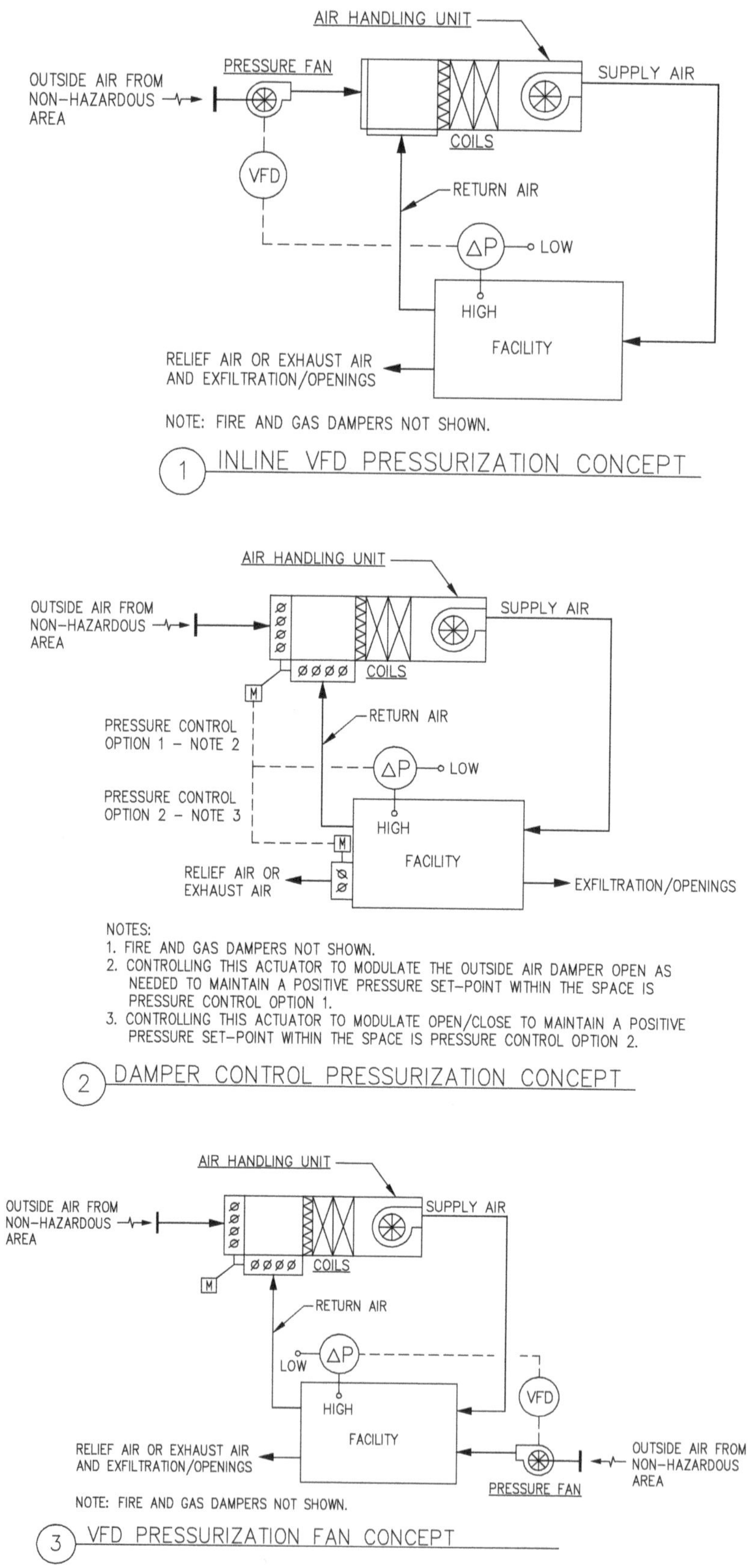

Figure 13.12 Pressurization control concepts.

ambient kit to allow for operation at lower temperatures. If the condensing unit is required for operation at temperatures below the allowable operating temperatures of the low ambient kit, then an indoor location should be considered. The condensing unit location needs additional consideration to either ensure that it is located outside of a hazardous electrical area classification boundary or that it is identified as being suitable for hazardous location operation (i.e., explosion-proof or intrinsically safe). Explosion-proof condensing units are not generally off-the-shelf items and a custom intrinsically safe unit may be needed. When specifying an intrinsically safe condensing unit make sure that all of the requirements in NFPA 70 are followed for intrinsically safe installations (NFPA 2020).

CONTROL SYSTEMS

It is important to identify the HVAC control system philosophy at the beginning of the project; the client may or may not already have a control philosophy identified for the facility. As with any facility, there are many control system options available depending on the scope, ranging from stand-alone control to integration into one of the process control systems. It is common to integrate the HVAC control either into a basic process control system, a fire and gas control system, or a separate building management system. Typicaloperation modes include:

- **Normal Operation**
  - Minimum outdoor air ventilation rate with heating/cooling system maintaining space temperatures.
- **Emergency Operation**
  - Gas detection: Increase to emergency (dilution) outdoor air ventilation rate as applicable. If a gaseous fire suppression or gas inerting system is present, shut down ventilation on release of agent. Space heating typically remains in operation if separate from the ventilation system.
  - Fire detection: Shutdown ventilation system as required.
  - Equipment failure: Alarms and start redundant equipment as applicable.

Although the fire and gas systems tend to have inputs into oil and gas HVAC control systems, it is important to differentiate that fire and gas detection must be provided by appropriately listed systems identified for fire and gas service. Fire and gas required shutdowns of the HVAC system must have a direct communication connection to the equipment and cannot be relayed through a secondary control system.

HOW COLD IS IT?

It is so cold ... we had to chop up the piano for firewood, but we only got one chord!

14 Commissioning

Commissioning of a building or facility is considered a best practice no matter the location or age of the facility. The commissioning process is a quality-focused process for enhancing the delivery of a project. The process focuses on verifying and documenting that the facility and all of its systems and assemblies are planned, designed, installed, tested, operated, and maintained to meet the owner's project requirements and the current facility requirements. ASHRAE has produced a family of guidance and standards documents to better define the concepts of commissioning (ASHRAE 2007, 2013d, 2019d).

Use of an effective commissioning process, as opposed to the all-too-common method of "commissioning by complaints" will increase the usability and efficiency of any building.

This chapter focuses on the application of commissioning principles in buildings located in cold climates and discusses the elements of the commissioning process, which manages project quality by observing and reviewing the inspection and testing processes, and technical commissioning, which is based primarily on hands-on inspections and functional measurement and verification of building systems performance.

GENERAL

Commissioning is accomplished by a team led by the commissioning authority, who is "an entity identified by the owner who leads, plans, schedules, and coordinates the commissioning team to implement the commissioning process" (ASHRAE 2019d). This can be anyone with a sufficient understanding of buildings and building science to direct the activities of the team or a skilled manager who directs the activities of the subject matter experts to optimize the commissioning process. This is true of any building, facility, or assembly independent of location or climate.

The commissioning process will differ between new construction and existing building commissioning. One of the fundamental differences is that the occupants become a critical part of the resources to be used by the existing building commissioning team. Similarly, an occupied building will have a well-defined set of operating needs or current facility requirements that will guide the decisions of the existing building commissioning team. The current facility requirements can include goals, measurable performance criteria, cost considerations, benchmarks, success criteria, and supporting information to meet the requirements of occupants, users, and owners of the facility.

In cold climates, the commissioning team will investigate all aspects of the facility that the owner directs them to examine to ensure the long-term success of the building as a whole. It should be a multidisciplinary team that considers the interaction of the various systems and assemblies, not just the narrow scope definitions of construction contracts.

Cold-climate commissioning could include ensuring the building envelope is practical for the application, heating systems are appropriate for the location, the operating staff are well trained, maintenance contractors are available, the maintainability of the building systems and assemblies, waste by-product removal is available, and the suitability of energy sources for the area.

There are two key elements of physical site investigation: condition evaluation and performance evaluation. Condition evaluation includes an assessment of the installed condition of the equipment, systems, and assemblies, using criteria such as appearance, physical damage, deterioration of condition, upgraded features, and change of use. Performance evaluation includes an assessment of how the equipment, systems, and assemblies are operated and how they are performing relative to expected performance and the current facility requirements. Performance evaluation may include active and passive testing. Execute the tests and repeat as necessary to obtain conclusive information about the performance of systems.

Some commissioning work may need to be deferred for seasonal reasons. The commissioning team will need to schedule tests to be able to observe the operation of the systems and assemblies under some load. For example, test heating systems in the winter and cooling systems in the summer.

BUILDING ENVELOPE

The most critical aspect of the building envelope revolves around air and moisture movement through the components. The details are covered in Chapter 8. In new construction or renovations, ensuring that the insulation is installed in all required cavities and that there are no direct conductive thermal bridges is critically important.

Commissioning of the building envelope begins with ensuring that the integrity of the insulation and vapor barrier is maintained in all instances where there is an interface with the structural, mechanical, and electrical penetrations, and other architectural openings. For new buildings, this should begin at the drawing stage to make sure what is proposed is appropriate. Of primary importance is the requirement to seal around all penetrations in the vapor barrier. Penetrations of the envelope should be reviewed to ensure that any thermally conductive pipes or ducts are adequately insulated to reduce the conductive thermal losses.

The methodology and construction logistics must be considered to ensure that all penetrations are identified, ensuring a constructible solution is determined prior to the installation of the finish surfaces. It is very expensive to go back after construction to fix errors and omissions.

Once a building envelope is complete, it is highly recommended that the leakage of the entire assembly can be tested using a blower door test in accordance with a standard outlining appropriate protocols for whole-building air testing standards, such as ASTM E779 (ASTM 2019). Typically, the building is pressurized to predefined pressure with all exterior penetrations closed. The amount of leakage is measured and normalized in accordance with ASHRAE standards and LEED® rating system requirements. Whole-building air testing is a mechanism for quality control of installation, and the only mechanism to measure actual installed performance of mechanical dampers, closures, and overall envelope airtightness. Whole-building air testing should be done by both pressurizing and depressurizing the building measuring infiltration and exfiltration rates. The testing should be done at a point in construction when a large number of mitigation strategies are still available. In new construction, this generally means after windows, ducting, and air barriers are installed, but prior to cladding and interior finishes are in place. Alternatively, component testing of areas of concern can be done (e.g., at windows to determine effectiveness of window-wall junction sealing methods).

The consequences of a leaky vapor barrier in cold climates includes the accumulation of moisture in insulation due to the movement of moist indoor air through the vapor barrier into a zone where the material is below the dew point. That moisture will freeze inside the insulation in wall systems. This can also include freeze-thaw damage in susceptible materials such as masonry and some wood products and can also cause mold and mildew problems in the building envelope. In attic spaces, the moisture can freeze on the roof structure or the insulation. When the attic warms up, this can result in major water problems for the building. Venting of attic spaces is therefore critical and there must be good coordination between the installation of the attic insulation, the roof structure, and the flow of vent air into and out of the attic space.

For envelope assemblies using a combined air/vapor barrier system, it is the air barrier commissioning that would be applied.

STRUCTURAL

The primary function of the foundation is to support the building superstructure. Challenges in building design include the elimination or mitigation of thermal bridges between the foundation and the superstructure. Care must be taken to ensure that the thermal envelope is maintained and that the vapor barrier is continuous around the various structural components. This can be especially challenging when working around the floor joists on multistory buildings.

Most building foundations are required by code to be a defined distance below the design frost penetration depth. Foundation walls in cold climates can include basement spaces that are normally heated. The degree of insulation on the basement (foundation) wall will have an impact on the soil temperature adjacent to the building and therefore the depth of frost. Addition or removal of insulation inside the structure will have an impact on adjacent frost depth. Similarly, soil is an insulator, so the removal or addition of soil will have an impact on building energy consumption and comfort.

Landscaping and irrigation will affect foundation walls in a number of ways. The biomass can act as an insulator; the irrigation, if improperly installed, can saturate the soil and deteriorate the foundation wall in time. Frost requires some ground moisture to form and irrigation can exacerbate the amount of water available. Ensure that there is proper weeping tile at the base of the foundation wall and that there is a drain or sump pit for the water to drain away from the foundation. The commissioning team must consider these and other factors when examining these structural aspects of the building.

Foundations in extreme cold climates that are installed into the permafrost may require the use of heat pipes or similar thermosiphons to maintain the integrity of the permafrost and therefore the building base. If the permafrost below a building melts, it could cause uneven settlement of the underlying soil and severely damage the building. There are a number of successful strategies available to counteract the effect of the building on the underlying permafrost that should be investigated when buildings are situated in these conditions. The commissioning team must ensure that the effect on the permafrost is considered when coordinating the provision of utilities from the site into the building.

PLUMBING AND DRAINAGE

The installation of domestic water pipes in the exterior wall should be avoided. Pipes should not be installed in cold attic spaces above the insulation layer. These situations involve the pipe penetrating the thermal envelope and the vapor barrier. The consequences of doing so include the possible freezing and rupture of water pipes that could flood a building or cause a drain line to freeze and plug up. Ideally, all piping should be installed in pipe chases inside the thermal envelope of the building. These can be pipe or duct shafts in larger buildings or wall cavities for smaller piping systems. In all cases, ensure there is sufficient cross-sectional area in the shaft to accommodate the pipes and the required insulation.

Drainage lines (sanitary and storm sewer) must be protected from freezing. Where these lines are located in unheated basements or crawlspaces, the lines should be heat taped, insulated, enclosed, or all of the preceding in order to protect the traps and drainage lines from freezing and causing problems for the occupants. The commissioning team should ensure that there is a planned and dependable way to direct the waste water out of the building and safely to the drainage point.

Sanitary sewer drain line incline is calculated using code slope values. But higher-than-minimum slopes may be desirable to prevent freeze up in buildings that may have sporadic or little use for long periods of time (such as schools during winter holidays). If using low-flush plumbing fixtures, consider using a full-flow water closet at the end of line to provide good flow down the line. Designers may prefer to use the minimum slope in order to avoid the need for a lift station in cases where the building drainage system must connect to an existing aboveground utilidor. The commissioning team should ensure that an informed decision is made when these competing priorities exist and that the decisions are well documented and well communicated.

PIPE INSULATION

This section applies equally to plumbing, drainage, heating, cooling, and sprinkler piping systems. The design, configuration, and installation of piping systems and related fittings and supports should not interfere with the installation of the pipe insulation thicknesses per ANSI/ASHRAE/IES Standard 90.1, Tables 6.8.3A and 6.8.3B (ASHRAE 2019b).

Ensure that the shutoff valves, check valves, and all other fittings are accessible and operable under the applied insulation.

MECHANICAL

This section will discuss the coordination of the heating and cooling systems with the building and the surrounding environment for the benefit of the people and systems that inhabit the facility.

Mechanical systems are generally dealing with the flow of fluids and the flow of heat energy. In cold climates, when mechanical systems fail, the risks are severe and can be costly.

The installation of heating pipes in the exterior wall should be avoided. These situations involve the pipe penetrating the thermal envelope and the vapor barrier. The consequences of doing so include rendering the heating system inoperable if a heating line breaks due to water freezing inside the pipe.

Ducts should not be installed in cold attic spaces above the insulation layer. If it is necessary to install ductwork outside the thermal and vapor barrier, the duct must be insulated and be sealed on the warm side of the insulation. When ductwork is exposed to the elements it may also be necessary to design a way to shed precipitation from penetrating the insulation. Similar to building envelope systems, the integrity of the thermal, air, and vapor barriers is critical and must be carefully coordinated between the sheet metal and building envelope trades.

Any hydronic heating system that is exposed to ambient air that potentially could be below the freezing temperature of water (32°F [0°C]) should be protected by using a thermal antifreeze solution suitable for the application. Thermal antifreeze solutions designed for hydronic heating and cooling systems contain water-treatment additives along with the antifreeze solution to protect the piping from corrosion and protect the system from freeze-ups when properly applied. It is critical to follow the manufacturer's installation instructions when adding these solutions because incorrect concentrations of the fluid in the system can actually increase the corrosion of piping and provide a false security that the system can be exposed to low-temperature air. High concentrations of thermal fluids reduces the heat transfer coefficient for hot- and cold-water systems—this must be allowed for when sizing systems or determining net capacity. Automotive antifreeze solutions are not suitable for hydronic heating systems.

Increasingly, closed-combustion systems are being used in new or replacement small-capacity combustion heating appliances. Condensing combustion appliances should be vented to a location that is accessible to the maintenance staff during peak heating seasons. The staff may need to deice the exhaust tubing to restore proper operation.

Older heating appliances require combustion air inlets and relief openings in accordance with the applicable building codes; the total area of inlets can create significant problems for the system designer and the building operator in cold weather. The total volume of ambient air required for good combustion can cause critically cold temperatures in the mechanical room. The system designer and commissioning team must ensure that there is sufficient heating capability in the mechanical room to offset the ventilation requirement of the combustion air at winter design conditions.

Conversely, in naturally vented combustion appliances, the volume of flue gases can put the mechanical room under a severe negative pressure if the combustion air inlet is undersized or blocked by snow or hoarfrost. If there are multiple chimneys, any appliances that are shut down could start to back draft and fill the machine room with combustion gases until the natural stack effect is reestablished. The commissioning team should ensure that there is a strategy in place to warn operators of the presence of carbon monoxide,

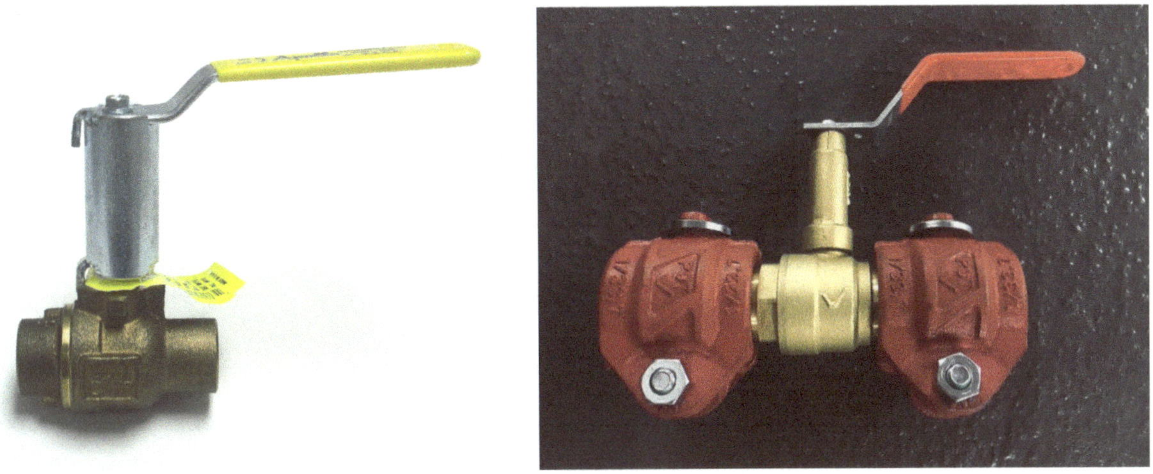

Figure 14.1 Valve stem extensions.

Courtesy Steve Clayman

carbon dioxide, or similar conditions in the mechanical room, and that there is a plan for regular testing.

In any cold climate that has below freezing temperatures that extend for more than one week, all critical heating components should have redundant backups and manual control overrides. The designer and owner should complete a risk analysis to ensure that there is sufficient heating capacity, heating equipment and heating systems to protect all critical parts of the facility when a critical component malfunctions. Most commonly, heating circulation pumps and the primary heating system will have redundant elements in an N+1 configuration. It is less common to have duplicate circulation piping or ducting on heating systems but that is where the designer and owner must agree on the degree of duplication.

The trend to seal the building envelope in cold climates produced buildings so tight that ventilation was required to offset the buildup of humidity and CO_2 as a by-product of occupancy. Heat recovery ventilators were added to provide the necessary ventilation and still maintain a high degree of energy efficiency by recapturing energy from the exhaust stream.

The commissioning team should ensure that there is a heat recovery ventilator in the building and that it is sized correctly for the application, and that ducts to and from the heat recovery ventilators are correctly connected in accordance with the manufacturer's instructions. The defrost control strategy is critically important to the successful operation of this equipment in below-freezing ambient conditions when the cold ventilation air extracts heat from the exhaust stream, takes that flow below the dew point, and then below the freezing point of water. System designers and equipment manufacturers have proven defrost and operating strategies that should be implemented to ensure long-term successful and energy-efficient operation.

If rooftop HVAC equipment is used, refer to Chapter 9. Control cabinets exposed to ambient temperatures may require heaters to operate correctly below 14°F (–10°C). The commissioning team may need to ensure that the controls and operators are suitable for cold-temperature operation.

Lubricants used in any system component exposed to cold air must be suitable for use at the design winter temperature for the location.

The commissioning team should work with the owner to ensure that there is a ready supply of critical parts in the local community or in stock at the building. This would include valves, belts, pulleys, pumps, fans, pipes, motors, compressors, wiring, and various control components.

ELECTRICAL

Electrical systems in cold climates are generally similar to those in all other climates. The possible exception is lighting systems that are sensitive to light output reductions at low temperatures. Lighting manufacturers do design fixtures that attempt to keep the lamps at optimal temperatures. The commissioning team should ensure that any exterior lighting systems are suitable for the temperature extremes.

Electric heat is often used to supplement other heat systems where redundancy is critical for a facility.

Electric motors and the equipment they drive, when exposed to winter design temperatures, should be lubricated with lubricants suitable for the extremes, particularly when the motor can cycle off for any reason during the cold weather.

Cold climates are predominately nearer the north and south poles. As such, the hours of daylight vary greatly between summer and winter. The exterior lighting system control systems should have the capability to sense minimum ambient light levels to trigger the on and off commands where exterior lighting is required for safety or security reasons. If time clocks are part of the control system they should be easily accessed to update the on/off times throughout the year.

In remote locations, generators are usually required to provide electrical power to the facility. In these circumstances, the security of the fuel supply becomes critically important. Ensure that there is a way to keep liquid fuel at a suitable temperature. For example, diesel fuel freezes around 17.5°F (–8.1°C) and biodiesel freezes at higher temperatures. The viscosity of diesel increases as temperatures drop and it will not flow in the fuel delivery system when it becomes a gel.

The commissioning team will need to carefully consider the facility electrical loads and future changes to the load to ensure that the fuel delivery and storage systems are adequate to provide the necessary power for a defined period of time as established by the owner.

CONTROLS

A general review of control systems is covered in Chapter 10. The commissioning of control systems in cold climates is primarily the same as for similar facilities in other climates. In any building with substantial seasonal differences in the building loads, it will be critical to test the operation of systems under load.

The control and automation system integration with the heating and ventilation equipment is a critically important aspect of cold-climate building operation. The controls interface with the people, the environment, and the equipment to produce a habitable building or operational facility.

The control systems should be no more complex than is necessary to successfully operate the facility. This is particularly important in remote locations where transportation challenges may limit the availability of replacement parts or where the sophistication of the system exceeds the capability of the operator.

The control system's ability to switch between heating and cooling modes, turning equipment on or off, and adjusting air and water set points in response to the loads on the building is another critical factor. Many buildings can experience heating needs in the morning and cooling needs in the afternoon.

The commissioning team should ensure that the building control systems are operating as intended or as required to meet the owner's current facility requirements. Trends should be established to monitor all important control functions. In simple systems, this may require the use of stand-alone temperature, pressure, humidity, and amperage meters. In facilities with BASs, it can include all of the same analog sensors and many binary functions such as start-stop commands. Trend data can then be analyzed to troubleshoot systems and optimize operations.

Fault detection and diagnostics (FDD) and or smart building controls (SBC) systems/software perform advanced data analytics that make ongoing adjustments to facility HVAC&R systems. FDD systems continuously analyze operational and energy data to provide real-time fault detection and diagnostics. SBC systems extract data, analyze, measure and verify energy use patterns and provide ongoing adjustments to maintain persistence in equipment performance and the associated energy utilization. These systems are typically installed on complex buildings with access to software support. Ensure that the system and operators are capable of operating in the context of the building environment.

FIRE AND SAFETY

Building fire suppression systems may include wet sprinkler systems. They require the same level of care as was discussed in the Plumbing and Drainage and the Mechanical sections. All water-filled piping should be installed in areas that will remain above freezing at all times. When systems are knowingly installed in areas that will be subjected to below-freezing conditions, there are proven ways to provide protection such as dry sprinkler systems filled with compressed air in the cold zone.

The commissioning team, in conjunction with the building owners and managers, should ensure that the building exit strategy is feasible during normal winter conditions. Similarly there should be provision of a safe haven for building occupant if the building must be evacuated in severely cold weather.

Under some specific conditions, moist indoor air exposed to cold outside air will create a fog that can resemble smoke to some smoke detectors. Caution should be used when placing the smoke detectors in ventilation plenums.

Rate-of-rise heat detectors can also be tripped when fans get turned off but heating coils remain active, which is a normal freezestat operating strategy. The operations and maintenance teams should be advised to look at their maintenance shutdown procedures to caution against similar occurrences.

SITE SERVICES

All utilities that connect to the building site must be adequately protected from the elements. In cold climates, water and sewer lines are buried at a depth below the historic frost penetration depth to reduce the risk of freezing. In areas where the ground conditions make that impractical, insulated and possibly heated utility conduits (or utilidors) can be built to deliver utilities.

Propane has a boiling point of –44°F (–42°C). As propane in the tank vaporizes to satisfy fuel demand, the liquid propane remaining in the tank is cooled below ambient temperatures, so when the demand for gaseous propane is at its peak, the fuel in the tank may remain as a liquid even though ambient temperatures are above the boiling point. Electric heaters are typically added to propane storage tanks in areas where the ambient temperature can approach the boiling point of propane to ensure the heating fuel source is available when needed.

Fuel oil is another fuel source in remote areas. At extremely low temperatures, the viscosity of the fuel increases and in some cases the fuel may gel, so preheating may be required to ensure continued operation of heating appliances.

Remote sites may have wells, cisterns, and septic tanks to handle the water and waste from the occupants. These are primary occupancy systems for the facility and must remain functional under any operating conditions. The commissioning team must ensure that the designers and operators understand how these systems will operate successfully at winter design temperatures.

Of equal importance is the repair of those systems at extreme temperatures, because that tends to be the time of maximum system stress and failure. Be sure there are adequate means to access equipment and controllers and proven methods to complete repairs on components at extreme temperatures.

TRAINING

The training of the current and future building operators in the correct operation of the facility is fundamental to the long-term health of the facility. ANSI/ASHRAE/IES Standard 202 (ASHRAE 2018b) is promoting the development of the facility guide. The facility guide is the equivalent of the owner's manual for a car or other complex piece of equipment. It will give the operator enough information to understand the intended operation of the facility. This will include maintenance information and frequency but not necessarily the details of repair or replacement of all components. Those details will remain in the systems manual for the facility. If high turnover of operations and maintenance staff or contractors is anticipated, consider creating operator training videos for training future staff. The commissioning team should make recommendations to the owner about the type, method, and frequency of training for the operations and maintenance staff and occupants of the building.

Train O&M personnel on operating the systems and equipment as well as how to use and update the systems manual. Build on any training that occurred throughout the construction and operation of the building to solidify the operation methods that ensure the longevity of the design. Document which staff receives the training provided.

Educate occupants on the systems and assemblies they will typically use and confirm they understand the effects of their actions on those systems and the facility. Training should cover not only the building systems and how to track performance with the BMS system, but should also include training on the following:

- Continually improving building performance by saving energy, reducing operational costs, and improving occupant comfort.
- Identifying and improving indoor environmental comfort and air quality.
- Extending equipment life cycles. Include any new maintenance procedures that need to be implemented to maximize the service life of both existing and new equipment.
- Ensuring the persistence of improvements over the building's life cycle.
- Using energy tracking protocols, dashboards, and benchmarking to track the facility's energy performance over time.

POST-CONSTRUCTION

Cold climates are well suited to the use of infrared thermography as a diagnostic tool for determining heat loss and airflow through buildings. The potentially high temperature differences between indoor and outdoor air temperatures will produce high-contrast images using currently available systems and equipment. This equipment can be used in a large number of applications, but is particularly good for assessing roof insulation and vapor barrier integrity.

Cold climates are home to hibernating animals that will be looking for warm, secure places to spend the coldest winter months. Care should be taken to ensure that the building envelope is secured against the most common pests to maintain a safe building for the intended occupants. Strategies will include physical barriers in vents, chimneys, and intakes.

EXTREME COLD

On water lines into the building, both heat trace and insulation may be required. Calculation guidelines and examples may be found in the ASCE Cold Regions Utilities Monograph, 3rd Edition (ASCE 1996).

For piped water system, a cold-water recirculation pump may be required. Run the recirculation line back to the main water supply. To prevent pipe freeze-up, water temperature should not drop below 33°F (0.5°C), which allows a small safety factor above the actual freezing temperature. Check ground temperature curves before composing this section of the facility guide. The minimum ground temperatures may occur weeks or months after the coldest ambient air temperatures of year. The control of the recirculation pump should not be based on outdoor temperature.

Ensure there are environmental protections for those that must service equipment or systems located outdoors, on roofs, or in unheated spaces. These protections can be permanent structures such as penthouses, or temporary tent-like shelters with lighting and appropriate heat for the duration of the service. Similarly, many jurisdictions require railings adjacent to rooftop equipment to prevent slip and fall accidents when ice and snow are present.

Sloped metal roofs with snow can be very slippery and caution must be exercised when accessing those roof structures.

REMOTE SITES

Firewater tank capacity must provide the duration of sprinkler water supply required by code or local authority. Ensure that the water in the tank can be supplied under fire conditions, including a loss of power at winter design temperatures.

HOW COLD IS IT?

It is so cold ... I chipped a tooth eating my chicken noodle soup!

15 Operation and Maintenance

INTRODUCTION

Refer to *ASHRAE Handbook—HVAC Applications*, Chapter 40, "Operation and Maintenance Management," for basic information on the subject (ASHRAE 2019e). This chapter focuses on issues that are more important for building operations in cold climates.

Operation and maintenance (O&M) of systems and equipment needs to be incorporated into the planning and layout of a building. It must ensure adequate access to mechanical and electrical equipment during extreme weather conditions including extreme low temperatures, cold, and significant snow with wind.

ACCESS TO SYSTEMS AND EQUIPMENT

Along with the selection of equipment and systems, the design of mechanical systems must consider how location and access can affect the simplicity and reliability of mechanical systems.

For example, the quality and frequency of servicing can be adversely affected if it has to be carried out in cramped and uncomfortable spaces, especially when heavy winter clothing is worn. Using a ladder in the winter to access equipment can be dangerous when snow and ice are present. Roof access with blowing snow and ice is very challenging.; maintenance personnel should have footwear that will ensure good traction under those conditions based on the type of roof surface or walkway materials used to allow access to the equipment. Staff will normally carry tools and replacement parts while traveling to the equipment and will often be working alone. The design of the facility should take those circumstances into account to ensure safe access to the building systems.

Ideally the building design team will work together to locate building equipment spaces that are inside the building envelope. This will allow for convenient access to both daily inspections and periodic maintenance of equipment without the need to go outside the facility. It will also allow replacement and spare parts to be stored adjacent to the equipment where practical. Additionally, it is usually easier to seal water barriers, vapor barriers, and thermal envelopes around architectural building systems than it is around rooftop unit penetrations.

Equipment rooms should be of sufficient dimensions to allow for regular maintenance to be accomplished without disrupting other systems and equipment. All heating, cooling, and ventilating equipment will eventually fail and/or need to be replaced. Think about how one would pull a large fan, coil, tube, or shaft from a component and fit the replacement through the machine room door. If this is impossible, there is another prob-

Figure 15.1 HVAC unit at grade allows safer access for maintenance.
Erich Binder

Figure 15.2 Insulated crawlspace access door.
Erich Binder

Figure 15.3 Frosty access platform.
Erich Binder

lem that must be solved. The thought of punching a large hole in the exterior wall in the middle of winter illustrates the complexity of the situation.

Equipment concealed in ceiling spaces, crawlspaces below floors, or mechanical rooms should be located in such a way that access and adequate space for servicing or replacement of all equipment is easily possible. If a room's only heating source is located in a ceiling or crawlspace that is difficult to access, the repair and maintenance of that system will take longer and will negatively impact the productivity of that space.

PREVENTATIVE MAINTENANCE

Design facilities and select equipment requiring maintenance levels appropriate to the intended function of the facility. Ideally equipment will be located inside the building, but that can increase the first cost of the facility. The total cost of ownership includes the cost of regular operations, maintenance, and utilities. Those metrics tend to favor buildings that are easier to operate and maintain.

If heating systems cannot be located inside the building, the level and frequency of service must be balanced with the system costs and the ability of the equipment exposed to ambient conditions to be maintained by individuals wearing bulky cold-weather clothing and insulated gloves. Is it reasonable to expect the maintenance work be accomplished with those constraints? Will the delay of maintenance work adversely impact the productivity of the facility? If the answer to either question is yes, the plan for exterior equipment needs to be reconsidered.

Predictive Maintenance

Determine and track the predicted life expectancy of facility and equipment components to replace them at the optimum time. Implement a predictive maintenance schedule for all systems and equipment. The majority of outdoor preventative maintenance work should be performed during milder temperatures during spring, summer, or fall.

When equipment is at the end of its expected life, any major equipment replacements should be scheduled around the expected weather and building operations. The timing will include the selection of equipment, budgeting procedures, procurement processes, manufacturing time, transportation to the site, contractor availability and equipment handling logistics. Remote sites may also have time-of-year transportation windows of opportunity that must be met. See the Remote Sites section at the end of this chapter.

Spare Parts

In new construction, facility maintenance and operators should determine, in consultation with designers, contractors and system vendors, what spare parts will be provided. In existing buildings a good preventative maintenance program will establish parts that require periodic replacement and the frequency of replacement. Parts should be ordered in sufficient time to ensure they arrive prior to the scheduled installation date. In most communities the lead time will be less than a week for common parts. Equipment age, specialization, or other unique factors can cause the lead time for parts to extend to weeks or months. Procurement of those replacement parts will need to take all of these factors into account.

Standardization

In the interest of maintenance simplicity, the equipment for any one particular function should be of one manufacturer and be compatible with any existing O&M parts inventories. Even with a given manufacturer, different ages of equipment could have different replacement parts requirements. Maintain a list of replacement parts within the building's Operations and Maintenance Systems Manual.

Redundancy

All primary heating systems should be designed and installed with a fully redundant backup. When analyzing a design or existing system installation, perform a risk analysis to identify aspects of the heating system where a single component, pipe, or duct could thwart the intended redundancy of the heating system. In cases where it is not practical to install a fully redundant system, such as unitary rooftop equipment or furnaces, the owner or operator should have a backup plan to provide adequate heat at design conditions when the primary system is disabled for repair. If a critical heating system requires maintenance, schedule that work during milder weather or unoccupied hours.

OPERATIONS AND MAINTENANCE SYSTEMS MANUAL

ASHRAE Guideline 1.4 outlines the recommended contents of a Facility Systems Manual (ASHRAE 2019c). *ASHRAE Handbook—HVAC Applications*, Chapter 40, Section 4 has a good summary of O&M documentation (ASHRAE 2019e). The O&M Systems Manuals are to be prepared in accordance with good engineering practice and should be readily available to the operators on site. The O&M Systems Manual is a living document that should be updated regularly by adding any new systems or equipment and deleting any equipment that is removed.

Identification

Nameplates, tags, and arrows can all be used to assist quick identification of equipment and piping. Consistent identification in all buildings is required so that maintainers and operators can easily orient and familiarize themselves with any building under their control. Operating and maintaining mechanical systems requires an understanding of their systems components, including movement of fluids, air and mechanical parts. Locate pipe markers and direction arrows on piping systems where they are visible from the floor of the usual operating areas or readily accessible points. This could be helpful to identify potential problems when water lines are located near combustion air inlets that could cool them well below freezing temperatures. When labels are located on systems and equipment located in areas subject to ambient conditions, ensure that labels can be visible when snow or ice is on the units.

Operational Set Points

Heating systems will usually be set up to ensure the best operation of the system under a wide range of normal conditions to optimize the total cost. This may include varying the supply water temperature on boiler systems. Care must be taken by the designer, contractor, and building operators to ensure that the set points and other control sequences are set and adjusted to ensure the correct operation of the terminal heating units in the occupied spaces, the piping and pumping system, and the boiler units. All piping that will be exposed to below-freezing ambient temperatures should maintain some flow at all times. Prevent dead heading of heating pumps during mild weather and ensure that boiler return water temperatures are within the manufacturers specifications.

Ventilation systems, especially makeup air units, have a wide range of inlet conditions but a very narrow range of supply air temperatures. Ensure that the control valve range matches the full range of supply air temperatures. In some very cold climates, gas or steam valves selected at design winter conditions will not control outputs within acceptable ranges when ambient temperatures are just below freezing because they begin

Figure 15.4 Proper Identification of equipment.
Erich Binder

Figure 15.5 Ensure labels are visible in winter.
Craig Fredeen

Figure 15.6 Snow adjacent to building.

Erich Binder

to hunt or cycle on and off. Generally, multiple smaller valves with different capacities are instead selected and installed to ensure that the comfort conditions are maintained for the total range of ambient air conditions.

Unoccupied temperature setback and warm-up cycles are often used to reduce a building's energy consumption. As ambient temperatures decrease, the total heat loss of a facility will increase and the warm-up cycle time will increase. The operators should monitor the building temperatures when occupants begin to arrive to ensure that the night setback and warm-up cycle times are optimized for building occupant comfort and energy efficiency.

SITE MAINTENANCE

Site work should be planned to ensure work can be completed using existing local equipment and operators where possible. Consider the cold-weather events (snow, ice, or low temperature) that the site could be subjected to and the need for the facility to be accessible and operational. Then, look at the resources that would be needed to accomplish the requirement. If there is a gap in the available resources, then the gap needs to be closed. For example, if a hospital is located in a location that rarely gets snow, there will not be a lot of snowplows available if snow falls. That site may need to invest in a suitable snowplow, even though it will only be used a few times each decade, to insure against closures.

Snow Clearing

The presence of snow and the need to clear snow and ice from a site is the norm in most cold-climate locations. A site that does not function well when covered in snow is unacceptable, depending on the normal number of days per year that snow is present.

Figure 15.7 Snow fence controls drifting.
Erich Binder

Consider how the snow will be removed (hand shovels, snow blowers, skid steer loader, tractor, grader, and so on), where snow will be seasonally piled or temporarily stored and the snow drifting patterns that may be affected by the snow pile. The protection of the building, vegetation, and fixed-site improvements from snow removal equipment is important to the successful operation of any cold-climate site.

Wind breaks and snow fences should be located to maximize their effectiveness without compromising the flow of people, vehicles, and equipment at the site. Similarly, they should not compromise vehicle parking or outdoor storage for the site.

In areas where high snowfall amounts can accumulate, it is wise to identify the location of fire hydrants, curbs, and walkways so that they can be accessed if necessary and protected from damage by snow removing equipment.

Spring Runoff

In most cold-climate communities, snowmelt can occur suddenly. The meltwater must be directed away from the building and into acceptable drainage courses on site to avoid

- flooding of basements, crawlspaces, and mechanical rooms;
- water cistern or sewage holding tanks floating and breaking piping connections; and
- granular or concrete pads being severely eroded by water seeping under or through the pad, resulting in structural damage.

These issues can be compounded when warm days are followed by subfreezing nights that refreeze the meltwater and form dams that impede the next warm day's meltwater. Snow piles can take months to melt and should be located to avoid blocking natural site drainage paths.

Ice dams can form on sloped roofs under similar freeze thaw cycles or when there is an accumulation of snow on a roof (which can be a good insulator) and poor insulation of

Figure 15.8 Snow drift over windows.

Bert Phillips

the ceiling below that roof. When a dam forms, trapping meltwater on the roof material, the water can back up under shingles and leak into the walls, attic, and ceiling of the building below, causing structural and IEQ problems for the building. Mitigation strategies include

- ensuring there is sufficient insulation in the attic,
- ventilating the attic to keep the roof at ambient temperatures,
- installing heat tape on the roof,
- providing a channel for the meltwater to escape, and
- designing a roof that will not accumulate snow.

To maintain a roof that does accumulate snow, it may be necessary to remove the amassed snow with a long-handled snow rake.

Planted Areas

Maintaining vegetation will vary depending on staff levels and location. Ideally, little effort will be necessary to maintain planted areas. The use of carefully selected indigenous and climate-hardy plants and grasses can play an important role in protecting slopes from erosion. Similarly, stormwater retention ponds will delay the release of stormwater and meltwater from the site to mitigate potential flooding in areas downstream from the property.

Pedestrian Access

Public buildings should be easily identifiable, with prominent, clearly visible entrances. All pathways, ramps, and stairs leading to entranceways should be barrier-free, easy to keep clear of snow and ice, and protected from vehicle traffic. Ensure that public access points have a place to pile snow without impeding access.

Whenever possible, eliminate the need for ramps and stairs by shaping the site. Grade elevation at the entrance should be as close to finished floor elevation as possible. Be sure to account for the site drainage, spring melting, and freeze-thaw events when designing access paths.

Locate entrances and major windows where snowdrifts will not normally form. Radiant slab heating is a viable option, especially for handicap access ramps and at door entrances. Refer to *ASHRAE Handbook—HVAC Applications*, Chapter 52, for guidance on these subjects (ASHRAE 2019e).

Vehicular Access

Facility sites must accommodate the type and size of emergency service vehicles and personnel to ensure they are able to approach fire standpipe connection points and principal entrances year-round with a minimum of difficulty. It is critical that there is no obstruction by snow, standing water, or steep slopes.

Cold-climate parking lots for staff, long-term, and visitor parking may require exterior electrical outlets to power block heaters and interior car warmers. Such systems should be placed on temperature-activated controllers or time clocks depending on the nature of the traffic parking pattern.

Lighting

The site lights are normally controlled to match ambient light cycles. In regions nearer the north and south poles the day lengths will vary substantially through the year. Ideally, site lighting should be actively controlled to match the length of the night to optimize energy consumption and occupant safety.

REMOTE SITES

Suppliers

Equipment specifications for new or replacement units should not unduly restrict local or regional suppliers. Consideration should be given to incorporating locally available products in buildings if they are suitable for the application.

Equipment

The use or repair of existing equipment benefits the community and can reduce owning and operating costs. Replacement equipment and parts are often difficult to transport to remote communities. Items with long delivery lead times should be kept on hand based on the risk associated with operating without that piece. Transporting heavy or bulky equipment and parts into most remote communities is extremely costly. Similarly, transportation schedules may be dependent on open waterways for barges in the summer or ice roads for trucks in the winter. If parts are needed when those options are not available then air freight may be the only other option, which will usually be substantially more expensive.

Labor and Training

To facilitate maximum local involvement, materials and methods used in buildings should be appropriate for the capabilities of the facility management team that will be operating the systems. In remote areas, there may be a limited number of experienced tradespeople available to properly operate the facility. There may be a both a need and an opportunity to train and develop building maintainers in these communities. Incorporating future maintenance personnel into the construction and commissioning team estab-

Figure 15.9 Snow-blocked exhaust hood.
Erich Binder

Figure 15.10 HVAC unit under building.
Erich Binder

lishes a training opportunity for those same O&M personnel and is a best practice. ASHRAE Standard 202 (2018b) and the numbers 0 and 1 series of ASHRAE guidelines detail the commissioning of new and existing buildings, including staff training.

Training modules for operations and maintenance personnel should be created. The training should start at a very basic level and progress to the highest level that operators will require to successfully operate the facility. Training should be recorded using technologies that are appropriate for current and future staff. This will be especially important in areas that are subject to high staff turnover. Establish a training log to document the education of all staff over time.

Maintenance Contractors

Due to the challenges of traveling to remote sites, the regularly scheduled visits of specialty contractors should be arranged well in advance. This may allow the contractor to visit multiple sites on a single visit, thereby sharing the travel costs between multiple owners and providing a higher level of service to all.

Services and Utilities Access

With winter conditions lasting from six to eight months of the year, it is important that building utility service points can be easily accessed by trucks and personnel and are protected from snow and ice build-up. In remote locations, heating fuels, (wood, heating oil, and propane), potable water, and sewage pumping are delivered exclusively by truck. For example, most municipal services to buildings in Nunavut and rural Alaska are delivered by vehicle.

Figure 15.11 Portable propane heater.
Erich Binder

Figure 15.12 Flex duct from portable heater.
Erich Binder

EXTREME COLD

Services and Utilities Access

Power and telephone services have traditionally been provided by overhead lines. If this is the case, ensure that there are no large trees above the lines that could cause a snow-laden branch to fall and disrupt power service to a building during a heavy snowfall or ice storm.

Propane liquefies at very low ambient temperatures. Prior to an expected cold spell, ensure that the tank heaters and controllers are operational. Confirm that there is sufficient fuel in the tank to last the expected duration of extreme weather. Fuel suppliers will be hard-pressed to deliver most fuels at extreme conditions, including high wind.

Equipment Service

In extreme climates, access to equipment that is subject to ambient conditions may require temporary heated service enclosures to protect the maintenance staff from the elements and allow them to perform repairs and maintenance over many hours. As noted previously in this chapter, the service enclosure should be large enough to accommodate all of the personnel required to do the work at any one time, including room for parts, tools, and sufficient elbow room while wearing cold-weather clothing. The enclosure will also need a heat source with enough capacity to maintain a temperature suitable for working without gloves but without too much heat at the source that could be uncomfortable or cause a fire risk.

Lighting

Some lighting types have substantially reduced light output when the fixtures are operated at very low ambient temperatures. Ensure that the selected or replacement lights are suitable for low ambient temperatures when used outdoors.

HOW COLD IS IT?

It is so cold ... the squirrels in the park were throwing their nut supply at an electric fence to thaw them!

Appendix: Case Studies

INSTITUTIONAL BUILDING, EDMONTON, AB, CANADA

A new facility at the University of Alberta in Edmonton will provide approximately 305,695 ft^2 (28,400 m^2) of space to support expanded educational and research activities in the Faculty of Engineering.

The building will house faculty members, staff, and students from every engineering discipline and be a hub of activity providing greater opportunities for interdisciplinary collaboration among professors and students.

Custom air-handling units (AHUs) were provided, complete with heating, cooling, and a heat recovery system.

Figure A.1 Institutional building, Edmonton, AB, Canada.
Courtesy University of Alberta

Unit Specifications

AHU 1
12 wide × 2 high = 24 fan cubes
120,000 cfm (56 634 L/s)
10 hp (7.46 kW) each = 240 hp (179 kW)
Total static pressure = 8 in. w.g. (1990 Pa)
Heat recovery, heating, and cooling coils

AHU 2
8 wide × 2 high = 16 fan cubes
94,500 cfm (44 599 L/s)
7.5 hp (5.59 kW) each = 120 hp (89.5 kW)
Total static pressure = 4.8 in. w.g. (1194 Pa)
Heat recovery coil

AHU 3
2 wide × 1 high = 2 fan cubes
12,000 cfm (5663 L/s)
7.5 hp each (5.59 kW) = 15 hp (11.2 kW)
Total static pressure = 4.3 in. w.g. (1070 Pa)
Heat recovery coil

Heat Recovery Hydronic Module
Prepackaged skid consisting of the following:

- Pumps
- Cooling/heat plate heat exchangers
- Control panel
- Expansion tank

Figure A.2 Piping of energy recovery coils so that each coil can be isolated.

Result

The hydronic skid recovers as much energy as possible from the exhaust air streams and transfers it to the supply air stream through the use of a closed-pumped glycol loop.

The system is able to meet 72% of the building's annual heating requirements through energy recovery only. When energy recovery alone is not enough to satisfy the building's heating or cooling demands, additional heating and cooling is brought into the system through use of plate heat exchangers located on the hydronic module.

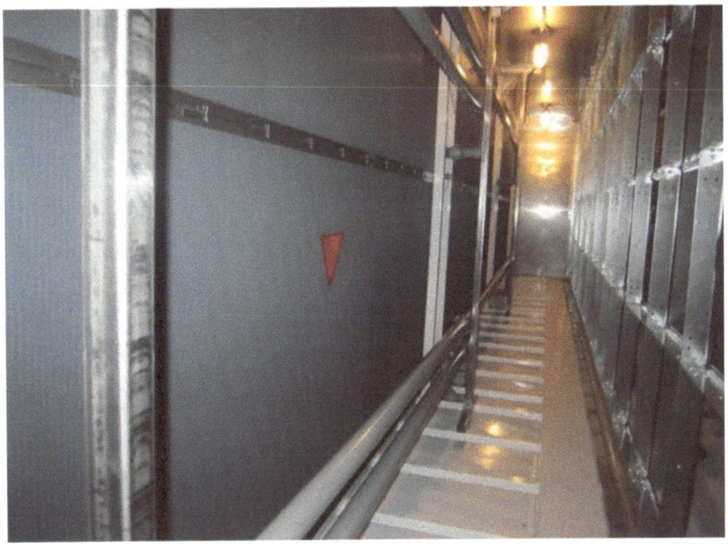

Figure A.3 Energy recovery coil surface.

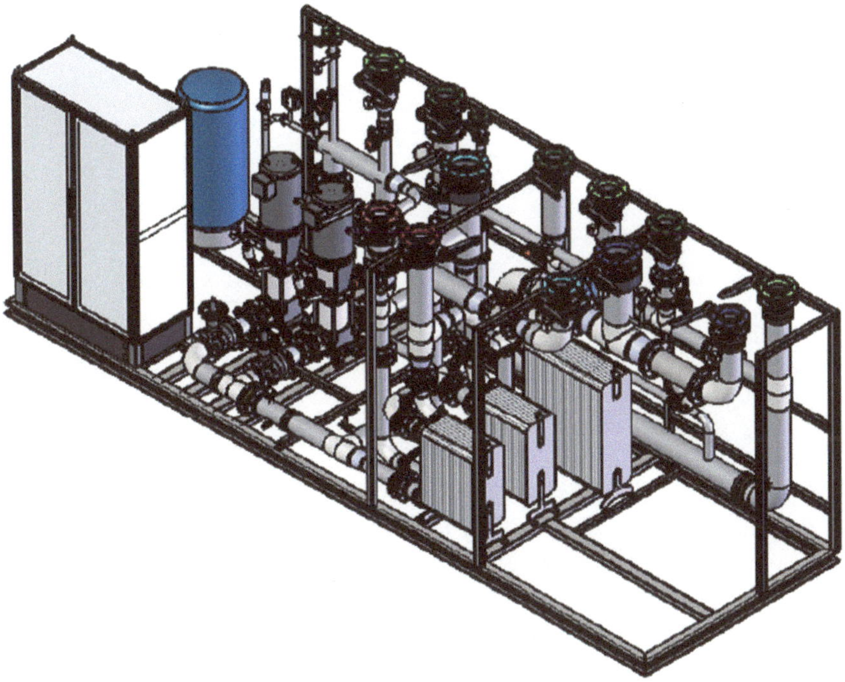

Figure A.4 Skid drawing.

Courtesy Konvekta USA Inc.

The energy recovery system provides a temperate glycol medium to meet all the seasonal requirements, saving labor and minimizing the risk of coils freezing in the winter. Its low-airflow pressure coil design with thick, strong fins allow 2600 psi (17.9 MPa) washing on the preheat coils, if necessary. The system is monitored year-round for the life of the equipment by engineers in Switzerland and can be an invaluable asset when recovering energy in cold climates.

SNOW AND ICE MELTING SYSTEMS

Snow and ice melting systems are hydronic systems intended to augment the removal of snow and ice by circulating a heat transfer liquid (inhibited glycol and water mixture) through cross-linked PEX pipe installed within the outdoor surface. Snow and ice melting systems work with most types of outdoor surfaces, such as concrete, paving stones, or

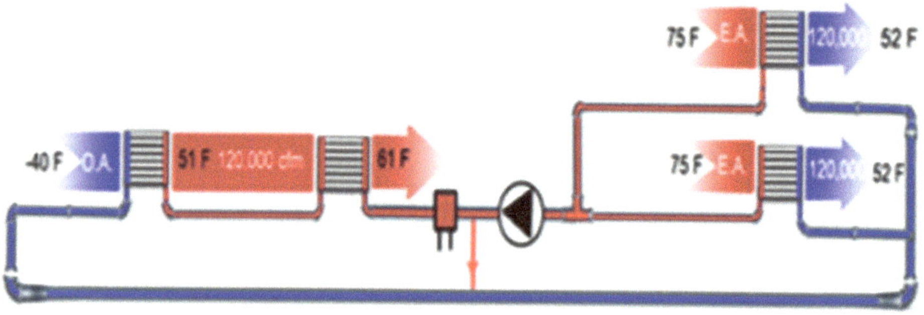

Figure A.5 System schematic.

Courtesy Konvekta USA Inc.

Figure A.6 Dual redundant pumps on skid.

asphalt, and the time to install the piping network is just before installation of the outdoor surfaces.

Design of a snow and ice melting system is a bit more complicated than design of a radiant heating system because the system output loads are based on several factors and not just the outdoor design temperature. *Loads* means the amount of heat required to operate the system.

Snow and ice melting system loads are based on melting snow into water and evaporating water into vapor. This leaves outdoor surfaces dry in some designs, while the majority of the water is expected to drain off. Factors that affect system loads include air temperature when snowing or melting, rate of snow fall, snow density, wind velocity, humidity level of the atmosphere, slab temperature at start of snowfall; and the apparent sky temperature.

There are a few options when it comes to controlling snow and ice melting systems.

- **Good: Semiautomatic Electronic Control**
  A semiautomatic control requires a person to activate the system manually, sometimes using a delay, and to set the runtime. It is an affordable type of control with good safety for many residential applications.

Figure A.7 These driveways and walkways are free of snow because of the snow and ice melting system. *Courtesy REHAU*

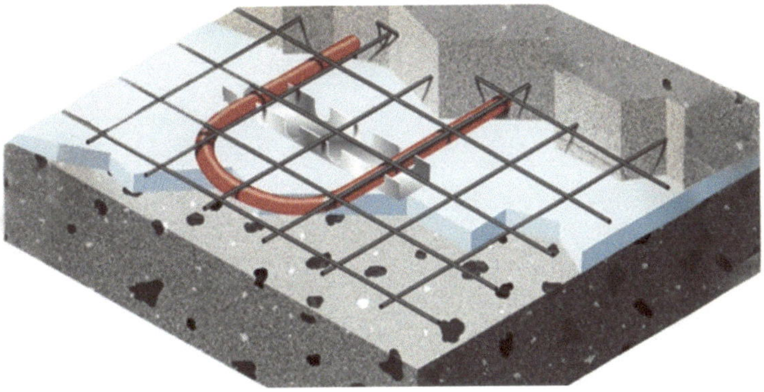

Figure A.8 Typical installation illustration of SIM pipes installed within concrete and over insulation. *Courtesy REHAU*

- **Better: Fully Automatic Control with Moisture Sensing**
  A fully automatic system uses a moisture sensor with an integrated outdoor temperature sensor placed in the melting area. When the sensor detects cold air temperature and moisture, it activates the melting system. The sensor detects the moisture in dry snow or ice by first melting the snow or ice with the integrated heating element. Then, the melted water conducts low-voltage electricity across the metal rings on the sensor surface. If it is cold and there is moisture on the sensor, then it must be snowing or icing. Automatic controls can improve safety and save energy for residential and commercial systems.
- **Best: Smart Controls System with Weather Anticipation**
  This is a computer-based system that accesses weather forecast data to predict incoming snow or ice and activate the system just in time. When installed, the latitude and longitude coordinates are used to get the closest weather information from the National Weather Service or Environment Canada. The system can be programmed to warm the melting area a certain amount of time before the snowstorm begins, so that it is ready to melt the first snowflake. With these capabilities, owners have ultimate control over safety, efficiency, and protection of heated areas. Smart controls offer the ultimate performance for critical outdoor areas.

In large facilities with distinct weather areas (e.g., east and west ramps), zoning may be achieved through separate sensors connected to separate circulators for each area. Some large SIM systems have three or more distinct zones.

In large residential projects, using separate manifolds with their own dedicated circulation pumps for each area can allow a different type of zoning flexibility. For instance, a homeowner with a large snow and ice melting system could have a manual switch to deactivate the circulator for guest parking, operating that area only when needed.

With correct zoning and controls, the system can be designed to ensure that all of the critical zones are cleared to provide safe surfaces for walking and driving all year round. In a cold climate, the ice and snow on walkways and sloped driveways cannot be ignored, because accidents occur frequently, sometimes causing lifelong consequences.

Figure A.9 Snow and ice melting system moisture sensor.
Courtesy REHAU

COMMERCIAL BUILDING, ALBANY, NY, USA

A commercial example is shown in Albany, NY.

- **Project:** Parking garage ramp, 50 ft long × 20 ft wide (15.2 m long × 6.1 m wide) for a total area of 1000 ft^2 (92.9 m^2)
  - Ramp is poured concrete 6 in. (152.4 mm) thick
  - With downward losses, the snow and ice melting system heat load = 150,000 Btu/h (158.3 MJ/h)
  - Typical temperature at start of snowfall is 18°F (–8°C)

Figure A.10 This snow and ice melting system has seven distinct zones, each with its own manifold (not shown) and circulator pump in a range of sizes based on size of zone.
Courtesy REHAU

Figure A.11 Snow and ice melting system heating materials installed on stairs.
Courtesy REHAU

- Weather data shows that it snows/ices 156 hours per year, and we assume this happens over 10 events per year with 15.6 hours of frozen precipitation at a time (10 × 15.6 = 156 hours annually)
- The designer selected a system capable of keeping a free area ratio (A_r) of 1.0 (100% snow free) during a 90% frequency distribution snowfall, which requires an output of 125 Btu/ft^2 (1,419 kJ/m^2) per hour

Some facilities generate waste heat that they are trying to reject using chillers or geothermal heat pumps during portions of the year. Some examples are office buildings, factories, hockey rinks, or car dealerships with waste-oil boilers. In these applications, an always-on snow and ice melting system can be an efficient and effective solution for rejecting the excess heat of the building or processes, while providing all the benefits of a hydronic snow and ice melting system (MacNevin 2015).

ENERGY ENVIRONMENT EXPERIENTIAL LEARNING (EEEL) PROJECT, UNIVERSITY OF CALGARY, CALGARY, AB, CANADA

The University of Calgary Energy Environment Experiential Learning (EEEL) Project is a 263,700 ft^2 (24 500 m^2) building that provides instructional space for expanded programs in energy and environments, new laboratories for biology and chemistry, and space for the Institute for Sustainable Energy, Environment and Economy Administrative Center. It is a LEED® Platinum facility that generates chilled water at night for use the next day, uses cascaded heating and cooling loops to save pumping energy, and uses earth tubes to precondition air.

Figure A.12 Fully automatic system, no idling.
Courtesy REHAU

The design team's goals were to minimize the use of energy, water, and material resources and maximize indoor environmental quality while also providing daylight to every regularly occupied space.

The HVAC system was designed mainly with dedicated outdoor air handlers and a decoupled sensible cooling system with radiant slab and panels. Alongside the DOAS ventilation, energy recovery systems were added to recover energy from 100% of the exhausted air to preheat the incoming outdoor air.

For the air handler supplying the lecture theatre of the EEEL Building, two 150 ft × 48 in. (45.7 m × 1219 mm) diameter concrete earth tubes are buried 14 ft (4.3 m) underground to help preheat or cool the air.

To further reduce the energy load, the system was designed to use a cascading heating and cooling system. With the following heating plant design, the heating system uses the return water of one load as the supply water for another. This allows the building to use a larger temperature differential throughout the system and save on heating equipment.

With the cascading cooling design, the system capitalizes on the cool nights to generate the required cool water for the facility's air-handling system and radiant cooling units. This is done with closed-circuit evaporative cooling towers. The system generates 50°F (10°C) water from 72°F (22.2°C) return water.

This also allows the cool water produced during the night to precharge the thermal mass of the building so that the load during the day is reduced and the cooling towers only operate during the night. The required cooling water for the next day is stored in an onsite, stratified chilled-water storage system located underground.

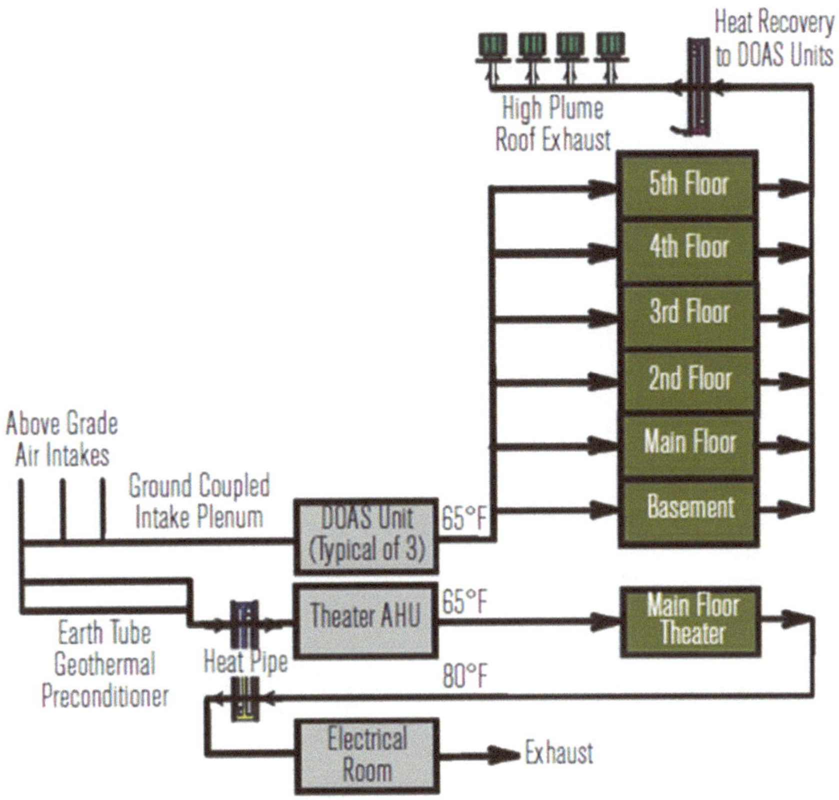

Figure A.13 Concept ventilation system (McGinn 2014).
Courtesy Tim McGinn

Figure A.14 Installation of an earth tube (McGinn 2014).
Courtesy Tim McGinn

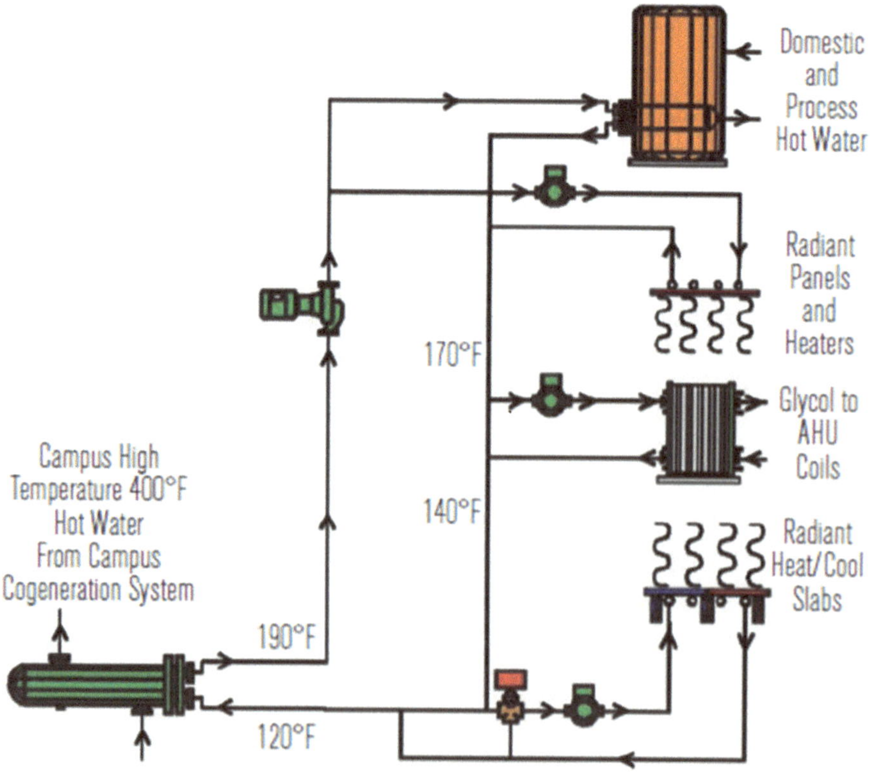

Figure A.15 Nested heating plant design (McGinn 2014).
Courtesy Tim McGinn

This system also incorporates the following strategies to minimize energy requirements:

- Low hot-water use plumbing fixtures
- Low-pressure-drop air-handling components, piping, and ducting
- Variable-speed pumps and fans
- Reuse of once-used air to condition service spaces
- Natural venting of general exhaust during mild weather to reduce fan power
- Demand-controlled ventilation

The efficient mechanical system is joined by a daylighting strategy to help make the space more conducive to teaching and learning.

A large light scoop at the top of the atrium drives light deep into the heart of the building while energy modeling ensured that the design minimizes peak solar loads to reduce mechanical cooling requirements. This combined with automated interior blinds ensure the building is not unnecessarily using artificial light.

In the end, the high-performance building envelope and HVAC system design of the EEEL produce a sustainable building that provides great indoor environmental conditions for working, taking a break, or meeting up with friends. For more information, please see the full article in the September 2014 *ASHRAE Journal* (McGinn 2014).

BULLITT CENTER, SEATTLE, WA, USA

The Bullitt Center, a commercial office building in Seattle, was designed to be the greenest commercial building in the world, aiming to be classified as a Living Building by the International Living Future Institute.

The six-floor, 52,000 ft^2 (4831 m^2) building has an expected lifespan of 250 years. Designed to be energy and carbon neutral, the building uses solar panels on the roof to dramatically reduce electricity grid consumption. This allows the building to use about one-third the electricity of comparably sized office buildings and, while electrical demand in the winter is higher than what the solar array can provide, the system should provide enough excess at other times to allow for recovery of the funds.

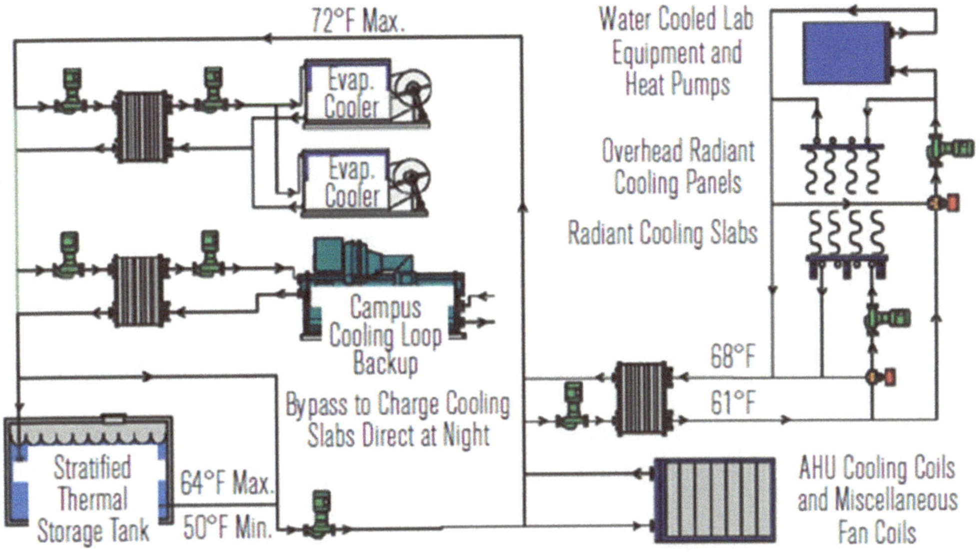

Figure A.16 Cooling plant concept (McGinn 2014).
Courtesy Tim McGinn

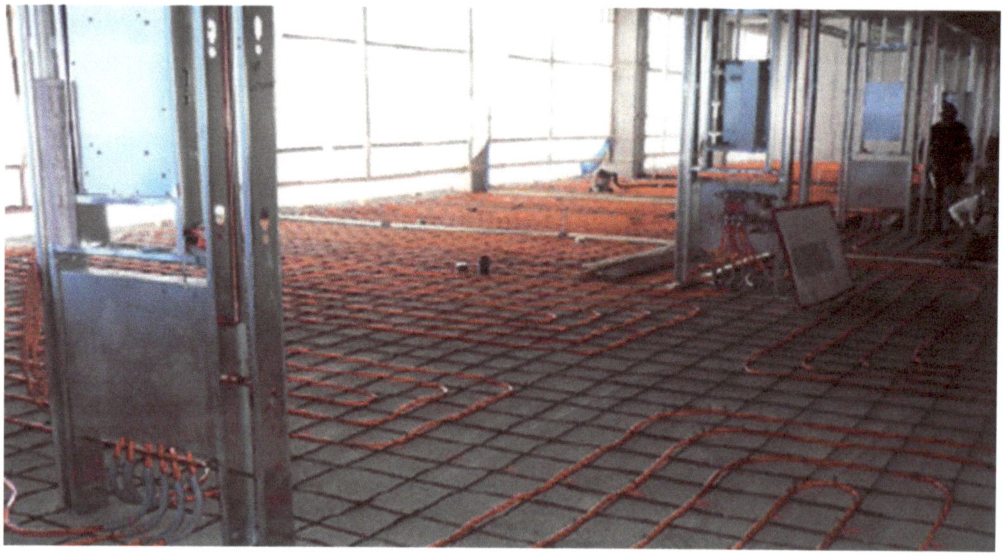

Figure A.17 Radiant tubing (McGinn 2014).
Courtesy Tim McGinn

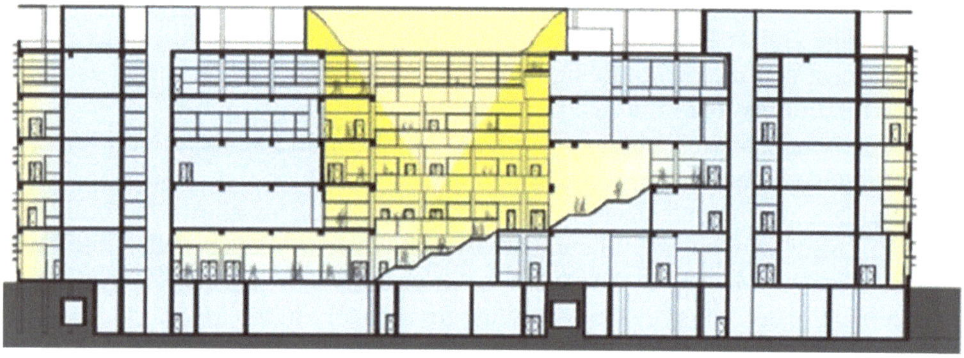

Figure A.18 A central atrium allows daylight into the building core (McGinn 2014).
Courtesy Tim McGinn

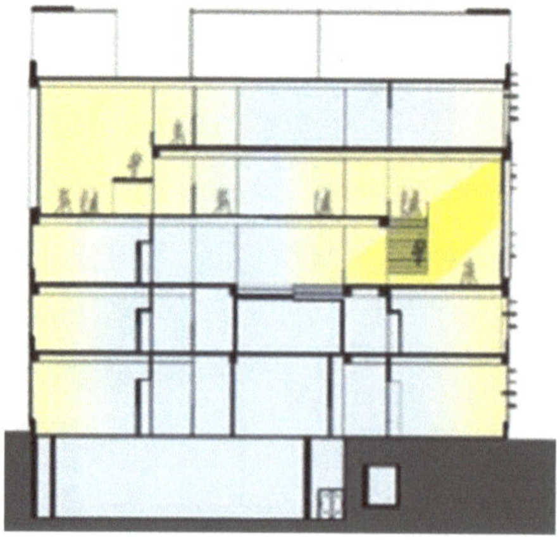

Figure A.19 Light also penetrates through window, atrium, and double-height lounges (McGinn 2014).
Courtesy Tim McGinn

To further reduce energy usage, the building has 26 geothermal wells that extend 400 ft (121.9 m) into the ground to produce 55°F (12.8°C) water for the building's heating system.

The geothermal heat exchange system is matched with a radiant-floor heating and cooling system to establish net zero energy status. An additional features that contributes to the building efficiency is high-efficiency windows with a retractable blind system that blocks heat before warming the building too much, and reduced plug loads.

The building was also designed to have net zero water. In order to do this, a 56,000 gal (211 983 L) rainwater collection system was put in place. Using ceramic water filters, ultraviolet light, activated charcoal, and chlorine, the center is able to provide its own water. However, until given final approval to use the rainwater process, the building water is provided by the municipal water supply. To complement the supply, the center also has its own sewage processing system that uses waterless toilets and a composting process in the basement, as well as a gray-water treatment system that allows for the water to be cleaned in a constructed wetland before being released back into the soil.

As the first of its kind, the building is meant to be an example and spark a change in the way buildings are designed and built in the future (Hayes n.d.; Bullitt Center 2014).

Figure A.20 Bullitt Center, Seattle.
Courtesy Nic Lehoux for the Bullitt Center

Figure A.21 Seattle skyline looking over the Bullitt Center's rooftop solar array.
Courtesy Nic Lehoux for the Bullitt Center

Figure A.22 Mechanical room.
Courtesy Nic Lehoux for the Bullitt Center

WALGREENS, EVANSTON, IL, USA

Established in 1901, Walgreens has gone on to become one of the largest retail brands in the world with more than 8000 stores. In furthering the growth, Walgreens wanted to create a scalable retail location that could be a model for innovative, sustainable, and high-performance building design in the retail industry.

The new 14,531 ft^2 (1350 m^2) building in Evanston, IL was to have the same functioning nature of most Walgreens buildings, so the building needed a conventional interior layout with at least 5356 operational hours annually. Additionally, while the average annual energy consumption of stores in the area was 1566 GJ (435,000 kWh), to achieve the net zero energy goal, the building had to consume 41% less energy, thereby requiring

Figure A.23 Walgreens, Evanston, IL, USA.
Courtesy Jason Robbins

Figure A.24 The roof of the Walgreens is entirely covered in solar panels.
Courtesy Camburas & Theodore Architects

the building to produce 921.6 GJ (256,000 kWh) per year. It was determined that this amount of energy could be produced with 840 microinverter panels on the roof along with two 5 kW wind turbines.

The two largest uses of energy in a typical store are the lighting and refrigeration, so careful attention was used in these areas. To maximize solar capacity, the store's design was changed from a typical rectangular design to a tired, multilevel, cantilevered roof with much more glass. While the roof and wall insulation were standard Walgreens specification, to control the window load, automatic shade controls were applied to west-facing fenestration below 14 ft (4.3 m) to limit sun penetration, while a light-directing film was used on the upper glass to direct 90% of the solar light and energy up and into the store and away from personnel. This was combined with LED lights to reduce energy use as much as possible.

To recover the energy dissipated from the refrigeration, a geoexchange well system was installed, similar to a previous Walgreens project in Chicago. Eight 499 ft (152 m) vertical wells were designed under the parking lot to store the rejected energy. With photovoltaic panels covering the roof and a zero lot line, the typical constant volume rooftop units used on previous Walgreens buildings was not an option—nearly all of the HVAC equipment had to be inside the building. In this manner, it was determined that the best method to reuse this stored energy was to match it with a chiller-heater central heat pump system. A secondary goal for Walgreens was to use all-natural refrigerants, so a CO_2 refrigerant heat pump was found in Europe. To use the system in the United States, it had to pass the Underwriters Laboratories tests requiring certification at five times the operating pressure, which varies from 1160 to 1450 psi (8000 to 10 000 kPa) for CO_2.

After passing the Underwriters Laboratories test, the system was finalized. The transcritical CO_2 heat pump uses three sets of variable-speed compressors. The low-temperature compressor is used for servicing the evaporators for the freezer, the medium-temperature compressor is used for servicing the evaporators for the cooler, and the high-temperature compressor is used for servicing the chilled-water system. The heat pump's dissipated energy provides heating for hot-water and service water preheat.

With low-temperature chilled water and heating hot water available, the building comfort system was designed with four-pipe fan coils, although a radiant heat system was also considered. The sales floor area was split into three separate zones, each with a variable-speed, single-zone AHU hanging in the space. Because of higher-than-anticipated noise levels, the fans were limited to 80% of maximum speed, which hindered the ability to provide heating. A variable-air-volume dedicated outdoor AHU was used to decouple the ventilation load and was designed to use demand control ventilation with CO_2 sensors to minimize the ventilation load, because occupancy is quite variable.

Performance

Walgreens recognized the opportunity to use this project to understand how these technologies perform in this environment. In this manner, a building management system

Figure A.25 The interior features eight lighting zones with daylight-sensing controls.
Courtesy Jason Robbins

Figure A.26 Additional daylight is possible through clerestory windows.
Courtesy Jason Robbins

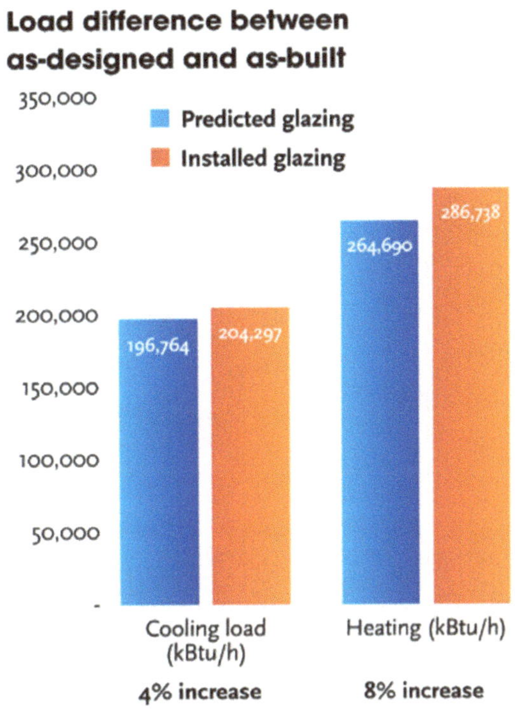

Figure A.27 The building is currently performing over its predicted load capacity.
Courtesy Cyclone Energy Group

was installed to monitor and control the system. An electrical submetering system was also used on nearly all the equipment to help verify the results.

This combination proved to be an invaluable tool during commissioning to determine where operation was deviating from the optimized design. With the store in operation and the metering systems calibrated, in the first year the store will fall short of the net zero energy consumption. Multiple elements contributed to this result. First, it was a severe winter and design-energy modeling was completed on typical weather conditions. As shown above, the curtain-wall system also underperformed relative to expectations. This was due to the framing system not being thermally broken, causing the assembly to underperform at peak conditions by 12%. The gas cooler to remove excess heat and prevent overheating of the geoexchange had electrical overloading issues, causing the geoexchange system to work less effectively. Additionally, the heat pump compressors were determined to be oversized, and a revolving door proved to be ineffective at providing energy savings. However, by changing out the compressors and resolving the other issues, the system is being tuned to reduce consumption by 18 GJ (5000 kWh) per month. With LEED Platinum status, Green Globes Certification, and a Green Chill Platinum rating, much can be learned from this project for similar buildings in the future (Skelton 2015; Bean 2014).

MULTIPURPOSE INDUSTRIAL FACILITY, BASSANO, AB, CANADA

This 55,000 ft^2 (5110 m^2) multipurpose industrial facility houses a mechanic shop, vehicle storage, truck wash, welding shop, carpentry shop, plumbing shop, field offices, and water treatment facility and is owned by a large-scale farming community. The building was designed for longevity while extracting the highest level of efficiency from the mechanical and electrical systems and ensuring thermal comfort and IAQ conditions for the workers.

Figure A.28 Multipurpose industrial facility.
Courtesy Robert Bean, Indoor Climate Consultants Inc.

The steel fabricated building is assembled with R-34 (RSI-5.99) batt insulation in the walls and roof and with R-10 (RSI-1.76) extruded polystyrene foam slab insulation under the entire building. Because of the large compressive loads from heavy equipment and machinery, the 6 to 8 in. (150 to 200 mm) slab was insulated with high-density extruded polystyrene foam insulation under the mechanic bays and truck wash. Motors and drives for cranes, ventilation, and water treatment pumping equipment were selected for maximum efficiency, as was the LED lighting throughout the buildings.

The 1.6 million Btu/h (469 kW) plant serves a hybrid mechanical system using low-temperature radiant floor heating with 100% dedicated outdoor air for ventilation. The heating system is comprised of 14 individually controlled zones regulated with space temperature thermostats, control valves, and pressure-regulated circuit setters.

For a variable-temperature and constant-flow approach, fluid temperature is reset based on weather conditions with target return temperatures of approximately 75°F to 90°F (24°C to 32°C) during 30% to 70% load profiles. The design for each zone was modified through radiant tube spacing and differential temperatures to achieve a single mean operating fluid temperature of approximately 105°F (41°C) at an outdoor design temperature of –40°F (–40°C). With the combination of reset and essentially constant low return flow temperatures, the boiler plant efficiency is maximized around 95% to 97%.

Because the distribution system for each zone originated from within the boiler room, an objective was to achieve similar differential pressure for each zone, within reason. This was accomplished by strategic placement of manifold locations, analysis of head losses in the radiant pipes, and diameter selection of the distribution pipe with a focus on maintaining acceptable velocities.

To facilitate the short construction season, all distribution lines were installed below the slab in sleeved PVC lines. Within a few days of the pour, the boiler plant was connected to the manifolds. Once the concrete cured, the radiant system could then be commissioned.

Figure A.29 Boiler room showing distribution piping, separator, boilers, and expansion tanks.
Courtesy Robert Bean, Indoor Climate Consultants Inc.

Additionally, special expansion loops had to be developed for those manifolds served by long distribution lines, because of the high coefficient of expansion for PEX pipe.

To improve motor efficiency and to handle the differential pressures from valves closing at low loads, pressure-controlled variable-speed circulators were used to maintain constant head at low flows.

Because efficiency and comfort were two key project objectives, low return fluid temperatures drove the enclosure and mechanical design. With the building and systems operating as intended, this makes for an excellent example of designing and commissioning multipurpose industrial facilities in cold climates.

LOUISE AND RICHARD GUY HUT AT MONT DES POILUS, BC, CANADA

In 2015 the Alpine Club of Canada (ACC) built a new hut on the Wapta Icefields near Mont Des Poilus, BC, Canada. The hut sleeps 18 people and operates in winter months, providing an alternate ski touring route for the Wapta traverse between the ACC's Bow and Stanley Mitchell Huts.

Challenges in designing the hut included the remote off-grid location, extreme low temperatures, strong gusty winds, heavy moisture-producing activities from cooking activities and drying wet ski gear, no live-in hut custodian to monitor mechanical systems during periods of occupancy and vacancy, periodic use, and high environmental protection standards. See Chapter 12 for in-depth discussion of these issues.

The vision for the hut's design team was to build a structure that enhances visitor experience and acts as a model for other similar off-grid facilities. Contemporary construction technologies have often failed, resulting in condensation and mold growth, energy inefficiencies, high operational costs, and environmental issues. A set of design goals for the hut included: 1) easy to operate and maintain technologies, 2) moisture control inside the hut, 3) superior energy efficiency, 4) excellent indoor air quality, and 5) environmental integrity.

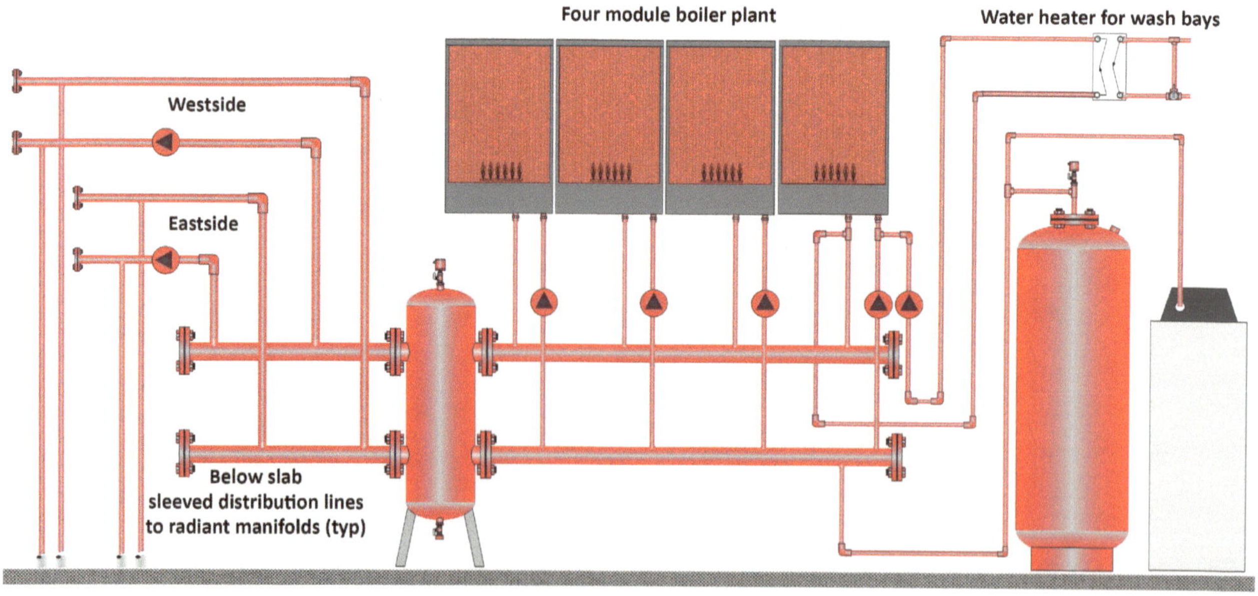

Figure A.30 Boiler room piping schematic.
Courtesy Robert Bean, Indoor Climate Consultants Inc.

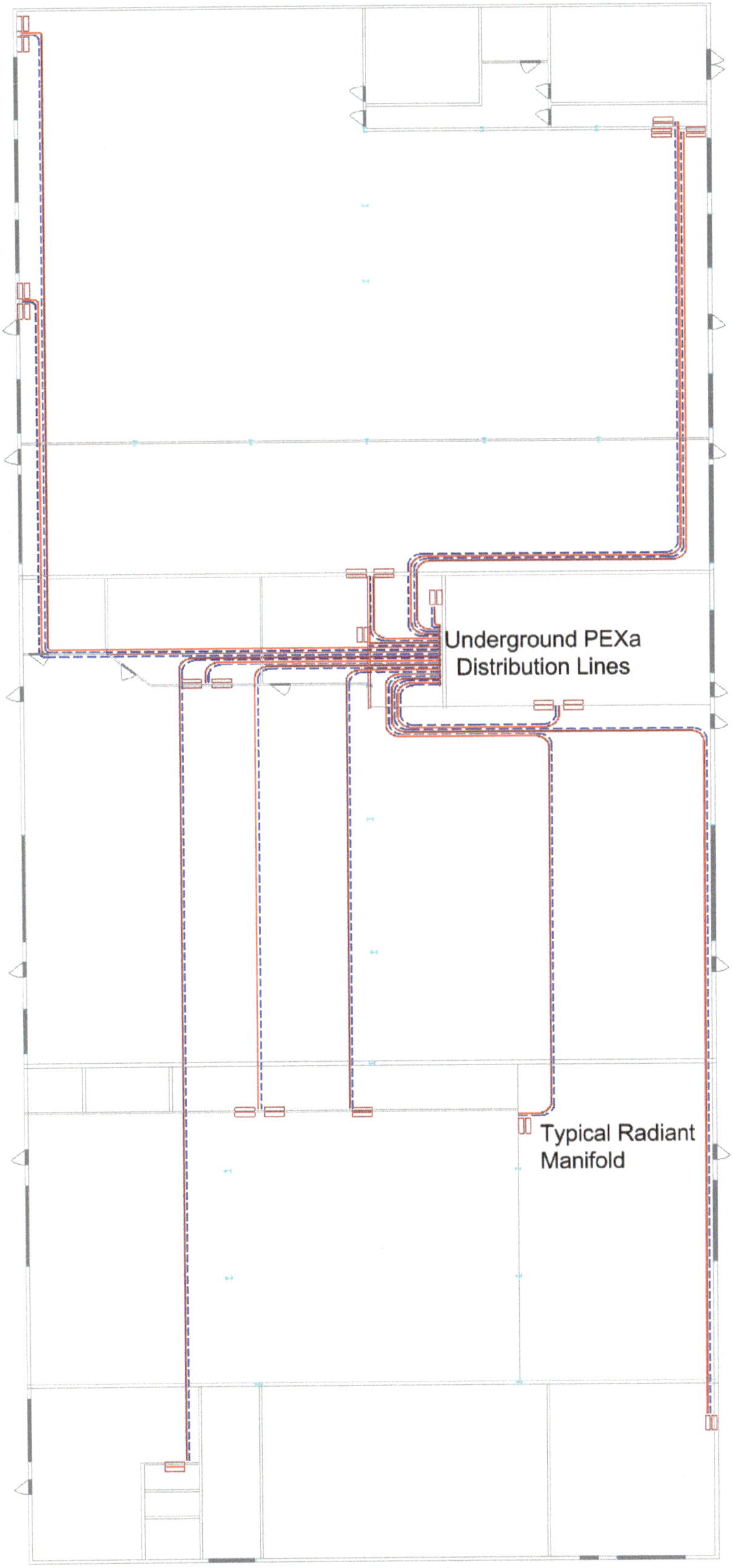

Figure A.31 Below-slab PEXa distribution lines to remote radiant manifolds.
Courtesy Robert Bean, Indoor Climate Consultants Inc.

Figure A.32 Below slab sleeves for PEXa distribution pipe.
Courtesy Robert Bean, Indoor Climate Consultants Inc.

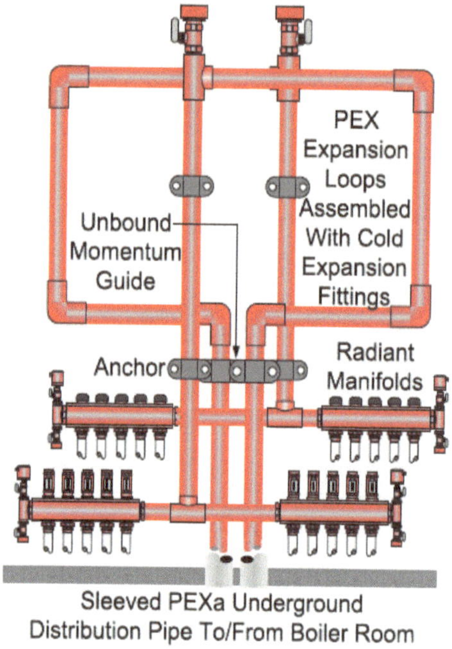

Figure A.33 Distribution expansion loop details at remote manifolds.
Courtesy Robert Bean, Indoor Climate Consultants Inc.

Figure A.34 East and west side radiant system circulators.
Courtesy Robert Bean, Indoor Climate Consultants Inc.

Table A.1 Design Goals and Design Solutions

Design Solutions	Design Goals: Easy to Operate and Maintain	Moisture Control	Energy Efficiency	Excellent Indoor Air Quality	Environmental Integrity
Superior construction	✓	✓	✓	✓	✓
Building layout	✓	✓	✓	✓	✓
Building materials: SIPs	✓	✓	✓	✓	✓
Building heat: sun and propane stove	✓	✓	✓	✓	✓
Ventilation: mechanical fans, operable windows	✓	✓	✓	✓	✓
Solar/wind hybrid renewable energy system	✓				✓
Potable water treatment system: boiling	✓				✓
Gray water dispersion field	✓				✓
Barrel fly-out system for human waste	✓				✓

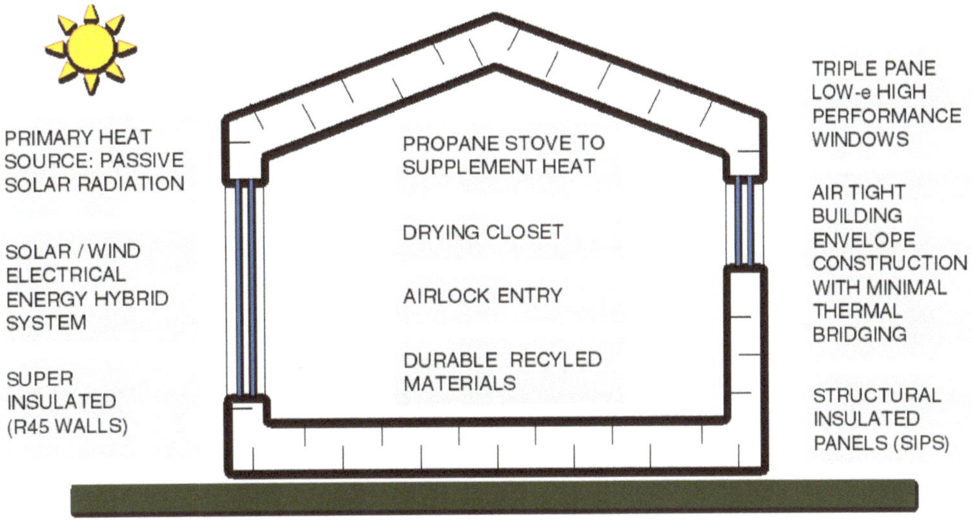

Figure A.35 Construction technologies and building features.
Courtesy DLB Associates

Solutions that satisfied the design goals include the following:

- **Construction:** sSuper insulation (R-45 walls), airtight construction, minimal thermal bridging, high-performance windows, ventilation
- **Building layout:** Compact design with main floor and sleeping loft above (small footprint, low visual impact), south wall elongated (to maximize solar energy potential, daylighting, and views), covered front entrance, airlock entry, drying room to isolate wet ski gear, furnishings kept away from exterior walls to promote free airflow across the inside surface of the exterior walls
- **Building materials:** Structural insulated panels (SIPs) for insulation value and shorter construction time, durable and resilient interior finishes, low-VOC-emitting building materials, recycled building materials content
- **Building heat source:** Passive solar supplemented by propane stove
- **Ventilation:** Moist air extracted from sleeping area and kitchen, outdoor air preheated by propane heaters, operable windows

- **Energy:** Solar/wind hybrid renewable energy system (12 solar panels of 250 W each; two wind turbines of 200 W each), deep cycle batteries, LED lighting and propane backup lights (energy for low-voltage ventilation fans, lighting, communications)
- **Potable water treatment system:** Boiling snowmelt
- **Gray-water management:** Dispersion field
- **Black-water management:** Barrel fly-out

Figure A.36 Aerial view of hut site.

Figure A.37 Twelve solar panels on the south face and one of the two wind turbines.

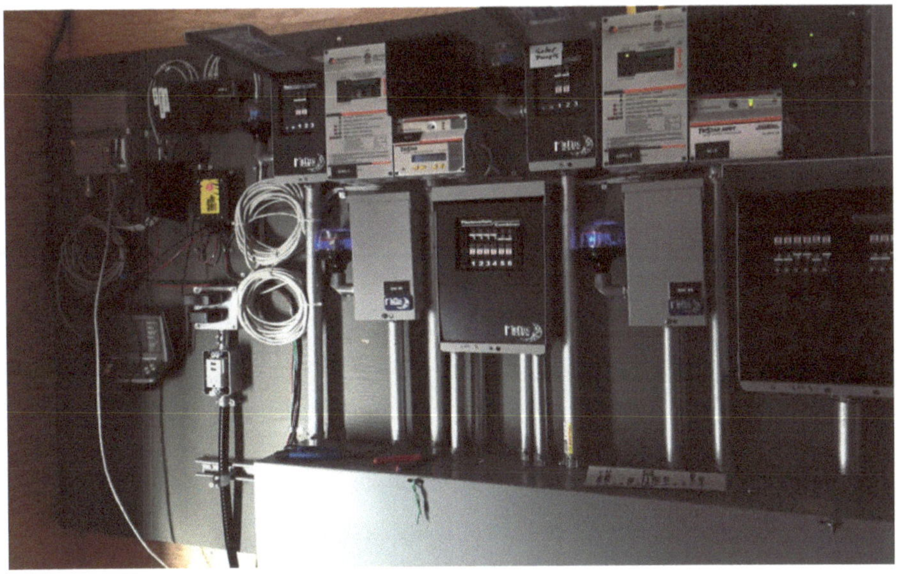

Figure A.38 Electrical system.

Figure A.39 Interior view of the hut's main floor.

Figure A.40 Sleeping area in loft.

Figure A.41 South-facing windows for solar energy gain, daylighting, and views.

Figure A.42 Louise and Richard Guy Hut at Mont Des Poilus.

MULTIFAMILY HOUSING

Description

Located on a large communal farm in ASHRAE Zone 7, this 88,000 ft^2 (8175.47 m^2) multifamily project is designed and built as utilitarian housing for multiple generations (Figure A.43). When finished, there will be 28 new homes in four buildings each with seven houses (Figure A.44). Two identical mechanical rooms are constructed, each serving 14 units. The below-grade enclosure is designed with 11 in. (280 mm) insulated concrete form (ICF) basement walls and a 4 in. (100 mm) slab bearing on 2 in. (50 mm) of rigid extruded polystyrene insulation on a gravel radon plenum. Interior wood frame

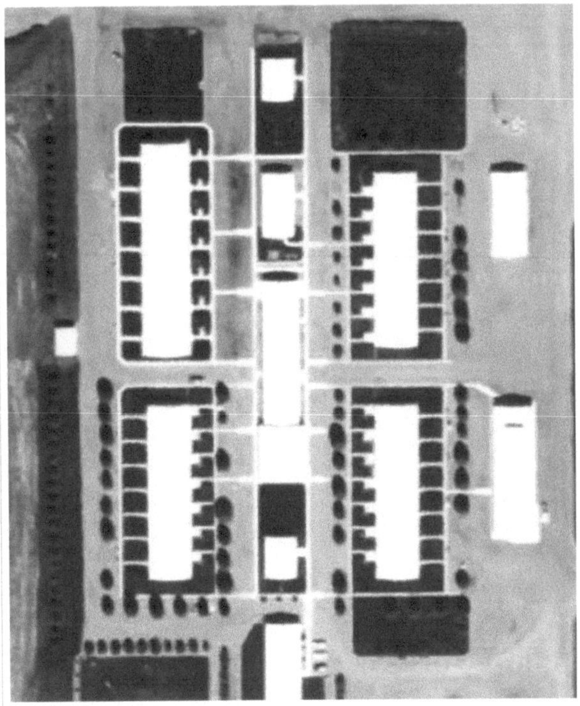

Figure A.43 Aerial view of the existing project. These older buildings being replaced by the new project.

Google Earth

Figure A.44 Phase 1 completed. Note access to central mechanical room, ground level left side.

Robert Bean

Figure A.45 Phase two under construction. Make note of the foundation drainage plane, steel roofing, ICF and foam walls. High performance system and simple geometries are ideal for resilient cold-climate homes.

Robert Bean

walls are sheathed with oriented strand board and then finished with drywall. Above grade walls are 6 in. (150 mm) expanded polystyrene panels with thermally isolated studs (Figure A.45). Walls are vapor barriered, sealed, and finished on the inside with drywall and outside with weather barrier, strapping, and PVC siding. The roof is assembled with engineered wood trusses, sheathed and finished with steel roofing. The attic is sealed with a polyethylene membrane, open-cell spray foam, and filled with cellulose. All PVC windows are triple-sealed, triple-pane, and argon-filled. Window glass with frame rated at $U = 0.14$ Btu/ft^2·h·°F (0.79 W/m^2·K).

Mechanical

Within this community each farm has their own tradespeople partially responsible for the design, assembly, and maintenance of the various building systems. In North America there is a long tradition of using hydronic-based HVAC systems with knowledge gained from one generation passed down to the next. On this particular property there are more than a dozen large structures, including a water treatment plant serving their industrial and agricultural businesses. As with most district energy systems it is advantageous for the owner/operator to maintain a few larger central plants rather than multiple smaller plants. The system eliminated the need to run 28 gas lines and regulators with associated venting and combustion air. This applies also to refrigeration lines, power, and outdoor condenser units. Additionally any service work on the heating and cooling plants can be done without having to enter each dwelling. Also considered was the ease to adopt future renewable energy sources to a centrally located plant. For this project, two identical hybrid systems were designed for the homes. Each system consists of a 1200 MBH (352 kW) six-stage boiler plant for heating (Figure A.46) and a 25 ton (88 k W) five-stage chilled-water plant for cooling (Figure A.47). The heating boilers serve the low temperature radiant floor systems (Figures A.48 and A.49) and makeup air heating coils (Figures A.50 and A.51). Both are designed to operate on reset with maximum supply of 120°F (49°C) and a 20°F (11°C) differential at design conditions of –35°F (–37°C).

Figure A.46 Space and makeup air heating and domestic hot-water plant for 14 homes.
Robert Bean

Figure A.47 Space and makeup air cooling for 14 homes.
Robert Bean

Figure A.48 Radiant floor heating installation; note also four of seven piping vaults and uninsulated sections for distribution pipe chase and cold storage areas.
Robert Bean

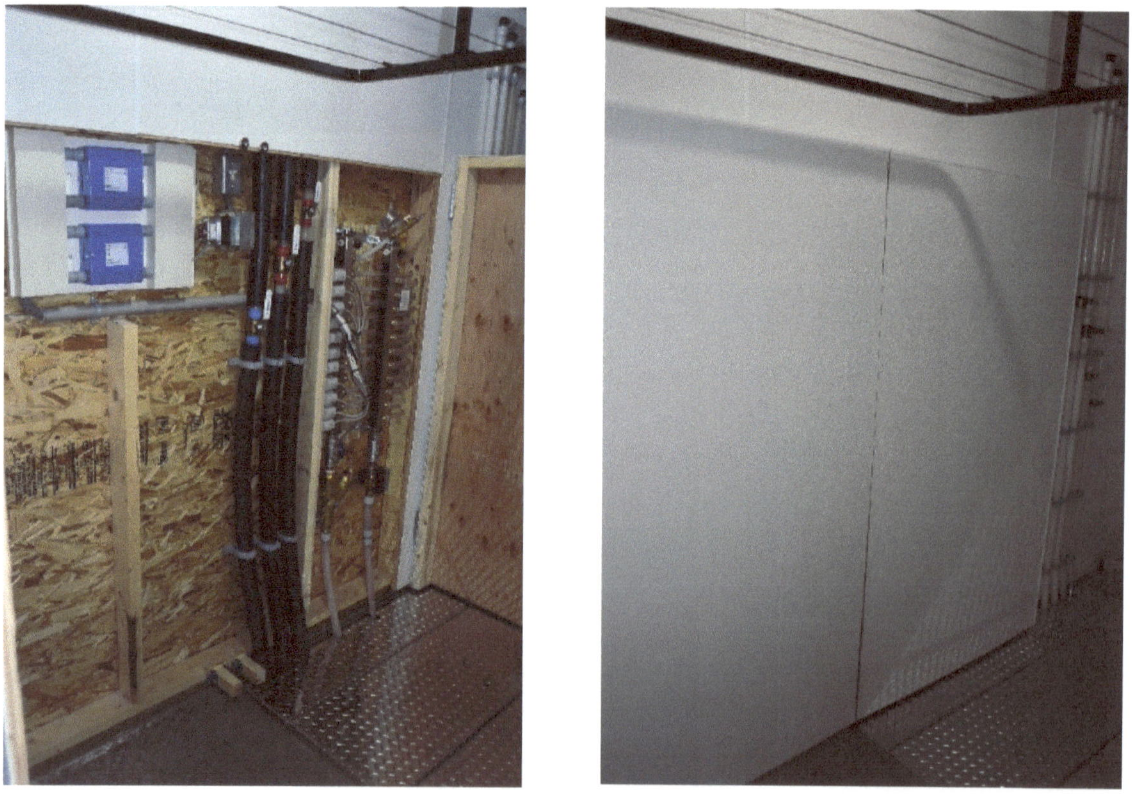

Figure A.49 Radiant distribution system and controls. Left, cabinet open. Right, closed.
Robert Bean

Figure A.50 Typical mechanical room in each home. Floor vault contains PEX distribution system, cabinet (left) contains radiant heating distribution and control systems, fan-coil above for space cooling and cooling/heating makeup air. The central plant at the far end of the building keeps these individual mechanical spaces tidy and accessible for service and maintenance.

Robert Bean

Figure A.51 Four-pipe heat/cool PEX distribution lines to fan-coil(s).

Robert Bean

Figure A.52 Heating, cooling, and domestic distribution pipe with expansion loops and system take offs: floor vault open (left) and closed (right).

Robert Bean

Boilers are also domestic-water prioritized, staged to switch to 180°F (82°C) for charging the four 120 gal (454 L) storage tanks through a preheat/boost/reheat brazed plate heat exchanger system for an 80°F (44°C) rise. The cooling coils were specified by the client for a 45°F (7°C) entering fluid temperature and a 10°F (6°C) rise for conditioning the space and makeup air. Cooling airflow per unit is 700 cfm (330L/s) at approximately 1 ton/1200 ft^2 (3.5 kW/112 m^2) based on 2200 ft^2 (204 m^2) of cooled space.

Ventilation is composed of a centrally located variable-speed in-line primary exhaust fan operating at 105 cfm (71 L/s) at 50% of total design exhaust flow. Ceiling-located bathroom fans provide additional exhaust capacity in conjunction with a unique odor exhaust systems connected to each toilet.

The exhaust system pulls makeup air through an outdoor duct connected into the return air plenum. This airflow passes across the heating and cooling coils of the fan/coil unit for distribution through the supply air system when the fan/coil blower is off. The temperature regulation of the makeup air and air for cooling is done with a unique non-electric auto heat/cool change over control. The control modulates fluid flow to each of the coils to maintain a client-controlled set point. A room thermostat is used to turn on the blower on call for cooling (or heating). Primary heating is radiant floors, but can be done in the shoulder seasons with the fan-coil if the occupants wish to use it for the occasional cold spell.

Heated, chilled, and domestic water from the central plants to each of the houses is done through an in-ground cross-linked polyethylene (PEX) distribution network running under the building (Figure A.52). Each home is fitted with a piping vault in the basement floor offering access to the pipes and expansions joints.

Discussion

The client's objectives were to replace an older housing project with four modern buildings designed around resiliency, indoor environmental quality, low maintenance, and energy efficiency. The homes must serve multiple generations of families living on the farm. The materials and methods of construction, HVAC, lighting, and interior finishes were all selected to meet these needs. The heating system was designed around the lowest possible common temperature profile that could serve both the needs of the radiant system and makeup air coils. By doing so, mixing valves, controls and circulators were eliminated, resulting in a simpler, higher-efficiency system. Domestic water was prioritized so that on call, one or more boilers could be switched to high temperature for charging and recharging the storage tanks. The load profile of the farm workers is similar to what a designer would find in a work camp on an industrial facility with high loads in the morning and evening. Due to the high performance of the enclosure, the morning and evening domestic charging times does not affect the thermal comfort of the occupants.

The underground PEX distribution systems facilitated speed and simplicity as continuous lengths could be fed down the entire length of the building through the sleeves joining the piping vaults (Figure A.52). From there expansion joints were made and takeoffs for systems easily assembled with the manufacturer's proprietary fittings.

Lessons Learned

This community consists of seasoned builders and professional project managers; as such, very few issues of consequence arose during the design and construction phases. However, due to the assignment of responsibilities within the community, it was not always possible to optimize the integration of each system during the early stages. For example, the project would be ideally suited for high-temperature radiant cooling as the flooring and embedded tubing was already being installed for low-temperature radiant heating. The dry climate zone, occupant load, and ventilation requirements offered a very low latent load. Had the homes been cooled with radiant, the chillers could have been operated with a 68°F (20°C) return temperature, increasing their efficiency. The radiant cooling would have enable further reduction in airflow down to ventilation rates only and likely the use of HRVs in lieu of the fan-coils. This would have resulted in reduced size of ductwork, registers, grills, supports, dampers, and filters.

References and Bibliography

AAC. 2017. Alaska Administrative Code. Title 13, Public Safety. Part 2: Fire Prevention. Chapter 50: Codes and Standards. Juneau, AK: State of Alaska. www.legis.state.ak.us/basis/folioproxy.asp?url=http://wwwjnu01.legis.state.ak.us/cgi-bin/folioisa.dll/aac/query=[JUMP:%2713+aac+50!2E020%27]/doc/%7B@1%7D?firsthit.

AIA. 2006. *Practicing architecture—Best practices.* Chapter 9. New York: The American Institute of Architects.

AMCA. 2016a. ANSI/AMCA Standard 99-16, *Standards Handbook.* Arlington Heights, IL: AMCA.

AMCA. 2016b. *ANSI/AMCA Standard 210-16 | ANSI/ASHRAE 51-07: laboratory methods of testing fans for certified aerodynamic performance rating.* Arlington Heights, IL: AMCA.

ANSI. 2019. CAN/CSA F326-M91 (R2019), *Residential Mechanical Ventilation Systems*. New York: American National Standards Institute.

API. 2012. API RP 500. *Recommended Practice for Classification of Locations for Electrical Installations at Petroleum Facilities Classified as Class I, Division 1 and Division 2*. Washington, DC: American Petroleum Institute.

API. 2018. API RP 505. *Recommended Practice for Classification of Locations for Electrical Installations at Petroleum Facilities Classified as Class I, Zone 0, Zone 1, and Zone* 2. Washington, DC: American Petroleum Institute.

Armstrong, R.S. 2012. *HVAC design considerations for cold climates*. Anchorage: RSA Engineering.

ASCE. 1996. *Cold regions utilities monograph*, 3rd ed. Daniel W. Smith, ed. Reston, VA: American Society of Civil Engineers.

ASHRAE. 2007. ASHRAE Guideline 1.1-2007, *HVAC&R technical requirements for the commissioning process.* Peachtree Corners, GA: ASHRAE.

ASHRAE. 2009. *The ASHRAE guide for buildings in hot and humid climates*, 2d ed. Peachtree Corners, GA: ASHRAE.

ASHRAE. 2011. *Advanced Energy Design Guide for Small to Medium Office Buildings: 50% Energy Savings.* Peachtree Corners, GA: ASHRAE

ASHRAE. 2015a. ASHRAE Guideline 24, *Ventilation and indoor air quality in low-rise residential buildings*. Peachtree Corners, GA: ASHRAE.

ASHRAE. 2015b. *Combined Heat and Power Design Guide*. Peachtree Corners, GA: ASHRAE.

ASHRAE. 2016. ASHRAE Guideline 10, *Interactions affecting the achievement of acceptable indoor environments*. Peachtree Corners, GA: ASHRAE.

ASHRAE. 2017. *ANSI/ASHRAE/IES Standard 90.1-2016 User's Manual*. Peachtree Corners, GA: ASHRAE.

ASHRAE. 2018a. ANSI/ASHRAE/IES Standard 90.2-2018, *Energy-efficient design of low-rise residential buildings*. Peachtree Corners, GA: ASHRAE.

ASHRAE. 2018b. ASHRAE/IES Standard 202-2018, *Commissioning process for buildings and systems*. Peachtree Corners, GA: ASHRAE.

ASHRAE. 2019a. ANSI/ASHRAE Standard 62.2-2019, *Ventilation and acceptable indoor air quality in residential buildings*. Peachtree Corners, GA: ASHRAE.

ASHRAE. 2019b. ANSI/ASHRAE/IES Standard 90.1-2019, *Energy standard for buildings except low-rise residential buildings*. Peachtree Corners, GA: ASHRAE.

ASHRAE. 2019c. ASHRAE Guideline 1.4-2019, *Preparing systems manuals for facilities*. Peachtree Corners, GA: ASHRAE.

ASHRAE. 2019d. ASHRAE Guideline 0-2013, *The commissioning process*. Peachtree Corners, GA: ASHRAE.

ASHRAE. 2019e. *ASHRAE Handbook—HVAC Applications*. Peachtree Corners, GA: ASHRAE.

ASHRAE. 2020a. *ASHRAE Handbook—HVAC Systems and Equipment*. Peachtree Corners, GA: ASHRAE.

ASHRAE. 2020b. ANSI/ASHRAE Standard 55-2020, *Thermal environmental conditions for human occupancy*. Peachtree Corners, GA: ASHRAE.

ASHRAE. 2021a. *ASHRAE Handbook—Fundamentals*. Peachtree Corners, GA: ASHRAE.

ASHRAE. 2021b. *ASHRAE Terminology*. Peachtree Corners, GA: ASHRAE. https://xp20.ashrae.org/terminology/index.php?letter=R.

ASHRAE. 2021c. ANSI/ASHRAE Standard 169-2020, *Climatic data for building design standards*. Peachtree Corners, GA: ASHRAE.

ASHRAE. 2021d. *Guide for Resilient Thermal Energy Systems Design in Cold and Arctic Climates*. Peachtree City, GA: ASHRAE. www.wbdg.org/ffc/army-coe/design-guides/guide-for-resilient-thermal-energy-systems-design-in-cold-and-arctic-climates.

ASME. 2018. B31.8 *Gas Transmission and Distribution Piping Systems*. New York: American Society of Mechanical Engineers.

ASME. 2021. *B31.3 - 2020 Process Piping*. New York: American Society of Mechanical Engineers.

ASTM. 2018. ASTM E2357-18. *Standard test method for determining air leakage of air barrier assemblies*. West Conshohocken, PA: ASTM International.

ASTM. 2019. ASTM E779-19. *Standard test method for determining air leakage rate by fan pressurization*. West Conshohocken, PA: ASTM International.

Bachu, S., and R.A. Burwash. 2012. Chapter 30: Geothermal regime in the Western Canada sedimentary basin. In *Geological atlas of the Western Canada sedimentary basin*, G.D. Mossop and I. Shetsen (comp.). Edmonton, AB: Alberta Geological Survey. www.ags.gov.ab.ca/publications/wcsb_atlas/atlas.html.

BC Hydro. 2020. *Building envelope thermal bridging guide*, version 1.4. Burnaby, BC: Morrison Hershfield. www.bchydro.com/content/dam/BCHydro/customer-portal/documents/power-smart/business/programs/BETB-Building-Envelope-Thermal-Bridging-Guide-v1-4.pdf?.

Bean, R. 2012. Built to code: What does it mean for consumer thermal comfort? *Healthy Heating*. www.healthyheating.com/Built-to-code.htm#.WZB8QFF95dg.

Bean, R. 2013. The Interaction and Connection between Buildings, HVAC System, and Indoor Environmental Quality. ASHRAE IAQ Conference, Vancouver, Canada.

Bean, R. 2014. Fairville farms. *Healthy Heating*. www.healthyheating.com/Commercial-Industrial-HVAC-Design-Services.htm#.VXnWhfnF92t.

Bean, R. 2015. Human Factors in HVAC: Thermal Comfort and IAQ for Lifetime Housing. Healthy Buildings America Conference. Boulder, CO. www.healthyheating.com/Healthy_Buildings_America_2015/Healthy_Buildings_America_2015.htm.

Bean, R. 2016c. Building Smart with Thermal Comfort, Solar Control and Passive Cooling. *Healthy Heating*. www.healthyheating.com/2016.BC.Housing/2016.BC.Housing.Comfort.htm#.YF37S51KhPY.

Bean, R. 2016a. The effect of heating coil performance on boiler efficiency and system complexity. *HPAC Magazine*. www.hpacmag.com/features/effect-heating-coil-performance-boiler-efficiency-system-complexity/

Bean, R. 2016b. Humidity: An Important Nexus in Indoor Environmental Engineering. *Healthy Heating*. www.healthyheating.com/Relative_Humidity_Therma-Stor.htm#.YFt9YJ1KhPY.

Bean, R. 2017. Integrated Design Curriculum. *Healthy Heating*. www.healthyheating.com/Integrated-Building-HVAC-Design-Course-Curriculum.htm#.YFpEzZ1KhPY.

Beate, B., and R. Salgado. 2005. *Geothermal country update for Ecuador, 2000–2005*. Proceedings of the World Geothermal Congress 2005. Antalya, Turkey, April 24–29.

Belzer, D.B. 2006. *Energy end-use flow maps for the buildings sector*. Oak Ridge, TN: U.S. Department of Energy, Office of Scientific and Technical Information. www.pnl.gov/main/publications/external/technical_reports/PNNL-16263.pdf.

Bernard, M. 1970. *Cold regions science and engineering monograph 3, section Blb: Ice pressure on engineering structures*. Hanover, NH: Cold Regions Research And Engineering Lab.

Bethune, D. 2015. *A novel urine evaporation and collection system for dry toilets*. Doctoral thesis. Calgary, AB: University of Calgary. doi: 10.11575/PRISM/24932.

Betts, J.G., et al. 2013. *Anatomy and physiology*. Houston, TX: Rice University.

Boecker, J., S. Horst, T. Keiter, A. Lau, M. Sheffer, B. Toevs, B. Reed, and S.R. Fredrizzi. 2014. *The integrative design guide to green building*. Hoboken, NJ: John Wiley & Sons.

BREEAM. 2015. *BREEAM communities*. Watford, UK: BRE Global. https://files.bregroup.com/breeam/briefingpapers/Introduction_to_BREEAM_Communities.pdf.

British Columbia Regulation. 2002. Organic Matter Recycling Regulation. Victoria, BC: Queens' Printer. www.bclaws.gov.bc.ca/civix/document/id/complete/statreg/15_18_2002a.

BSI. 2009. *Ergonomics of the thermal environment. Methods for the assessment of human responses to contact with surfaces. Hot surfaces*. BS EN ISO 13732-1:2008. London: British Standards Institution.

BSI. 2012. *Heating boilers. Heating boilers for solid fuels, manually and automatically stoked, nominal heat output of up to 500 kW. Terminology, requirements, testing and marking*. London: British Standards Institution.

BSI. 2019. *Energy performance of buildings. Ventilation for buildings. Indoor environmental input parameters for design and assessment of energy performance of buildings addressing indoor air quality, thermal environment, lighting and acoustics. Module M1-6*. BS EN 16798-1:2019. London: British Standards Institution.

BSRIA. 2021. *Soft landings*. Bracknell, Berkshire, UK: BSRIA Limited. www.bsria.com/uk/consultancy/project-improvement/soft-landings/.

Burkhard, E., N. Russell, B. Musil-Schläffer, A. McCarry, C. Schmidl, and W. Haslinger. European wood-heating technology survey. Albany, NY: NYSERDA. www.nyserda.ny.gov/-/media/Files/Publications/Research/Biomass-Solar-Wind/European-wood-heating-technology-survey.pdf.

Cabanac M. 1988. Alliesthesia. In *Sensory Systems: II. Readings from the Encyclopedia of Neuroscience*. Boston: Birkhäuser. https://doi.org/10.1007/978-1-4684-6760-4_1.

Carpenter, S. 2012. *Advances in modelling thermal bridges in building envelopes*. http://citeseerx.ist.psu.edu/viewdoc/download?doi=10.1.1.128.7094&rep=rep1&type=pdf.

CCHRC. n.d. What are the advantages of hot water in-floor heat versus baseboard heating? Alaska: Cold Climate Housing Research Center. www.cchrc.org/sites/default/files/docs/radiant_heat.pdf, accessed July 2017.

CENELEC. 2007. CENELEC Guide 29. *Temperatures of hot surfaces likely to be touched*. Brussels: CENELEC.

CIB. 2014. *Code of practice for project management for construction and development*, 5th ed. Hoboken, NJ: Wiley-Blackwell.

CIBSE. 2005. *CIBSE Climate Change and the Indoor Environment: impacts and adaption (TM36)*. London: CIBSE.

CIBSE. 2006. *CIBSE TM31 building log book toolkit*. London: CIBSE.

CIBSE. 2013. *TM54 evaluating operational energy performance of buildings at the design stage*. London: CIBSE.

CIBSE 2015. *Environmental design: CIBSE Guide A*. London: CIBSE.

Clayton, J. 2014 BlueSkyModel: 1 kilowatt-hour. BlueSkyModel.org. http://blueskymodel.org/kilowatt-hour#carbon-dioxide-from-one-kilowatt-hour.

Cornell University. 2010. How natural and built environments impact human health. *Design & Environmental Analysis: Outreach & Extension*. Ithaca, NY: Cornell University.

Craven, C. and R. Garber-Slagt. 2011. *Reflective insulation in cold climates*. Fairbanks: Cold Climate Housing Research Center. www.insulationstop.com/resources/TR_2011-01_Reflective_Insulation_in_Cold_Climates.pdf.

CSA Group. 2018. *AAMA/WDMA/CSA 101/I.S.2/A440, NAFS — North American Fenestration Standard/Specification for windows, doors, and skylights (NAFS)*. Alpharetta, GA: CSA Group.

CSA Group. 2021. *CSA B214:21, Installation code for hydronic heating systems*. Alpharetta, GA: CSA Group.

CSI. 2004. *The project resource manual: CSI manual of practice*. New York: McGraw-Hill Education.

CSI. 2011. *The CSI project delivery practice guide*. Hoboken, NJ: Wiley.

Danfoss. n.d. *The Heating Book: 8 steps to control of heating systems*. Part 2-Control of heating systems used in Western Europe. Nordborg, Denmark: Danfoss.

Danfoss. 2012. *District heating application handbook*. Nordborg, Denmark: Danfoss. www.danfoss.com/en/service-and-support/downloads/dhs/district-heating-application-guide/#tab-overview.

Decision Analyst. 2016. *American Home Comfort Study*. Arlington, TX: Decision Analyst.

DOE. 2009. *Vapor control layer recommendations*. BSC Information, Research Report. Washington, D.C.: U.S. Department of Energy.

Egger, C. 2012. *Transformation of the Austrian high-efficiency biomass heating market*. Washington, D.C.: Biomass Thermal Energy Council. www.biomassthermal.org/resource/pdfs/BTEC_ChristianeEgger.pdf.

EIA. 2021. *Short-term energy outlook*. Washington, D.C.: Energy Information Administration. www.eia.gov/outlooks/steo/.

Emmerich, S.J., T.P. McDowell, and W. Anis. 2005. *Investigation of the impact of commercial building envelope airtightness on HVAC energy use*. NIST Interagency/ Internal Report (NISTIR)—7238. Gaithersburg, MD: National Institute of Standards and Technology.

EPA. 1999. Water Efficiency Technology Fact Sheet: Incinerating Toilets. Washington, D.C: United States Environmental Protection Agency, Office of Water.

Ericsson, K. and P. Svenningsson. 2009. *Introduction and development of the Swedish district heating systems—Critical factors and lessons learned.* (RES-H/C Policy project report D2.3). Environmental and Energy Systems Studies. Lund, Sweden: Lund University.

FEMA. 2013. *Hurricane Sandy in New Jersey and New York: building performance observations, recommendations, and technical guidance: Mitigation Assessment Team report.* Washington, D.C.: FEMA.

FEMA. 2013. *Mitigation Assessment Team Report Hurricane Sandy in NJ and NY.* FEMA P-942. Washington, D.C.: FEMA.

Finch, G., J. Straube, and B. Hubbs. 2007. *Hygrothermal performance and drying potential of wood frame rainscreen walls in Vancouver's coastal climate.* Banff, AB: 11th Canadian Conference on Building Science and Technology.

Francescato, V., E. Antonini, L.Z. Bergomi, C. Metschina, C. Schnedl, N. Krajnc, K. Koscik, P. Gradziuk, G. Nocentini, and S. Stranieri. 2008. *Wood fuels handbook.* Legnaro, Italy: Italian Agriforestry Energy Association.

GNWT. 2004. *Good engineering practice for northern water and sewer systems.* Yellowknife, NT, Canada: Department of Public Works and Services, Government of the Northwest Territories.

GNWT. 2011a. *Biomass heating system: Wood pellet boilers*. Yellowknife, NT, Canada: Department of Public Works and Services, Government of the Northwest Territories.

GNWT. 2011b. Public works and services energy conservation projects—Annual report 2010. Yellowknife, NT, Canada: Department of Public Works and Services, Government of the Northwest Territories.

GNWT. 2013. *Good building practice for northern facilities*. Yellowknife, NT, Canada: Department of Public Works and Services, Government of the Northwest Territories.

Halley, G. 1998. Boiler/burner combustion air supply requirements and maintenance. *The National Board of Boiler and Pressure Vessel Inspectors Bulletin*. Columbus, OH: The National Board of Boiler and Pressure Vessel Inspectors. www.nationalboard.org/index.aspx?pageID=164&ID=236.

Hawkins, G. 2009. *Building services job book—A project framework for engineering services*. Bracknell, Berkshire, UK: BSRIA.

Hayes, D. n.d. Better, faster, more: Toward sustainable cities. *Bullitt Center.* www.bullittcenter.org/vision/message-from-denis-hayes.

Health Canada. 2011. Guidelines for Canadian drinking water quality—Chlorine guideline technical document, N-nitrosodimethylamine. Ottawa, ON: Water, Air and Climate Change Bureau, Healthy Environments and Consumer Safety Branch, Health Canada. www.canada.ca/content/dam/canada/health-canada/migration/healthy-canadians/publications/healthy-living-vie-saine/water-nitrosodimethylamine-eau/alt/water-nitrosodimethylamine-eau-eng.pdf.

Health Canada, 2017. Guidelines for Canadian Drinking Water Quality—Summary Table. Ottawa, ON: Water, Air and Climate Change Bureau, Healthy Environments and Consumer Safety Branch, Health Canada. www.canada.ca/en/health-canada/services/environmental-workplace-health/reports-publications/water-quality/guidelines-canadian-drinking-water-quality-summary-table.html.

Hill, G., and G. Henry. 2013. The Application and Performance of Urine Diversion to Minimize Waste Management Costs Associated with Remote Wilderness Toilets. *International Journal of Wilderness* 19(1).

HOT|COOL Magazine. 2013. *Focus: District heating in germany.* Frederiksberg, Germany: DBDH. http://dbdh.dk/download/Publications/HC2_13_LOWg.pdf.

ICC. 2007. *2007 supplement to the International Codes.* Washington, D.C.: International Code Council.

ICC. 2018. *International green construction code® powered by ANSI/ASHRAE/ICC/USGBC/IES Standard 189.1.* Washington, DC: International Code Council.

Institute of Medicine. 2011. *Climate Change, the Indoor Environment, and Health.* Washington, DC: The National Academies Press. https://doi.org/10.17226/13115.

International Geothermal Association. 2021. *What is geothermal energy?* Bonn, Germany: International Geothermal Association. www.geothermal-energy.org/what_is_geothermal_energy.html.

ISO. 2005. *ISO 7730:2005—Ergonomics of the thermal environment—Analytical determination and interpretation of thermal comfort using calculation of the PMV and PPD indices and local thermal comfort criteria.* Geneva, Switzerland: ISO.

ISO. 2019. *ISO 11855-7:2019—Building environment design—Design, dimensioning, installation and control of embedded radiant heating and cooling systems—Part 7: Input parameters for the energy calculation.* Geneva, Switzerland: ISO.

Jones, D., B. Brown, T. Thompson, and G. Finch. 2014. Building enclosure airtightness testing in Washington State—Lessons learned about air barrier systems and large building testing procedures. *ASHRAE Transactions* 120(2).

Kapedan, E., J. Herssens, E. Nuyts, G. Verbeeck. 2017. The importance of comfort indicators in home renovations: A merger of energy efficiency and universal design. *Proceedings, PLEA.* Edinburgh: Passive and Low Energy Architecture.

Karakash, J. and D.D. Richter. 2010. *Architects and energy professionals—The missing link in wood energy.* Princeton, WV: Wood Education and Resource Center. www.biomassthermal.org/resource/documents/8_ArchitectsEnergyProfessionals.pdf.

Kosinski, R.J. 2013. *A literature review on reaction time.* Clemson, SC: Clemson University. www.cognaction.org/cogs105/readings/clemson.rt.pdf.

Küpper, T., V. Schöffl, and J. Milledge. 2012. Consensus statement of the UIAA medical commission Vol. 6 "Water Disinfection in the Mountains." Berne, Switzerland: The International Mountaineering and Climbing Federation. www.theuiaa.org/documents/mountainmedicine/English_UIAA_MedCom_Rec_No_6_Water_Disinfection_2012_V3-1.pdf.

Lark, J.P. 2014. *BWL December 2013 ice storm outage report.* Lansing, MI: Lansing Board of Water and Light. http://docshare02.docshare.tips/files/20783/207832181.pdf.

LBNL. 2006. Fenestration Heat Transfer Basics, Condensation Resistance. In *THERM 5.2/WINDOW 5.2 NFRC Simulation Manual.* Berkeley, CA: Lawrence Berkeley National Laboratory.

LBNL. 2020. CBE Thermal Comfort Tool. Berkeley, CA: Lawrence Berkeley National Laboratory. https://comfort.cbe.berkeley.edu/.

Leinster, P., and E. Mitchell. 1992. *A review of indoor air quality and its impact on the health and well-being of office workers.* Luxembourg: Commission of the European Communities.

Lepage, R., C. Schumacher, and A. Lukachko. 2013. *Moisture management for high R-value walls.* Oak Ridge, TN: U.S. Department of Energy, Office of Scientific and Technical Information. www.nrel.gov/docs/fy14osti/60487.pdf.

Lewis, M. 2004. Integrated design for sustainable buildings. *ASHRAE Journal* 46(9): S22-S29.

Lstiburek, J. 2000. *Air barriers vs. vapor barriers.* Somerville, MA: Building Science Press. www.buildingscience.com/documents/reports/rr-0004-air-barriers-vs-vapor-barriers.

Lstiburek, J. 2004. *Vapor barriers and wall design.* Somerville, MA: Building Science Press. www.buildingscience.com/documents/reports/rr-0410-vapor-barriers-and-wall-design.

Lstiburek, J. 2008. BSI-001: The perfect wall. *Building Science Insights.* www.buildingscience.com/documents/insights/bsi-001-the-perfect-wall.

Lstiburek, J. 2015. BSI-062: Thermal bridges redux. *Building Science Insights.* www.buildingscience.com/documents/insights/bsi062-thermal-bridges-redux.

MacNevin, L. 2015. Snow and ice melting systems: Needed now more than ever (Part I). *Rehau Blog.*

Mathew, L. S.S. Purkayastha, A. Jayashankar, and H.S. Nayar. 1981. Physiological characteristics of cold acclimatization in man. *International Journal of Biometeorology* 25(3): 191–8.

McGinn, T. 2014. Learning by experiencing. *ASHRAE Journal* 56.

MIT. 2006. *The future of geothermal energy.* Chapter 2—Geothermal resource-base assessment. Idaho Falls, ID: Renewable Energy and Power Department Idaho National Laboratory. www1.eere.energy.gov/geothermal/pdfs/egs_chapter_2.pdf.

Morata, D. 2014. *South America region: Geothermal progress.* GRC 2014 38th Annual Meeting. Portland, OR, Sept. 28–Oct. 1.

NAS. 2011. *Climate Change, the Indoor Environment, and Health.* Washington, DC: National Academy of Sciences.

NFPA. 2020. *NFPA 70. National electric code.* Quincy, MA: National Fire Protection Association. www.nfpa.org/nec.

NFPA. 2021a. NFPA 496. Standard for Purged and Pressurized Enclosures for Electrical Equipment. Quincy, MA: National Fire Protection Association. www.nfpa.org/codes-and-standards/all-codes-and-standards/list-of-codes-and-standards/detail?code=496.

NFPA. 2021b. *NFPA 497. Recommended practice for the classification of flammable liquids, gases, or vapors and of hazardous (classified) locations for electrical installations in chemical process areas.* Quincy, MA: National Fire Protection Association. www.nfpa.org/codes-and-standards/all-codes-and-standards/list-of-codes-and-standards/detail?code=497.

Nicholls, D. 2009. Wood energy in Alaska—Case study evaluations of selected facilities. Portland, OR: U.S. Forest Service, Pacific Northwest Research Station. www.fs.fed.us/pnw/pubs/pnw_gtr793.pdf.

NMIC. 2016. Simple Calculators. *Whole Building Design Guide.* Washington, D.C.:National Institute of Building Sciences. www.wbdg.org/guides-specifications/mechanical-insulation-design-guide/simple-calculators.

NMIC. 2017. Estimate Time To Freezing For Water In An Insulated Pipe Calculator. *Whole Building Design Guide.* Washington, D.C.:National Institute of Building Sciences. www.wbdg.org/guides-specifications/mechanical-insulation-design-guide/design-objectives/time-freezing-fluid-insulated-pipe-calculator.

Norris, N., P. Ropell, and M. Lawton. 2012. *The concept of linear and point transmittance and its value in dealing with thermal bridges in building enclosures.* Atlanta: Morrison Hershfield. www.brikbase.org/sites/default/files/best3_norris.pdf.

NRC. 2006. *Building envelope, heating, and ventilating practices and technologies for extreme climates.* Ottawa, ON: National Resources Canada.

NRC. 2021. *Keeping the heat in.* Ottawa, ON: National Resources Canada. www.nrcan.gc.ca/energy-efficiency/homes/make-your-home-more-energy-efficient/keeping-the-heat/15768.

NRCAN. 2010. *National building code of Canada.* Ottawa, ON: National Research Council Canada.

NRCAN. 2013. *User's guide to the national energy code of Canada for buildings 2013.* Ottawa, ON: National Research Council.

NTPC. 2014. *Northern Canada set to realize benefits from liquefied natural gas.* Hay River, NWT, Canada: Northwest Territories Power Corporation.

OpenStax. 2013. *Anatomy and Physiology.* Houston: Rice University.

OSHA. 2011. OSHA Standard 1910.134, *Respiratory protection.* Washington, D.C.: Occupational Safety and Health Administration. www.osha.gov/laws-regs/regulations/standardnumber/1910/1910.134.

Panagoupoulos, T. 2008. Using microclimatic landscape design to create thermal comfort and energy efficiency. *Actas da 1st Conferência sobre Edifícios Eficientes*: Universidade do Algarve. January 25, 2008, 1–4. www.researchgate.net/publication/228505837_Using_microclimatic_landscape_design_to_create_thermal_comfort_and_energy_efficiency.

Passive House Canada. n.d. *About Passive House.* Victoria, BC: Passive House Canada. www.passivehousecanada.com/about-passive-house/.

PDE. 2021. FlexPDE, ver. 7.20. USA: PDE Solutions Inc. www.pdesolutions.com.

Peyton, H.R. 1969. *Thermal design in permafrost soils.* Third Canadian Conference on Permafrost. July 10–13, Edmonton, AB, Canada.

Pierce, M. 2008. The opportunity (and necessity) for new district energy systems in the Northeast United States. *Energy & Environment* 2:6–8.

Phillips, E.G., R.Chant, D. Fisher, and B. Bradley. 1989. Comparison of freezing control strategies for residential air-to-air heat recovery ventilators. *ASHRAE Transactions* 95(2):484.

Phillips, E.G., D.R. Fisher, R.E. Chant and B.C. Bradley. 1992. Freeze-control strategy and air-to-air energy recovery performance. *ASHRAE Journal* 34(12):44.

Pine Tree Legal Assistance. 2010. How much heat and other basic utilities does my landlord have to provide? Portland, ME: Pine Tree Legal Assistance. http://ptla.org/how-much-heat-and-other-basic-utilities-does-my-landlord-have-provide.

PMI. 2014. *Implementing organizational project management: A practice guide.* Newtown Square, PA: Project Management Institute, Inc.

PwC. 2013. Rising sick bill is costing UK business £29bn a year. *PwC research.* London: PricewaterhouseCoopers LLP https://pwc.blogs.com/press_room/2013/07/rising-sick-bill-is-costing-uk-business-29bn-a-year-pwc-research.html.

RDH Building Engineering. 2013. *Air leakage control in multi-unit residential buildings.* Vancouver: RDH Building Engineering, Ltd. www.rdh.com/resource/air-leakage-control-in-multi-unit-residential-buildings/.

Rice, E. 2008. *Building in the north, fifth edition.* Anchorage, AK: University of Alaska Press. www.uaa.alaska.edu/civilengineering/arctic/upload/Building _in_the_North.pdf.

Richter, D.D., D.H. Jenkins, J.T. Karakash, J. Knight, L.R. McCreery, and K.P. Nemestothy. 2009. Wood energy in America. *Science* 323(5920):1432–1433.

Robinson, J. 2017. *The effects of stress on your body.* New York: WebMD. www.webmd.com/balance/stress-management/effects-of-stress-on-your-body.

Rodolfo, K. 2000. What is homeostasis? *Scientific American.* www.scientificamerican.com/article/what-is-homeostasis/.

Rollins, J., and C. Young. 1992. Personal Communication with F. Zahl. in *An Inventory of Hut Sites in the Canadian Rockies and Northern Selkirks*. Prepared for the Alpine Club of Canada for Waste Management. Ottawa: Environment Canada.

Rollins, K. 2014. *Best environmental practices for managing energy, water, and wastewater at off-grid mountain facilities.* Backcountry Energy Environmental Solutions (BEES) and Alpine Club of Canada. www.slideshare.net/sustainablesummitsinitiative/karen-rollins-best-environmental-practices-for-managing-energy-water-and-wastewater-at-offgrid-mountain-facilities.

Roppel, P. and W. Marif. 2011. *ASHRAE research project report RP-1365. Thermal performance of buildings envelope details for mid- and high-rise buildings.* Peachtree Corners, GA: ASHRAE.

Rosenlund, H. 2000. Climatic design of buildings using passive techniques. *Building Issues* 10(1): 3–26.

SCC. 2011. CAN/ULC-S742-11, *Standard for air barrier assemblies—Specification.* Ottawa: Standards Council of Canada.

Siegenthaler, J. 2014. Hydronics for high efficiency biomass boilers. Holland Patent, NY: Appropriate Designs. www.nyserda.ny.gov/-/media/Files/EERP/Renewables/Biomass/biomass-hydronics-training.pdf.

Skelton, B.A. 2015. Holding out for a zero. *CIBSE Journal*. January 2015.

Smegal, J. and J. Straube. 2010. *High-R walls for the Pacific Northwest—A hygrothermal analysis of various exterior wall systems.* Somerville, MA: Building Science Press. www.buildingscience.com/documents/reports/rr-1014-high-r-walls-pacific-northwest-hygrothermal-analysis/view.

Smegal, J. and J. Straube. 2011. *Hygrothermal analysis of exterior Rockwool insulation.* Somerville, MA: Building Science Press. www.buildingscience.com/documents/reports/rr-1104-hygrothermal-analysis-exterior-rockwool-insulation.

Spielman, F. 2015. City moves to protect tenants at five South Side buildings with inadequate heat. *Chicago Sun-Times*. https://chicago.suntimes.com/city-hall/2015/2/19/18558951/city-moves-to-protect-tenants-at-five-south-side-buildings-with-inadequate-heat.

Stevens, V., C. Craven, and B. Grunau. 2013. *Thermal storage technology assessment.* Fairbanks, AK: Cold-Climate Housing Research Center. http://cchrc.org/thermal-storage-technology/.

Straube, J. 2009. *Thermal metrics for high performance enclosure walls: The limitations of R-value*. Somerville, MA: Building Science Press. www.buildingscience.com/documents/reports/rr-0901-thermal-metrics-high-performance-walls-limitations-r-value.

Straube, J.F. 2012. *High performance enclosures*. Somerville, MA: Building Science Press.

Straube, J. and J. Smegal. 2009. *Building America special research project: High-R walls case study analysis.* Somerville, MA: Building Science Press. www.buildingscience.com/documents/bareports/ba-0903-building-america-special-research-project-high-r-walls/view.

Toilet Tech. Solutions. n.d. Seattle: ToiletTech.com

Ueno, K. 2010. Building energy performance metrics. *Building Science Digest* 152.

UL. 2006a. UL 555. *Fire dampers.* Northbrook, UL: UL.

UL. 2006b. UL 823. *Electric heaters for use in hazardous (classified) locations.* Northbrook, IL:UL.

UL. 2014. UL 555S, *Smoke dampers.* Northbrook, UL: UL.

UL. 2015. UL 1995. *Heating and cooling equipment.* Northbrook, IL: UL.

University of Washington. 2021. Cold regions engineering. Short course. Seattle: University of Washington Professional and Continuing Education. www.pce.uw.edu/courses/cold-regions-engineering.

Urban Green. 2013. *Building resiliency task force*. New York: Urban Green Council. http://issuu.com/urbangreen/docs/brtf_executive_summary.

USGBC. n.d. *LEED for neighborhood development*. www.usgbc.org/articles/getting-know-leed-neighborhood-development.

WHO. 1948. Preamble to the Constitution of the World Health Organization as adopted by the International Health Conference, New York, 19-22 June, 1946; signed on 22 July 1946 by the representatives of 61 States (Official Records of the World Health Organization, no. 2, p. 100) and entered into force on 7 April 1948.

WHO. 1987. *Health impact of low indoor temperatures*. Copenhagen: World Health Organization.

WHO. 1997. *Guidelines for drinking-water quality,* second edition. Volume 3.Geneva: World Health Organization.

WHO. 2010. World Health Organization international workshop on housing, health and climate change: Developing guidance for health protection in the built environment - mitigation and adaptation responses. Geneva: World Health Organization. www.bridport-tc.gov.uk/wp-content/uploads/2019/04/Housing-and-Health-for-HNA.pdf.

WHO. 2017. 10 facts on ageing and health. Geneva: World Health Organization. www.who.int/features/factfiles/ageing/en/.

Williams, C.F., M.J. Reed, R.H. Mariner, J. DeAngelo, and S.P. Galanis. 2008. Assessment of moderate- and high-temperature geothermal resources of the United States. Menlo Park, CA.: U.S. Geological Survey. http://pubs.usgs.gov/fs/2008/3082/pdf/fs2008-3082.pdf.

Wolf, J. 2013. The future of refrigerants. *The Anthracite Chapter NEWS.* Scranton, PA: The Anthracite ASHRAE Chapter. http://anthraciteashrae.weebly.com/uploads/2/2/8/5/22852420/anthracite_chapter_news-_october_2013.pdf.

Yellow Wood Associates and Richmond Energy Associates. 2011. Perliminary feasibility report: Biomass heating analysis for Indiana area senior high school. Indiana, PE: U.S. Department of Agriculture. www.yellowwood.org/biomass-heating-analysis-for-indiana-area-senior-high-school-resource.html.

Zarr, R.R., D.M. Burch, T.K. Faison, and C.E. Arnold. 1986. Thermal resistance measurements of a well-insulated wall using a calibrated hot box. *Journal of Building Physics* 10(1): 54-75.

Zmeureanu, R., and X.Y. Wu. 2006. Energy and exergy performance of residential heating systems with separate mechanical ventilation system. *Energy, The International Journal* 32(3):187–95.

Index

A

B

C

D

E

F

G

H

I

K

L

M

N

O

P

R

S